AF577190

# Seismic Applications of Acoustic Reciprocity

# Seismic Applications of Acoustic Reciprocity

**J.T. Fokkema**

*Department of Petroleum Engineering and Technical Geophysics,*
*Faculty of Mining and Petroleum Engineering,*
*Delft University of Technology,*
*Delft, The Netherlands*

and

**P.M. van den Berg**

*Laboratory for Electromagnetic Research,*
*Faculty of Electrical Engineering,*
*Delft University of Technology,*
*Delft, The Netherlands*

1993

ELSEVIER
AMSTERDAM • LONDON • NEW YORK • TOKYO

ELSEVIER SCIENCE PUBLISHERS B.V.
Sara Burgerhartstraat 25
P.O. Box 211, 1000 AE Amsterdam, The Netherlands

ISBN 0-444 89044 0



This book is printed on acid-free paper.

Printed in The Netherlands

Dedicated to A.T. de Hoop

# Preface

This book is dedicated to Adrianus T. de Hoop in honor of his sixty-fifth birthday. Every student who has studied under Professor de Hoop knows the egg-shaped figure that plays such an important role in his theoretical description of acoustic, electromagnetic and elastodynamic wave phenomena. Among the students, including the present authors, this figure is affectionally known as the "de-Hoop's egg". On the one hand this figure represents the domain for the application of a reciprocity theorem in the analysis of a wavefield and on the other hand it symbolizes the power of a consistent wavefield description. In the present book we concentrate on the acoustic formulation and application of this theorem.

It will not come as a surprise that the seismic applications of the reciprocity theorem, developed in this book, are based on lecture notes and publications from Professor de Hoop. For the roots of the theorem we revert to Green's theorem for Laplace's equation and Helmholtz's extension to the wave equation. In 1894, J. W. Strutt, who later became Lord Rayleigh, introduced in his book *The Theory of Sound* this extension under the name of Helmholtz's theorem. Nowadays, it is known as Rayleigh's reciprocity theorem.

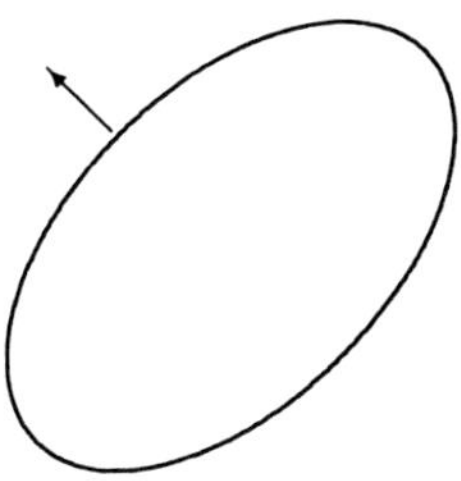

The egg-shaped domain.

Progress in seismic data processing requires the knowledge of all the theoretical aspects of the acoustic wave theory. We choose the reciprocity theorem as the central theme of this book because it constitutes the fundaments of the seismic wave theory. In essence, two states are distinguished in this theorem. These can be completely different, although they share the same time-invariant domain of application and they are related via an interaction quantity. The particular choice of the two states determines the acoustic application. This makes it possible to formulate the seismic experiment in terms of a geological system response to a known source function.

In linear system theory, it is well known that the response to a known input function can be written as an integral representation where the impulse response acts as a kernel and operates on the input function. Due to the temporal invariance of the system, this integral representation is of the convolution type. In seismics, the temporal behavior of the system is dealt with in a similar fashion; however, the spatial interaction needs a different approach. The reciprocity theorem handles this interaction by identifying one state with the spatial impulse function, which is also known as the Green's function, while the other state is connected with the actual source distribution. In general, the resulting integral representation is not a spatial convolution. Moreover, the systematic use of the reciprocity theorem leads to a hierarchical description of the seismic experiment in terms of increasing complexity. Also from an educational point of view this approach provides a hierarchy. The student learns to decompose the seismic problem into consistent partial solutions, in contrast with disconnected ad-hoc solutions.

We hope that this book contributes to the understanding that the reciprocity theorem is a powerful tool in the analysis of the seismic experiment. We are very grateful that we are able to pass on the scientific philosophy of Professor A.T. de Hoop.

We are indebted to Professor A.J. Berkhout for initiating the writing of this book. We acknowledge the assistance of Jan Thorbecke, for the computations of the strip dataset; Roald G. van Borselen, for the computations of the multiple elimination and the domain imaging; Radmilla Tatalovic and Menno Dillen, for the computations of the boundary imaging; and Evert Slob for proofreading the manuscript. We are also grateful to Shell Research B.V., Rijswijk, The Netherlands, for their stimulating support with respect to the multiple elimination.

December 24, 1992

Jacob T. Fokkema
Peter M. van den Berg

# Contents

# Introduction

In mathematical physics, the description of a particular experiment involves the postulation of a model by which the measurement obtains an operational status. In the conceptual view of the model roughly three elements can be recognized: the measurement itself, the mathematical framework which is known as the theory, and the relational parameters that constitute the material existence of the medium. It is important to note that these constitutive parameters only have significance in the realm of the proposed model. Different theoretical models yield complementary views of the medium constitution. It is obvious that the theory, formulated as a mathematical deductive system, has the key role in this conceptual model. Basically there are two ways to use this model: one way is going from the material parameters to the simulated measurements, this process is known as (forward) modeling. The other way is going from the measurement to the material parameters and this process is usually denoted as inversion. The two information streams of these processes are shown in Fig. 1 and clearly illustrate the central role of the theory.

In this book we concentrate on the seismic problem, where the measurement is a sampled version of the acoustic wavefield. The theory is based on the acoustic wave equations and the constitutive parameters are the mass density and the compressibility. The determination of these parameters and their spatial distribution from seismic measurements is the objective of an inversion process. Of course, the physical relation between these parameters obtained by different models is important, especially where it concerns the geological validation. For example, a geological discontinuity does not necessarily coincide with discontinuities of mass density and compressibility. This physical relation between the geological model and the seismic model falls outside the scope of the present book.

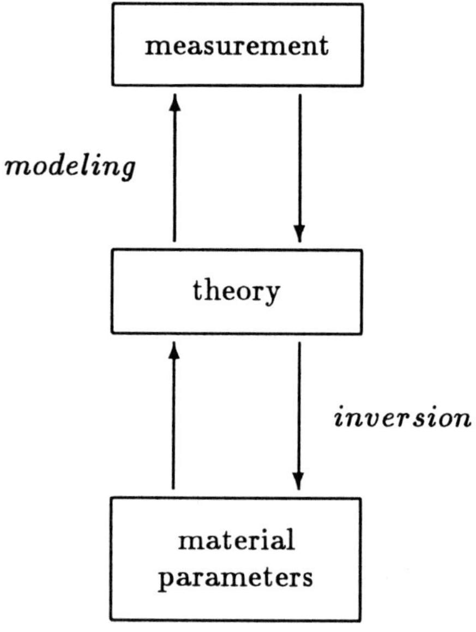


Figure 1. Conceptual view of the model.

The modular structure of the conceptual view, Fig. 1, suggests an approach that is similar to the linear system theory. However, due to the spatial distribution of the material parameters, the higher dimensionality of the acoustic problem requires a careful analysis to arrive at the system approach. We argue that the appropriate vehicle for obtaining this approach is furnished by the acoustic reciprocity theorem, where the spatial-temporal impulse response plays a central role. We show that an analysis along these lines allows for a hierarchical description that controls in an optimal way the complexity of the model. Moreover, the different analyses or processing steps are only then treated in a consistent way.

In the seismic experiment, the measurement consists of the registration of the temporal and the spatial distribution of the acoustic wavefield, conveniently represented in the so-called seismogram. In the direction of the time coordinate, this process is physically constrained by the property of causality. By the latter we mean that changes in the time behavior of the sources that generate the wavefield may only manifest themselves in changes in the time behavior of the wavefield after some elapse of time. Moreover, we consider the physical parameters that characterize the medium shift-invariant in time. But, in the spatial direction, the notion of causality does not exist and the spatial shift-invariance only applies for simple configurations.

In Chapter 1 we discuss the mathematical tools we employ in the analysis of the acoustic wavefield. In particular, we concentrate on the use of integral-transformation methods. In view of the aforementioned causality and shift-invariance in the time direction, the most suitable temporal transformation is the one-sided Laplace transformation. The commonly used temporal Fourier transformation is treated as a special case of the Laplace transformation. In cases where it is admissible, we also define an integral transformation of the Fourier type in the spatial direction. To honor the wavefield aspects, we modify the exponential behavior of the transformation kernel in such a way that it is linearly dependent on the Laplace transform parameter. This is a so-called Fourier transformation of the Radon type and decomposes the wavefield in a superposition of generalized rays. We conclude this chapter by briefly reviewing the essence of the numerical implementation of the integral transformation, focussing on the conservation of the symmetry properties of the continuous transform and on the restrictions imposed by the discretization.

Many characterization problems in seismics are associated with the solution of an integral equation. The method of solution of an integral equation is the subject of Chapter 2. In general, apart from some canonical examples, an analytic solution does not exist. This implies that we have to be satisfied with an approximate solution, obtained by numerical means. The degree of resemblance to the exact solution is a measure that has to be defined. We opt for the root-mean-square error in the equality sign of the pertaining integral equation that has to be satisfied by the exact solution. In a discretized version of a realistic seismic problem, the minimization of the error leads to a large system of linear algebraic equations which is intractable for a direct numerical implementation. Therefore, we advocate the iterative solution by means of recursive minimization. In the analysis of this method we offer several operational schemes to control the successive decline of the error. In addition, the ideas behind the recursive minimization are also applicable to the non-linear problem of the inversion process, where the reconstruction of the material parameters is updated iteratively.

In Chapter 3 we present the partial differential equations that govern the dynamical state of matter on a macroscopic scale and hence the acoustic wave propagation in the medium. The first equation is Newton's law of motion which interrelates locally the spatial gradient of the acoustic pressure and the temporal change of the mass flow density. The second equation is the deformation equation, which translates how the spatial gradient of the particle velocity is locally related to the temporal change of the volume. These equations determine the theoretical framework of the conceptual view of the model. The constitutive relations define how the mass flow density and the volume change are interrelated to the acoustic pressure and the particle velocity. The relational parameters are the mass density and the compressibility. The constitutive relations are established by a physical experiment, which must include the concepts of the seismic wave model. In those areas where these material parameters change abruptly, the partial differential equations no longer hold and have to be supplemented by the boundary conditions. These conditions relate the acoustic wavefield quantities at either side of the discontinuity. Since we are dealing with a time-invariant medium, it is convenient to consider the acoustic wave equations in the Laplace-transform domain.

In Chapter 4 we derive expressions for the acoustic wavefield that is causally related to the action of sources of bounded extent in an unbounded homogeneous medium. The spatial invariance of the medium permits us to carry out a spatial Fourier transformation on the Laplace-transformed wavefield quantities. Then the acoustic wave equations reduce to a system of two algebraic equations. In this process, the algebraic inverse of the transformed wave equations is identified as a spectral representation of the so-called Green's function. In the space-time domain, this function represents the space-time impulse response of the medium and is related to the action of a point source. We present two alternatives to arrive at the space-time domain expressions of this Green's function. Then, after the introduction of the scalar and vector potential, the final representation of the wavefield is given as a spatial convolution of the Green's function and the source strengths.

In Chapter 5 we introduce the concepts of acoustic states. The acoustic states are defined in a time-invariant, bounded domain. They encompass the set of circumstances which completely describes the wave motion in the domain of consideration. We distinguish three constitutive members of the set: the material state, which relates to the parameter distribution, the source state, which corresponds to the source distribution, and the field state, which represents the induced wavefield quantities. The central theme of this chapter is the reciprocity theorem. This theorem relates two non-identical acoustic states that can occur in the domain of interest. It directly relates the spatial divergence of the wavefield interaction quantity to the differences between the material and the source distributions of the two states. In the Laplace-transform domain, we present two forms of this theorem: the field reciprocity theorem and the power reciprocity theorem. In the time domain we obtain reciprocity theorems of the convolution type and of the correlation type, respectively. In the subsequent chapters, we show that both forms of the reciprocity theorem are important for a consistent wavefield analysis. Many seismic processing methods can be viewed as the result of a proper application of the reciprocity theorems.

The first application of the field reciprocity theorem with a direct physical appeal is discussed in Chapter 6. In particular, we consider these experiments where the transducers (source and receiver) interchange their position. We model the transducers as distributions of point transducers, surface transducers and volume transducers. The resulting relations are important for understanding the redundancy in the physical experiment. Moreover, these relations may serve as a numerical check in computational acoustics.

The analysis of the acoustic radiation generated by known volume sources in a known inhomogeneous medium, the so-called direct source problem, is discussed in Chapter 7. The representations for the acoustic wavefield are obtained by the application of the field reciprocity theorem. As mentioned before, two acoustic states are distinguished. In this particular application, one state is associated with the actual wavefield; the other state is taken as a suitable Green's state. This latter state is the point-source solution of the acoustic wavefield in the actual medium. Mathematically, the integral representation for the acoustic wavefield quantities is recognized as a spatial convolution of the Green's state and the source distribution. Now, the resemblance with the linear system theory is evident. In this system approach we have a one-dimensional temporal convolution of the system response and the input signal. In the seismic case, the Green's state plays the role of the (three-dimensional) configurational system response. In Fig. 2, we depict the conceptual view of the direct source problem.

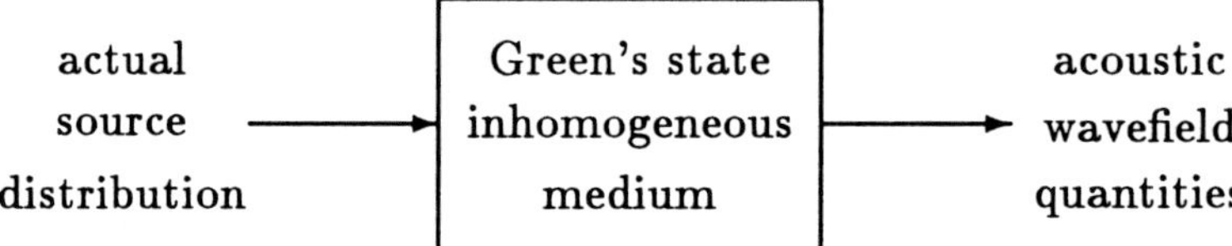


Figure 2. Conceptual view of the direct source problem.

Further analysis shows that, by using the reciprocity theorem again, we can replace the action of the sources by equivalent surface sources at simply closed surfaces. This surface completely encloses the domain of active sources. In the same fashion, we arrive at the acoustic wavefield in a bounded subdomain, under the condition that we know the acoustic wavefield on the confining surface of the subdomain.

The scattering of acoustic waves by a contrasting domain of finite extent, present in an inhomogeneous embedding, the background medium, is the topic of Chapter 8. This subject is known as the direct scattering problem and is also denoted as forward modeling. We show that, by reordering the governing acoustic wavefield equations and by employing their linearity, the total wavefield splits into an incident wavefield and a scattered wavefield. The incident wavefield is the wavefield that originates from the actual sources and would be present in the background medium in absence of the contrasting domain. The pertaining wavefield equations for the scattered wavefield are formulated such that they describe the wave motion in the background medium, originating from the volume sources located in the scattering domain. These contrast sources are originated by the total wavefield and their strengths are determined by the contrast parameters and the total wavefield quantities. This procedure reduces the pertaining scattering problem to a direct source problem and the conceptual view is depicted in Fig. 3. This approach allows us to decide upon the hierarchy of the solution of the pertaining scattering problem. The background medium can be chosen in

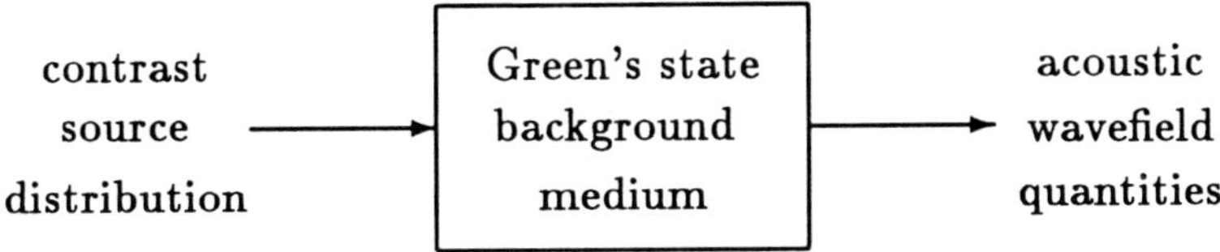


Figure 3. Conceptual view of the direct scattering problem.

such a way that the spatial support and the amplitudes of the contrast parameters of the scatterer are optimally chosen. This optimum is constrained by the complexity of the Green's states and the complexity of the scatterer. For example, in the seismic problem of the earth, the horizontal layering is predominant and it is advantageous to determine the Green's state of a horizontally layered background medium; the scatterer is then determined by the contrast with respect to this chosen background medium.

Determination of the total acoustic wavefield and the related contrast sources is not trivial. The total acoustic wavefield follows from a linear system of domain-integral equations. The numerical treatment of these equations is based on the theory of Chapter 2. Further, we show that a formulation in terms of contrasting surface sources is also possible. This formulation leads to a linear system of boundary-integral equations.

As an example of scattering by a contrasting domain, we discuss in Chapter 9 the problem of acoustic wave scattering by an infinitely thin disk. The disk is impenetrable: either perfectly compliant or perfectly rigid and immovable. The background is either homogeneous or inhomogeneous. As an example of an inhomogeneous embedding, we consider the homogeneous halfspace. For these problems, we derive integral equations of the convolution type. The numerical solution is obtained iteratively and is based on the theory of Chapter 2. In the computational procedure, we employ standard Fourier transformation techniques to calculate the operator expressions involved in the various iterative schemes. We restrict ourselves to the two-dimensional problem of scattering by a strip. The numerical results are collected in a dataset and will be used as synthetic input data to demonstrate different seismic-processing operations in the subsequent chapters.

The decomposition of the acoustic wavefield in a downgoing and upgoing part is important in certain seismic applications, such as redatuming. In Chapter 10 we formalize this procedure with the aid of the reciprocity theorems. We consider a homogeneous subdomain of infinite lateral extent, bounded vertically by two interfaces. The wavefield decomposition is realized in a horizontal plane of this region. With the aid of the field reciprocity theorem and the causal Green's function we show that the downgoing wavefield is associated with an integral contribution of time-retarded surface-source distributions over the upper interface. Using the power reciprocity theorem and the anti-causal Green's function we obtain an integral expression for the upgoing wavefield in terms of time-advanced surface-source distributions over the upper interface. The fact that the downgoing and upgoing parts are both related to the upper interface, makes the decomposition feasible in surface seismics.

In the marine case, the reflections against the water surface generate shadow features in the recorded signal that are known as receiver and source ghosts. These events are removed by an operation which is denoted as deghosting. In Chapter 11 we develop a deghosting procedure. The method is based on the fact that the ghosts are the only downgoing wave constituents in the recorded signal. The decomposition theory outlined in Chapter 10 allows us to isolate the ghosts effectively, and as a next step, to remove them. The performance is illustrated by deghosting the dataset generated in Chapter 9.

The surface-related wave phenomena such as the water-surface multiples in the marine case blur the recorded signal. The multiples only prove that the water surface is a strong reflector. In further processing of the seismic data, the presence of these multiples hinders the good performance of e.g. imaging and inversion. For that reason they need to be removed. However, this removal is conditional: it has to be effected without changing any relevant subsurface information present in the data. The latter requirement implies that the removal procedure must be independent of the subsurface information. In Chapter 12 we show that the field reciprocity theorem provides the most suitable mathematical framework for formalizing the removal procedure. It allows us to formulate the requirement in a natural way. One state is associated with the actual situation, while the other state is the desired multiple-free situation. In the domain of application, both states share the same unknown geology, but in the desired state the water surface is absent. After some consistent applications of the field reciprocity theorem, we arrive at an integral equation of the second kind for the reflected wavefield of the desired case. The kernel of this integral equation and the known term are both related to the deghosted wavefield. The precise knowledge of the acquisition parameters and the source wavelet is prerequisite. This shows that the success of the multiple removal validates our knowledge of these input data. In this sense, it is a necessary condition for seismic inversion. Further, the causality in the time domain admits that the integral equation can be solved by means of a Neumann expansion. We illustrate the performance of the multiple removal with the deghosted dataset of Chapter 11 as input.

The last chapters of this book are devoted to the subject of medium reconstruction. When we observe the structure of the recorded data, we note that the velocity distribution in the earth is solely responsible for the fact that the depth-dependent layering is directly related to the arrival times of the corresponding events in the seismograms. Due to causality, we know that the wavefields reflected from the deeper layers arrive at later times. This observation suggests that we first should try to determine the velocity distribution, in other words to delineate the velocity discontinuities. This process is known as imaging. Then, inversion as the final reconstruction process determines the material parameters by including the knowledge from the imaging process in the construction of the background medium.

In Chapter 13 we discuss the imaging procedure of a single boundary. We start our analysis with the boundary-integral representation. We assume that the wavefield reflected at the boundary is linearly related to the incident wavefield, through a frequency-independent reflection factor. Performing a sequence of spatial Fourier transforms with respect to the source and receiver coordinates separately and using a high-frequency approximation, we obtain a representation that is suitable to image the boundary as the envelop of the arrivals of the reflected causal source-wavelet function. This procedure clearly reveals how the acquisition parameters determine the quality of the image. The image procedure is tested on the synthetic dataset constructed in the previous chapters.

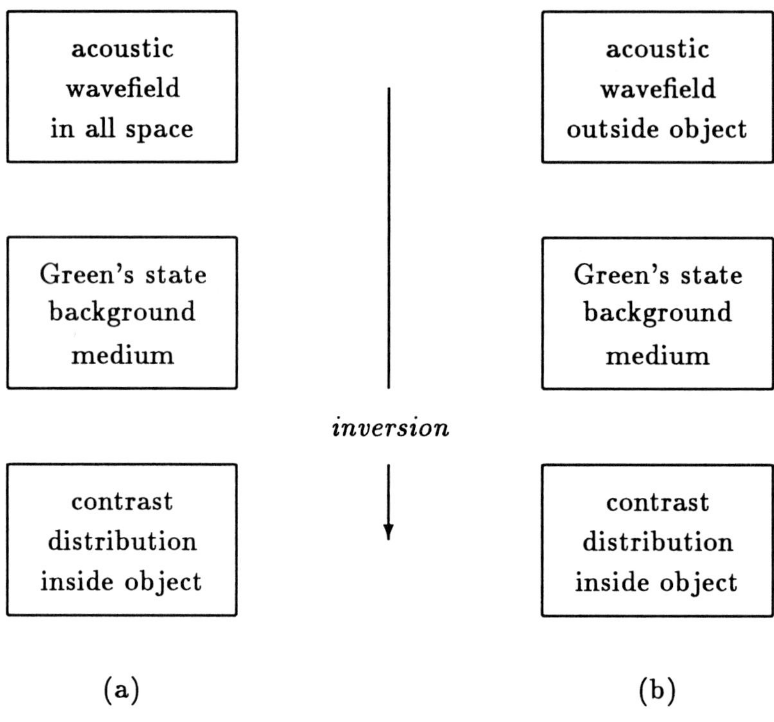


Figure 4. Conceptual views of the ideal seismic model (a) and the actual seismic model (b).

In Chapter 14 we present an alternative imaging procedure based on the domain-integral equation of the reflected field. Here, the first Born approximation is used for arriving at an expression, in which the contrast sources in the domain are linearly related to the incident wavefield, through a frequency-independent contrast factor. Then, an analysis similar to that in Chapter 13 leads again to an imaging procedure, where the velocity discontinuities manifest themselves as a temporal convolution of the source wavelet and the contrast function at the time depth. The relation with the boundary imaging in piecewise homogeneous domains is established. We conclude this chapter by showing the resulting image of a circular cylinder.

In Chapter 15 of this book we discuss the problem of seismic inversion. What we first need is a proper formulation of the problem or to be more precise, an operational context, in which the mathematical relation between the knowns and unknowns is precisely stated. Again the reciprocity relation serves this purpose. In Chapter 8 we argued that the forward modeling is equivalent to the computation of the scattered wavefield due to contrast sources. In their turn the contrast sources are related to the deviation of the actual parameter distribution from the background medium. In the case of forward modeling, the latter distribution is known and via the field reciprocity theorem using the Green's state of the background medium, the acoustic wavefield is calculated. In the case of inversion, the same formulation can be used. Then the knowledge of the acoustic wavefield in all space leads via the field reciprocity theorem and the background Green's state to the contrast sources, and consequently to the constitutive parameters. In both cases, the solution follows from a domain-integral equation with the scattering object as supporting domain. This situation is depicted in Fig. 4a, which we denote as the conceptual view of the ideal seismic model. A comparison with the model of Fig. 1 is relevant. However, in the actual situation, the seismic experiment is used as a diagnostic tool, where the wavefield probing is essentially outside the scattering object. The conceptual view of the actual model is depicted in Fig. 4b. In this case, the relevant integral equation lacks data support in the object. Therefore, we do not have an integral equation, but an integral representation of the scattered acoustic wavefield outside the object, which results in an ill-posed inverse problem. In Chapter 15 we propose handling the inversion problem by the usual minimization of the error between the integral representation of the scattered field and the data observed, but the domain-integral equation inside the scattering object is used as a consistency constraint. This constraint is included in the pertaining minimization problem. In fact, we propose a non-linear iterative scheme that minimizes the error in both the data domain and the object domain simultaneously. The concepts of the iterative schemes presented in Chapter 2 are incorporated in this inversion scheme.

# Chapter 1

# Integral Transformations

The seismic quantities that describe the acoustic waves, depend on position and on time. Their time dependence in the domain where the seismic source is acting is impressed by the excitation mechanism of the source. The subsequent dependence on position and time is governed by propagation and scattering laws. To register the position we employ a Cartesian reference frame with an origin $O$ and three base vectors $\{\boldsymbol{i}_1, \boldsymbol{i}_2, \boldsymbol{i}_3\}$ that are mutually perpendicularly oriented and are of unit length each. The property that each base vector specifies geometrically a length and an orientation, makes it a vectorial quantity, or a vector; notationally, vectors will be represented by bold-face symbols. Let $\{x_1, x_2, x_3\}$ denote the three numbers that are needed to specify the position of an observer, then the vectorial position of the observer $\boldsymbol{x}$ is the linear combination

$$\boldsymbol{x} = x_1\boldsymbol{i}_1 + x_2\boldsymbol{i}_2 + x_3\boldsymbol{i}_3 \, . \tag{1.1}$$

The numbers $\{x_1, x_2, x_3\}$ are denoted as the orthogonal Cartesian coordinates of the point of observation. The time coordinate is denoted by $t$. We employ the International System of Units (Système International d'Unités), abbreviated to SI, for expressing the physical quantities of the acoustic wave motion.

In this chapter we discuss the integral transformations that serve as mathematical tools for the analysis of the acoustic wavefield. Specifically, we introduce the Laplace transformation with respect to the time coordinate and the Fourier transformation with respect to the spatial coordinates.

## 1.1. Cartesian vectors

The mathematical framework of the theory of acoustic waves is furnished by vector calculus. For this reason, this section summarizes those properties of Cartesian vectors (JEFFREYS, 1974, p. 3) that are needed in our further analysis.

*The summation convention*

The summation convention is a shorthand notation to indicate the sum of products of arithmetic arrays. The arrays under consideration have either the same or different dimensions, but the bounds on their subscripts are all the same. In the present acoustic wave theory the arithmetic arrays are the arithmetic representation of acoustic wave motion quantities. The subscripts $\{k, l, p, q\}$ are then to be assigned the values 1, 2 and 3. The convention prescribes that to these lowercase subscripts in a product of arrays the values 1, 2 and 3 are successively to be assigned, while after each assignment the result is added to the previous one. Let, for example, $a_k$ and $b_k$, with $k \in \{1, 2, 3\}$, denote one-dimensional arrays and let $c_{k,l}$, with $k \in \{1, 2, 3\}$ and $l \in \{1, 2, 3\}$, be a two-dimensional array. Then

$$a_k b_k \text{ stands for } \sum_{k=1}^{3} a_k b_k \, , \tag{1.2}$$

$$a_k c_{k,l} \text{ stands for } \sum_{k=1}^{3} a_k c_{k,l} \, , \tag{1.3}$$

$$a_k b_l c_{k,l} \text{ stands for } \sum_{k=1}^{3} \sum_{l=1}^{3} a_k b_l c_{k,l} \, . \tag{1.4}$$

*Addition, subtraction and multiplication of vectors*

Vectors can be subjected to the algebraic operations of addition, subtraction and multiplication. Let the components of $\boldsymbol{\sigma}$ be given by $\sigma_k$ and those of $\boldsymbol{\tau}$ by $\tau_k$, then the components of the sum (difference) of $\boldsymbol{\sigma}$ and $\boldsymbol{\tau}$ is given by

$$(\sigma \pm \tau)_k = \sigma_k \pm \tau_k \, . \tag{1.5}$$

The product of the constant $\gamma$ and $\boldsymbol{\sigma}$ is given by

$$(\gamma\sigma)_k = \gamma\sigma_k \, . \tag{1.6}$$

The inner product of $\boldsymbol{\sigma}$ and $\boldsymbol{\tau}$ is given by

$$\boldsymbol{\sigma} \cdot \boldsymbol{\tau} = \sigma_k \tau_k \, . \tag{1.7}$$

*Differentiation of a vector*

As regards the differentiation of a vector, two cases have to be distinguished: differentiation with respect to a parameter, and differentiation with respect to the spatial (Cartesian) coordinates of the space in which the vector is defined.

Let $\boldsymbol{\sigma}$ be a vector function and assume that $\boldsymbol{\sigma}$ is a differentiable function of the parameter $t$ (in seismics often the time coordinate). Let $\sigma_k$ denote the components of $\boldsymbol{\sigma}$, then the derivative $\partial_t \boldsymbol{\sigma}$ of $\boldsymbol{\sigma}$ with respect to $t$ is a vector whose components are given by $\partial_t \sigma_k$.

Let $\boldsymbol{\sigma}$ be a vector function and assume that $\boldsymbol{\sigma}$ is a differentiable function of the spatial (Cartesian) coordinates $x_1, x_2, x_3$. Let $\sigma_l$ denote the components of $\boldsymbol{\sigma}$, then for each $k$ ($k = 1, 2, 3$), the derivative $\partial_k \boldsymbol{\sigma}$ of $\boldsymbol{\sigma}$ with respect to the spatial coordinate $x_k$ is a vector function. For each $k$, its components are given by $\partial_k \sigma_l$, where $\partial_k$ denotes the partial derivative with respect to $x_k$. Derivatives of a higher order are defined in a similar manner.

In three-dimensional Euclidean space, the gradient of a scalar function $\phi$ of position is introduced as grad $\phi = \partial_k \phi$, the divergence of a vector function $\boldsymbol{v}$ of position as div $\boldsymbol{v} = \partial_k v_k$.

*Gauss' integral theorem for vectors*

Let $\boldsymbol{\sigma}$ denote a continuously differentiable vector function of position defined in some bounded domain $\mathbb{D}$ of a three-dimensional Euclidean space. Let, further, $\partial\mathbb{D}$ denote the boundary of $\mathbb{D}$ (Fig. 1.1). Then, Gauss' integral theorem states that

$$\int_{\boldsymbol{x} \in \mathbb{D}} \partial_k \sigma_l \mathrm{dV} = \int_{\boldsymbol{x} \in \partial\mathbb{D}} \sigma_l \nu_k \mathrm{dA} \, , \tag{1.8}$$

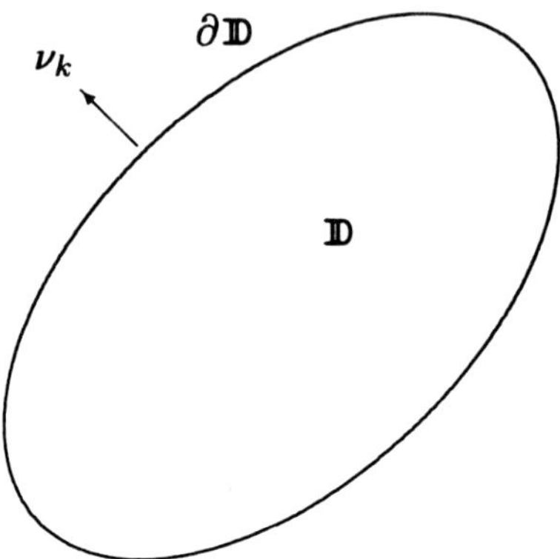


Figure 1.1. Configuration for the application of Gauss' integral theorem.

where $\mathrm{dV} = \mathrm{d}x_1\mathrm{d}x_2\mathrm{d}x_3$ is the elementary volume in three-dimensional Euclidean space and dA is the elementary area of $\partial\mathbb{D}$ and $\nu_k$ is the unit vector normal to $\partial\mathbb{D}$ and oriented away from $\mathbb{D}$.

## 1.2. Integral-transformation methods

In the analysis of physical problems, often integral-transformation methods are employed, see e.g., TRANTER (1966). Their application proves to be most useful in problems associated with configurations whose properties (though not the wavefields occurring in them) are shift invariant in time and/or in one or more of the spatial coordinates. As far as the time coordinate is concerned, we have, in addition, to take into account the property of causality. By the latter we mean that changes in the time behavior of the sources that generate the wavefield may only manifest themselves in changes of the time behavior of the wavefield after some elapse of time. The causality condition can mathematically most easily be accounted for by the use of the one-sided Laplace transformation. For this reason, we shall employ the one-sided Laplace transformation as the integral transformation with respect to time. A similar argument does not apply to the variations of the wavefield in space. Here, it is of importance that we shall be able to handle wavefields in unbounded domains. From this point of view, the Fourier transformation is the most appropriate one, and hence we shall employ the (one-, two-, or three-dimensional) Fourier transformation as the integral transformation with respect to one or more of the spatial variables.

### 1.2.1. Laplace transformation of a causal time function

Let us assume that the seismic sources that generate the wave motion are switched on at the instant $t_0 > 0$ . In view of the causality condition we are then interested in the behavior of the wavefield in the interval (see Fig. 1.2)

$$\mathrm{T} = \{t \in \mathbb{R}; t > t_0\}. \tag{1.9}$$

Further, we shall denote by $\mathrm{T}'$ the complement in $\mathbb{R}$ of the union of T and the instant $t_0$. Hence,

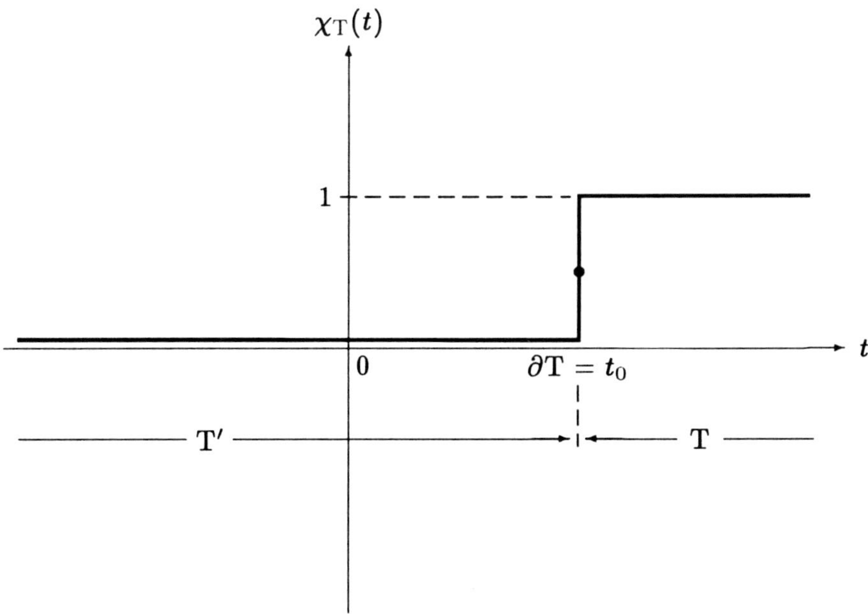


Figure 1.2. Time interval T and its characteristic function $\chi_{\mathrm{T}}$.

$$\mathrm{T}' = \{t \in \mathbb{R}; t < t_0\} . \tag{1.10}$$

Occasionally, we shall denote the instant $t_0$ by $\partial\mathrm{T}$, i.e.,

$$\partial\mathrm{T} = \{t \in \mathbb{R}; t = t_0\} . \tag{1.11}$$

The one-sided Laplace transform of some seismic space-time quantity $u = u(\boldsymbol{x}, t)$, defined in $t > t_0$ and in some as yet unspecified domain in space, is then given by

$$\hat{u}(\boldsymbol{x}, s) = \int_{t \in \mathrm{T}} \exp(-st) u(\boldsymbol{x}, t) \mathrm{d}t . \tag{1.12}$$

Equation (1.12) is considered as an integral equation with known function $\hat{u}(\boldsymbol{x}, s)$, unknown function $u(\boldsymbol{x}, t)$ and kernel $\exp(-st)$. Now, causality is enforced by extending the range of $u$ by the value zero when $t < t_0$, and requiring that this integral equation to be solved for $u(\boldsymbol{x}, t)$, has a unique solution, viz. the value zero when $t < t_0$ and the reproduction of the function that we started with when $t > t_0$. It can be shown, see for example WIDDER (1946, p. 243), that this requirement can be met by a proper choice of the transform parameter $s$. Because of the practical reason that in seismics all quantities have bounded values, we shall restrict ourselves to functions $u$ that are bounded. Then, the right-hand side of Eq. (1.12) is convergent, and Eq. (1.12) considered as an integral equation has a unique solution, if $s$ is either real and positive (which choice has its advantages in the theory of a number of wave propagation problems), or complex with $\mathrm{Re}(s) > 0$. The latter choice leads in the limiting case when $s = j\omega$, where $j$ is the imaginary unit and $\omega$ is real and positive, to the well-known frequency-domain analysis with complex time factor $\exp(j\omega t)$, $\omega$ being the circular frequency of the relevant frequency component. Introducing the characteristic function of the set T,

$$\chi_{\mathrm{T}}(t) = \{1, \frac{1}{2}, 0\} \text{ when } t \in \{\mathrm{T}, \partial\mathrm{T}, \mathrm{T}'\} , \tag{1.13}$$

we can also write

$$\hat{u}(\boldsymbol{x}, s) = \int_{t \in \mathbb{R}} \exp(-st) \chi_{\mathrm{T}}(t) u(\boldsymbol{x}, t) \mathrm{d}t . \tag{1.14}$$

In elucidating the properties of the Laplace transformation it is often advantageous to use Eq. (1.14) rather than Eq. (1.12).

In a number of cases encountered in the theory of the propagation of impulsive waves, the transformation from the $s$-domain back to the time domain is carried out either by direct inspection or by inspection after having applied some elementary rules of the Laplace transformation. For this reason, some of these rules are discussed below.

*Differentiation with respect to time*

Let $u = u(\boldsymbol{x}, t)$ denote a function that is defined when $t \in \mathrm{T}$ and that is equal to zero when $t \in \mathrm{T}'$. Then, the (one-sided) Laplace transform of the derivative $\partial_t u$ of $u$ is found via partial integration as

$$\int_{t_0}^{\infty} \exp(-st)\partial_t u(\boldsymbol{x}, t)\mathrm{d}t = -\exp(-st_0) \lim_{t \downarrow t_0} u(\boldsymbol{x}, t) + s\hat{u}(\boldsymbol{x}, s) . \tag{1.15}$$

Further, the Laplace transform of the time derivative of $\chi_{\mathrm{T}} u(\boldsymbol{x}, t)$ is found as

$$\begin{aligned}
&\int_{t \in \mathbb{R}} \exp(-st)\partial_t[\chi_{\mathrm{T}}(t)u(\boldsymbol{x}, t)]\mathrm{d}t \\
&= \int_{t \in \mathbb{R}} \exp(-st)[\partial_t \chi_{\mathrm{T}}(t)]u(\boldsymbol{x}, t)\mathrm{d}t + \int_{t \in \mathbb{R}} \exp(-st)\chi_{\mathrm{T}}(t)\partial_t u(\boldsymbol{x}, t)\mathrm{d}t \\
&= \exp(-st_0) \lim_{t \downarrow t_0} u(\boldsymbol{x}, t) + \int_{t_0}^{\infty} \exp(-st)\partial_t u(\boldsymbol{x}, t)\mathrm{d}t \\
&= s\hat{u}(\boldsymbol{x}, s) ,
\end{aligned} \tag{1.16}$$

where Eq. (1.15) has been used. The term $\exp(-st_0) \lim_{t \downarrow t_0} u(\boldsymbol{x}, t)$ accounts for the presence of an impulse function (Dirac distribution) at $t = t_0$, whose strength equals the jump in $u$ when passing the instant $t = t_0$ in the direction of increasing $t$. Upon incorporating the latter contribution in the definition of the time derivative of $u$, the rule applies that the $s$-domain equivalent of the operation of time differentiation is the multiplication by a factor of $s$.

Equation (1.16) exemplifies that the transformation rules find their simplest expression when $\mathbb{R}$ is taken as the domain of $u$.

*Asymptotic behavior as* $|s| \to \infty$

Performing a partial integration in the right-hand side of Eq. (1.12), the asymptotic behavior of the Laplace transform as the transform parameter goes to infinity is found to be

$$\lim_{|s|\to\infty} \exp(st_0)\hat{u}(\boldsymbol{x}, s) = 0, \quad \mathrm{Re}(s) > 0 . \tag{1.17}$$

Furthermore, we have (cf. Eq. (1.15))

$$\lim_{|s|\to\infty} \exp(st_0)s\hat{u}(\boldsymbol{x}, s) = \lim_{t\downarrow t_0} u(\boldsymbol{x}, t), \quad \mathrm{Re}(s) > 0 . \tag{1.18}$$

This latter relation is known as the initial-value theorem.

*Temporal convolution*

Let $u = u(\boldsymbol{x}, t)$ and $v = v(\boldsymbol{x}, t)$ denote two functions that are defined on $\mathbb{R}$. Then, the temporal convolution $C_t\{u, v\}(\boldsymbol{x}, t)$ of $u$ and $v$ is defined as

$$\begin{aligned} C_t\{u, v\}(\boldsymbol{x}, t) &= \int_{t'\in\mathbb{R}} u(\boldsymbol{x}, t - t')v(\boldsymbol{x}, t')\mathrm{d}t' \\ &= \int_{t'\in\mathbb{R}} u(\boldsymbol{x}, t')v(\boldsymbol{x}, t - t')\mathrm{d}t' = C_t\{v, u\}(\boldsymbol{x}, t) . \end{aligned} \tag{1.19}$$

Equation (1.19) shows that the convolution is a symmetrical functional of the two constituent functions. Taking the Laplace transformation of Eq. (1.19), we arrive at

$$\hat{C}_t\{u, v\}(\boldsymbol{x}, s) = \hat{u}(\boldsymbol{x}, s)\hat{v}(\boldsymbol{x}, s) . \tag{1.20}$$

For Eq. (1.20) to be valid, there must exist a value of $s$ for which the two definition integrals for $\hat{u}$ and $\hat{v}$ converge simultaneously.

*Temporal correlation*

Let $u = u(\boldsymbol{x}, t)$ and $v = v(\boldsymbol{x}, t)$ denote two functions that are defined on $\mathbb{R}$. Then, the temporal correlation $C_t'\{u, v\}(\boldsymbol{x}, t)$ of $u$ and $v$ is defined as

$$\begin{aligned} C_t'\{u, v\}(\boldsymbol{x}, t) &= \int_{t'\in\mathbb{R}} u(\boldsymbol{x}, t + t')v(\boldsymbol{x}, t')\mathrm{d}t' \\ &= \int_{t'\in\mathbb{R}} u(\boldsymbol{x}, t')v(\boldsymbol{x}, t' - t)\mathrm{d}t' = C_t'\{v, u\}(\boldsymbol{x}, -t) . \end{aligned} \tag{1.21}$$

Equation (1.21) shows that the correlation is not a symmetrical functional of the two constituent functions. Taking the Laplace transformation of Eq. (1.21), we arrive at

$$\hat{C}_t'\{u,v\}(\boldsymbol{x},s) = \hat{u}(\boldsymbol{x},s)\hat{v}(\boldsymbol{x},-s)\,. \tag{1.22}$$

For Eq. (1.22) to be valid, there must exist a value of $s$ for which the two definition integrals for $\hat{u}(\boldsymbol{x},s)$ and $\hat{v}(\boldsymbol{x},-s)$ converge simultaneously. In wave-field problems dealing with lossless media, this only occurs for the limiting case of imaginary values of the transform parameter $s$.

*Inverse Laplace transformation*

The inverse Laplace transformation can be carried out explicitly by evaluating the following inversion integral (also denoted as the Bromwich integral) in the complex domain

$$\frac{1}{2\pi j}\int_{s_0-j\infty}^{s_0+j\infty} \exp(st)\hat{u}(\boldsymbol{x},s)\mathrm{d}s = \chi_{\mathrm{T}}(t)u(\boldsymbol{x},t)\,, \tag{1.23}$$

where the path of integration is along the line $s = s_0$, $\mathrm{Re}(s_0) > 0$, parallel to the imaginary axis of the complex $s$-domain.

*Temporal Fourier transformation*

In this subsection we consider the consequences for the limiting value when $s \to j\omega$, where $\omega$ is real. Then, the Laplace transform is equivalent with the temporal Fourier transform

$$\hat{u}(\boldsymbol{x},j\omega) = \int_{t\in\mathbb{R}} \exp(-j\omega t)\chi_{\mathrm{T}}(t)u(\boldsymbol{x},t)\mathrm{d}t\,. \tag{1.24}$$

A sufficient condition for the convergence of the integral of Eq. (1.24) is the absolute integrability of $u(\boldsymbol{x},t)$ over the domain T (VAN DER POL and BREMMER, 1950, p. 8). The function $\chi_{\mathrm{T}}(t)u(\boldsymbol{x},t)$ is retrieved from $\hat{u}(\boldsymbol{x},j\omega)$ as

$$\frac{1}{2\pi}\int_{\omega\in\mathbb{R}} \exp(j\omega t)\hat{u}(\boldsymbol{x},j\omega)\mathrm{d}\omega = \chi_{\mathrm{T}}(t)u(\boldsymbol{x},t)\,. \tag{1.25}$$

From Eq. (1.24) we observe that $\hat{u}(\boldsymbol{x},-j\omega) = \hat{u}^{\star}(\boldsymbol{x},j\omega)$, where the star denotes the complex conjugate. Hence, Eq. (1.25) may be written as

$$\frac{1}{\pi}\mathrm{Re}\left[\int_0^{\infty} \exp(j\omega t)\hat{u}(\boldsymbol{x},j\omega)\mathrm{d}\omega\right] = \chi_{\mathrm{T}}(t)u(\boldsymbol{x},t)\,. \tag{1.26}$$

In this book we only consider seismic wave problems. We assume that $\hat{u}(\boldsymbol{x}, j\omega) = 0$ for $\omega = 0$. From Eq. (1.26) we observe that only positive values of $\omega$ have to be considered. Therefore, we restrict our analysis in this book to positive values of $\omega$.

*Asymptotic behavior as $\omega \to \infty$*

The asymptotic behavior of the temporal Fourier transform as the transform parameter goes to infinity is found to be

$$\lim_{\omega \to \infty} \hat{u}(\boldsymbol{x}, j\omega) = 0 \,. \tag{1.27}$$

The result is based on the Riemann-Lebesgue lemma (WHITTAKER and WATSON, 1927, p. 172).

The initial-value theorem becomes now

$$\lim_{\omega \to \infty} \exp(j\omega t_0) j\omega \hat{u}(\boldsymbol{x}, j\omega) = \lim_{t \downarrow t_0} u(\boldsymbol{x}, t) \,. \tag{1.28}$$

Note that the initial-value theorem, previously given by Eq. (1.18), holds not only for $\mathrm{Re}(s) > 0$, but also for $s = j\omega$.

### 1.2.2. Spatial Fourier transformation of a localized function

Let us consider the scalar wavefield quantity $u = u(\boldsymbol{x}, t)$ that is defined in some bounded domain $\mathbb{D}$ in space and let $\hat{u} = \hat{u}(\boldsymbol{x}, s)$ denote its time Laplace transform. The spatial Fourier transform of this localized function $\hat{u}$ over the domain $\mathbb{D}$ is then defined as

$$\tilde{u}(js\boldsymbol{\alpha}, s) = \int_{\boldsymbol{x} \in \mathbb{D}} \exp(js\alpha_q x_q) \hat{u}(\boldsymbol{x}, s) \mathrm{dV} \,, \tag{1.29}$$

where dV is the elementary volume in $\mathbb{R}^3$. We have put the factor $s$ in the exponential function, because this is very convenient for seismic wave problems. Let, further, $\partial\mathbb{D}$ denote the boundary surface of $\mathbb{D}$ and let $\mathbb{D}'$ denote the complement of $\mathbb{D} \cup \partial\mathbb{D}$ in $\mathbb{R}^3$. By introducing the characteristic function $\chi_{\mathbb{D}}(\boldsymbol{x})$ of the set $\mathbb{D}$ as (Fig. 1.3)

$$\chi_{\mathbb{D}}(\boldsymbol{x}) = \{1, \tfrac{1}{2}, 0\} \quad \text{when} \quad \boldsymbol{x} \in \{\mathbb{D}, \partial\mathbb{D}, \mathbb{D}'\} \,. \tag{1.30}$$

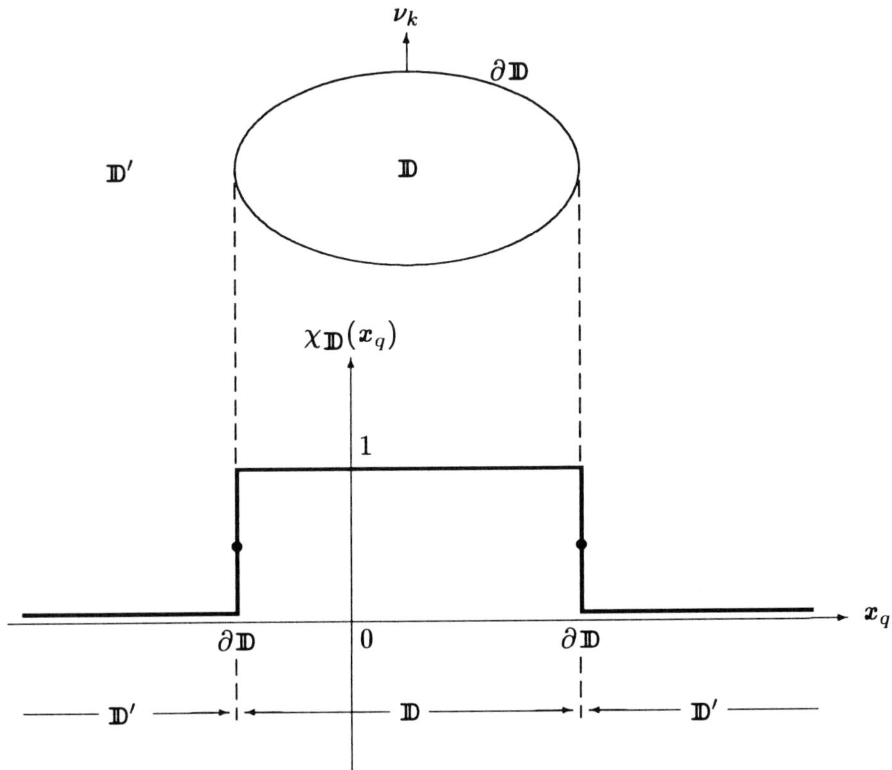


Figure 1.3. Domain $\mathbb{D}$ and its characteristic function $\chi_{\mathbb{D}}$.

Equation (1.29) can also be written as

$$\tilde{u}(js\boldsymbol{\alpha}, s) = \int_{\boldsymbol{x}\in\mathbb{R}^3} \exp(js\alpha_q x_q)\chi_{\mathbb{D}}(\boldsymbol{x})\hat{u}(\boldsymbol{x}, s)\mathrm{dV} . \tag{1.31}$$

In Eqs. (1.29) and (1.31), $\boldsymbol{\alpha}$ is denoted as the angular-slowness vector; in terms of its Cartesian components we have

$$\boldsymbol{\alpha} = \alpha_1\boldsymbol{i}_1 + \alpha_2\boldsymbol{i}_2 + \alpha_3\boldsymbol{i}_3 . \tag{1.32}$$

Although $\boldsymbol{\alpha}$ may be complex, we take $s\boldsymbol{\alpha}$ to be real, i.e., $s\boldsymbol{\alpha} \in \mathbb{R}^3$. In the complex $\alpha_q$-domain ($q = 1, 2, 3$), we take $\alpha_q$ always in the direction of the complex conjugate $s^\star$ of $s$ (see Fig. 1.4). Then, a sufficient condition for the convergence of the definition integral (TITCHMARSH, 1948) is the absolute integrability of $\hat{u}(\boldsymbol{x}, s)$ over the domain $\mathbb{D}$. Since then $\hat{u}(\boldsymbol{x}, s)$ necessarily goes to zero as $|\boldsymbol{x}| \to \infty$ when $\mathbb{D}$ is unbounded, we denote this class of functions also as localized (in space). The Fourier transform domain is also denoted as the angular-slowness-vector domain or the *spectral* domain.

The transformation from the angular-slowness-vector domain back to the spatial domain is carried out by employing the Fourier inversion integral

$$\frac{1}{(2\pi)^3}\int_{s\boldsymbol{\alpha}\in\mathbb{R}^3} \exp(-js\alpha_q x_q)\tilde{u}(js\boldsymbol{\alpha}, s)\mathrm{dV} = \chi_{\mathbb{D}}(\boldsymbol{x})\hat{u}(\boldsymbol{x}, s)\,. \tag{1.33}$$

The integration path in the complex $\boldsymbol{\alpha}$-domain is chosen in such a way that $s\boldsymbol{\alpha}$ is real valued (see Fig. 1.4). The result of the left-hand side of Eq. (1.33) for $\boldsymbol{x} \in \partial\mathbb{D}$ holds on the assumption that $\partial\mathbb{D}$ has a unique tangent plane, while the integral has to interpreted as a Cauchy principal-value integral around infinity (VAN DER POL and BREMMER, 1950, p. 8). In a number of cases the integrations with respect to $s\alpha_1$, $s\alpha_2$ and/or $s\alpha_3$ can be evaluated by employing theorems of the theory of functions of a complex variable.

Next, some elementary rules for the spatial Fourier transformation will be discussed.

*Differentiation with respect to the spatial coordinates*

Let $u = u(\boldsymbol{x}, t)$ denote a function that is defined on $\mathbb{D}$ and that is equal to zero on $\mathbb{D}'$. Let, further, $\hat{u} = \hat{u}(\boldsymbol{x}, s)$ denote its Laplace transform. Then, the spatial Fourier transform of the spatial derivative $\partial_k\hat{u}$ is found as

$$\begin{aligned}
&\int_{\boldsymbol{x}\in\mathbb{D}} \exp(js\alpha_q x_q)\partial_k\hat{u}(\boldsymbol{x}, s)\mathrm{dV} \\
&\quad = \int_{\boldsymbol{x}\in\mathbb{D}} \{\partial_k[\exp(js\alpha_q x_q)\hat{u}(\boldsymbol{x}, s)] - [\partial_k\exp(js\alpha_q x_q)]\hat{u}(\boldsymbol{x}, s)\}\,\mathrm{dV} \\
&\quad = \int_{\boldsymbol{x}\in\partial\mathbb{D}} \exp(js\alpha_q x_q)\hat{u}(\boldsymbol{x}, s)\nu_k\mathrm{dA} - js\alpha_k\tilde{u}(js\boldsymbol{\alpha}, s)\,,
\end{aligned} \tag{1.34}$$

where Gauss' integral theorem has been used to arrive at the integral over $\partial\mathbb{D}$; $\nu_k$ denotes the unit vector along the normal to $\partial\mathbb{D}$ pointing away from $\mathbb{D}$, and the value of $\hat{u}$ on $\partial\mathbb{D}$ is the limiting value approaching $\partial\mathbb{D}$ via $\mathbb{D}$. Further, the spatial Fourier transform of the spatial derivative of $\chi_D(\boldsymbol{x})\hat{u}(\boldsymbol{x}, s)$ is found as

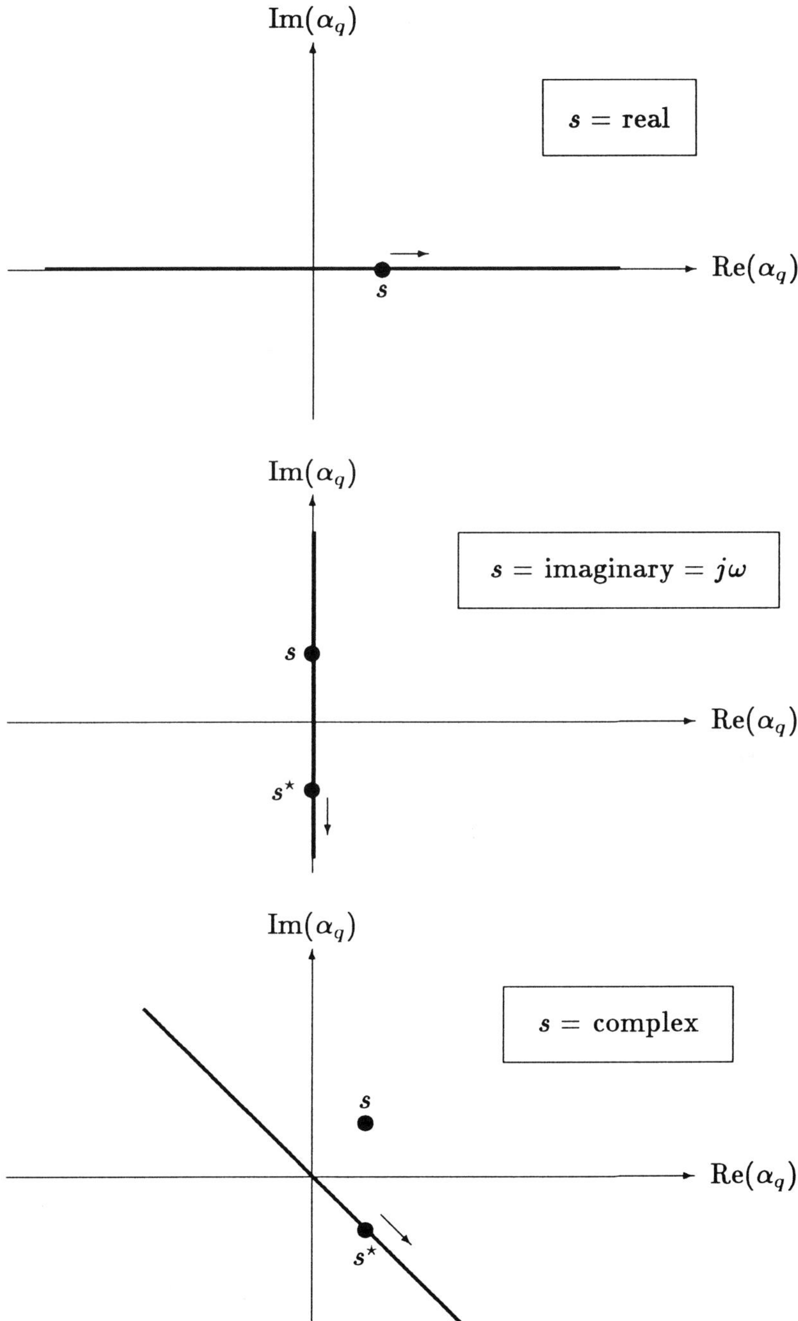


Figure 1.4. Integration path in the spectral domain.

$$\int_{\boldsymbol{x}\in\mathbb{R}^3} \exp(js\alpha_q x_q)\partial_k[\chi_{\mathbb{D}}(\boldsymbol{x})\hat{u}(\boldsymbol{x},s)]\mathrm{dV}$$

$$= \int_{\boldsymbol{x}\in\mathbb{R}^3} \exp(js\alpha_q x_q)[\partial_k\chi_{\mathbb{D}}(\boldsymbol{x})]\hat{u}(\boldsymbol{x},s)\mathrm{dV} + \int_{\boldsymbol{x}\in\mathbb{R}^3} \exp(js\alpha_q x_q)\chi_{\mathbb{D}}(\boldsymbol{x})\partial_k\hat{u}(\boldsymbol{x},s)\mathrm{dV}$$

$$= -\int_{\boldsymbol{x}\in\partial\mathbb{D}} \exp(js\alpha_q x_q)\nu_k\hat{u}(\boldsymbol{x},s)\mathrm{dA} + \int_{\boldsymbol{x}\in\mathbb{D}} \exp(js\alpha_q x_q)\partial_k\hat{u}(\boldsymbol{x},s)\mathrm{dV}$$

$$= -js\alpha_k\tilde{u}(js\boldsymbol{\alpha},s)\,, \tag{1.35}$$

where Eq. (1.34) has been used. In the derivation we have also used the property that $\partial_k\chi_{\mathbb{D}}(\boldsymbol{x})$ has a spatial unit impulse function (Dirac distribution) behavior in the opposite direction of the unit vector along the normal to $\partial\mathbb{D}$ pointing away from $\mathbb{D}$.

Equation (1.35) exemplifies that the transformation rules find their simplest expression when $\mathbb{R}^3$ is taken as the domain of $\hat{u}$.

*Asymptotic behavior as* $|s\boldsymbol{\alpha}| \to \infty$

The asymptotic behavior of the spatial Fourier transform as the transform parameter goes to infinity is found to be

$$\lim_{|s\boldsymbol{\alpha}|\to\infty} \tilde{u}(js\boldsymbol{\alpha},s) = 0\,, \quad |s\boldsymbol{\alpha}| \in \mathbb{R}^3\,. \tag{1.36}$$

The result of Eq. (1.36) is based on the Riemann-Lebesgue lemma (WHITTAKER and WATSON, 1927, p. 172).

*Spatial convolution*

Let $\hat{u} = \hat{u}(\boldsymbol{x},s)$ and $\hat{v} = \hat{v}(\boldsymbol{x},s)$ denote two $s$-domain functions that are defined on $\mathbb{R}^3$. Then, the spatial convolution $C_{\boldsymbol{x}}\{\hat{u},\hat{v}\}(\boldsymbol{x},s)$ of $\hat{u}$ and $\hat{v}$ is defined as

$$C_{\boldsymbol{x}}\{\hat{u},\hat{v}\}(\boldsymbol{x},s) = \int_{\boldsymbol{x}'\in\mathbb{R}^3} \hat{u}(\boldsymbol{x}-\boldsymbol{x}',s)\hat{v}(\boldsymbol{x}',s)\mathrm{dV}$$

$$= \int_{\boldsymbol{x}' \in \mathbb{R}^3} \hat{u}(\boldsymbol{x}', s)\hat{v}(\boldsymbol{x} - \boldsymbol{x}', s)\mathrm{dV} = C_{\boldsymbol{x}}\{\hat{v}, \hat{u}\}(\boldsymbol{x}, s) . \tag{1.37}$$

Equation (1.37) shows that the convolution is a symmetrical functional of the two constituent functions. Taking the Fourier transformation of Eq. (1.37), we arrive at

$$\tilde{C}_{\boldsymbol{x}}\{\hat{u}, \hat{v}\}(js\boldsymbol{\alpha}, s) = \tilde{u}(js\boldsymbol{\alpha}, s)\tilde{v}(js\boldsymbol{\alpha}, s) . \tag{1.38}$$

For Eq. (1.38) to be valid, there must exist a value of $s\boldsymbol{\alpha}$ for which the two definition integrals for $\tilde{u}$ and $\tilde{v}$ converge simultaneously. For absolutely integrable functions this is the case (for real values of $s\boldsymbol{\alpha}$).

*Spatial correlation*

Let $\hat{u} = \hat{u}(\boldsymbol{x}, s)$ and $\hat{v} = \hat{v}(\boldsymbol{x}, s)$ denote two $s$-domain functions that are defined on $\mathbb{R}^3$. Then, the spatial correlation $C'_{\boldsymbol{x}}\{\hat{u}, \hat{v}\}(\boldsymbol{x}, s)$ of $\hat{u}$ and $\hat{v}$ is defined as

$$\begin{aligned} C'_{\boldsymbol{x}}\{\hat{u}, \hat{v}\}(\boldsymbol{x}, s) &= \int_{\boldsymbol{x}' \in \mathbb{R}^3} \hat{u}(\boldsymbol{x} + \boldsymbol{x}', s)\hat{v}(\boldsymbol{x}', s)\mathrm{dV} \\ &= \int_{\boldsymbol{x}' \in \mathbb{R}^3} \hat{u}(\boldsymbol{x}', s)\hat{v}(\boldsymbol{x}' - \boldsymbol{x}, s)\mathrm{dV} = C'_{\boldsymbol{x}}\{\hat{v}, \hat{u}\}(-\boldsymbol{x}, s) . \end{aligned} \tag{1.39}$$

Equation (1.39) shows that the correlation is not a symmetrical functional of the two constituent functions. Taking the Fourier transformation of Eq. (1.39), we arrive at

$$\tilde{C}'_{\boldsymbol{x}}\{\hat{u}, \hat{v}\}(js\boldsymbol{\alpha}, s) = \tilde{u}(js\boldsymbol{\alpha}, s)\tilde{v}(-js\boldsymbol{\alpha}, s) . \tag{1.40}$$

For Eq. (1.40) to be valid, there must exist a value of $s\boldsymbol{\alpha}$ for which the two definition integrals for $\tilde{u}$ and $\tilde{v}$ converge simultaneously. For absolutely integrable functions this is the case (for real values of $s\boldsymbol{\alpha}$).

*Special case of imaginary s*

In the special case of $s = j\omega$ the slowness vector must be taken imaginary. We therefore introduce the real slowness vector

$$\boldsymbol{p} = p_1\boldsymbol{i}_1 + p_2\boldsymbol{i}_2 + p_3\boldsymbol{i}_3 , \tag{1.41}$$

as

$$\boldsymbol{p} = j\boldsymbol{\alpha} \,. \tag{1.42}$$

Hence, the transform pair of Eqs. (1.31) and (1.33) is rewritten as

$$\tilde{u}(j\omega\boldsymbol{p}, j\omega) = \int_{\boldsymbol{x}\in\mathbb{R}^3} \exp(j\omega p_q x_q)\chi_{\mathbb{D}}(\boldsymbol{x})\hat{u}(\boldsymbol{x}, j\omega)\mathrm{dV} \,, \tag{1.43}$$

$$\left(\frac{\omega}{2\pi}\right)^3 \int_{\boldsymbol{p}\in\mathbb{R}^3} \exp(-j\omega p_q x_q)\tilde{u}(j\omega\boldsymbol{p}, j\omega)\mathrm{dV} = \chi_{\mathbb{D}}(\boldsymbol{x})\hat{u}(\boldsymbol{x}, j\omega) \,. \tag{1.44}$$

Equation (1.43) represents the temporal Fourier transform of the Radon transform. To show this, we transform Eq. (1.43) back to the time domain. Using Eq. (1.25) we arrive at

$$\breve{u}(\boldsymbol{p}, t) = \int_{\boldsymbol{x}\in\mathbb{R}^3} \chi_{\mathbb{D}}(\boldsymbol{x})\chi_T(t + p_q x_q)u(\boldsymbol{x}, t + p_q x_q)\mathrm{dV} \,. \tag{1.45}$$

The latter represents the three-dimensional Radon transformation (DEANS, 1983; CHAPMAN, 1981). It is clear that the Radon transformation can conveniently be carried out via the temporal Fourier domain, using Eq. (1.43).

### 1.2.3. Spatial Fourier transformation with respect to the horizontal coordinates

It is common in seismic problems to assign the $x_3$-coordinate to the vertical depth position. Then, $x_1$ and $x_2$ represent the horizontal positions. Very often the spatial Fourier transform of a bounded function with respect to these horizontal coordinates is used. This transform pair is defined as

$$\bar{u}(js\alpha_1, js\alpha_2, x_3, s) = \int_{(x_1,x_2)\in\mathbb{R}^2} \exp(js\alpha_1 x_1 + js\alpha_2 x_2)\hat{u}(x_1, x_2, x_3, s)\mathrm{dA} \,, \tag{1.46}$$

$$\frac{1}{(2\pi)^2} \int_{(s\alpha_1,s\alpha_2)\in\mathbb{R}^2} \exp(-js\alpha_1 x_1 - js\alpha_2 x_2)\bar{u}(js\alpha_1, js\alpha_2, x_3, s)\mathrm{dA} = \hat{u}(x_1, x_2, x_3, s) \,. \tag{1.47}$$

*Special case of real s*

In the special case of real $s$, the slowness vector $\boldsymbol{\alpha}$ must be taken real as well. This case occurs in the generalized-ray theory, e.g., the Cagniard-de Hoop method (see Chapter 4). Then the inverse transform, defined by Eq. (1.47), may be written as

$$\left(\frac{s}{2\pi}\right)^2 \int_{(\alpha_1,\alpha_2)\in\mathbb{R}^2} \exp(-js\alpha_1 x_1 - js\alpha_2 x_2)\bar{u}(js\alpha_1, js\alpha_2, x_3, s)\mathrm{dA}$$
$$= \hat{u}(x_1, x_2, x_3, s)\,. \tag{1.48}$$

*Special case of imaginary s*

When $s$ approaches the imaginary value, the introduction of $s = j\omega$ and $p_1 = j\alpha_1$, $p_2 = j\alpha_2$ leads to the transform pair

$$\bar{u}(j\omega p_1, j\omega p_2, x_3, j\omega)$$
$$= \int_{(x_1,x_2)\in\mathbb{R}^2} \exp(j\omega p_1 x_1 + j\omega p_2 x_2)\hat{u}(x_1, x_2, x_3, j\omega)\mathrm{dA}\,, \tag{1.49}$$

$$\left(\frac{\omega}{2\pi}\right)^2 \int_{(p_1,p_2)\in\mathbb{R}^2} \exp(-j\omega p_1 x_1 - j\omega p_2 x_2)\bar{u}(j\omega p_1, j\omega p_2, x_3, j\omega)\mathrm{dA}$$
$$= \hat{u}(x_1, x_2, x_3, j\omega)\,. \tag{1.50}$$

Equation (1.49) represents the temporal Fourier transform of the Radon transform with respect to the horizontal coordinates. To show this, we transform Eq. (1.49) back to the time domain. Using Eq. (1.25) we arrive at

$$\check{u}(p_1, p_2, x_3, t) = \int_{(x_1,x_2)\in\mathbb{R}^2} \chi_T(t+p_1x_1+p_2x_2)u(x_1, x_2, x_3, t+p_1x_1+p_2x_2)\mathrm{dA}\,. \tag{1.51}$$

The latter represents the two-dimensional Radon transformation with respect to the horizontal coordinates $x_1$, $x_2$ (DEANS, 1983; MCCOWAN AND BRYSK, 1982). It is advantageous to carry out this Radon transformation via the temporal Fourier domain (FOKKEMA *et al.*, 1992), using Eq. (1.49).

## 1.3. Discrete Fourier-transformation methods

In seismic data processing we need the discrete counterparts of the pertinent Fourier transforms. In the numerical treatment that leads to the discrete Fourier transform, we require that the symmetry properties of the continuous Fourier transform are maintained. To that end we first identify these properties.

*Symmetry properties of the continuous Fourier transform*

The more-dimensional spatial Fourier transform, previously given by Eq. (1.29), is considered as a repeated version of a one-dimensional Fourier transformation. The temporal Fourier transform is already of the one-dimensional type (cf. Eq. (1.24)). This allows us to focus the discussion on the one-dimensional case. The Fourier transform of $u = u(x)$ is generally given by

$$U(\alpha) = F\{u\}(\alpha) = \int_{x\in\mathbb{R}} \exp(j2\pi\alpha x)u(x)\mathrm{d}x\,, \tag{1.52}$$

while the function $u(x)$ is retrieved from $U(\alpha)$ by employing the Fourier inversion integral with the result

$$F^{-1}\{U\}(x) = \int_{\alpha\in\mathbb{R}} \exp(-j2\pi\alpha x)U(\alpha)\mathrm{d}\alpha = u(x)\,. \tag{1.53}$$

The function $u(x)$ is considered as the sum of an even part $u^e(x)$ and an odd part $u^o(x)$,

$$u(x) = u^e(x) + u^o(x)\,. \tag{1.54}$$

The even part is given by

$$u^e(x) = \frac{1}{2}u(x) + \frac{1}{2}u(-x) \tag{1.55}$$

and the odd part is given by

$$u^o(x) = \frac{1}{2}u(x) - \frac{1}{2}u(-x)\,, \tag{1.56}$$

with the properties

$$u^e(x) = u^e(-x) \tag{1.57}$$

and

$$u^o(x) = -u^o(-x)\,. \tag{1.58}$$

Substituting this decomposition in even and odd parts in Eq. (1.52) we arrive at

$$U(\alpha) = U^e(\alpha) + U^o(\alpha)\,, \tag{1.59}$$

where the even and odd parts follow from

$$U^e(\alpha) = \frac{1}{2}U(\alpha) + \frac{1}{2}U(-\alpha) \tag{1.60}$$

and

$$U^o(\alpha) = \frac{1}{2}U(\alpha) - \frac{1}{2}U(-\alpha)\,. \tag{1.61}$$

Consequently, the even parts in the $x$-domain and $\alpha$-domain are related through

$$U^e(\alpha) = \int_{x\in\mathbb{R}} \exp(j2\pi\alpha x)u^e(x)\mathrm{d}x \tag{1.62}$$

and

$$\int_{\alpha\in\mathbb{R}} \exp(-j2\pi\alpha x)U^e(\alpha)\mathrm{d}\alpha = u^e(x)\,, \tag{1.63}$$

while the odd parts are linked through

$$U^o(\alpha) = \int_{x\in\mathbb{R}} \exp(j2\pi\alpha x)u^o(x)\mathrm{d}x \tag{1.64}$$

and

$$\int_{\alpha\in\mathbb{R}} \exp(-j2\pi\alpha x)U^o(\alpha)\mathrm{d}\alpha = u^o(x)\,. \tag{1.65}$$

The relation between even and odd parts are consequences of the fact that the Fourier transform and its inverse are symmetrical operators in the $x$-domain and the $\alpha$-domain, respectively. We require that these symmetry properties are conserved in the numerical discretization.

*Real functions in the $x$-domain*

In the case that $u(x)$ is real, $U^e(\alpha)$ is real and $U^o(\alpha)$ is imaginary valued, which allows us to rewrite Eqs. (1.63) and (1.65) as

$$2\,\mathrm{Re}\left[\int_{\alpha\in\mathbb{R}} \exp(-j2\pi\alpha x)\chi_{\mathbb{R}^+}(\alpha)U^e(\alpha)\mathrm{d}\alpha\right] = u^e(x) \tag{1.66}$$

and

$$2\,\mathrm{Re}\left[\int_{\alpha\in\mathbb{R}} \exp(-j2\pi\alpha x)\chi_{\mathbb{R}^+}(\alpha)U^o(\alpha)\mathrm{d}\alpha\right] = u^o(x)\,. \tag{1.67}$$

In Eqs. (1.66) and (1.67) we have used the definition (cf. Eq. (1.30)) of the characteristic function $\chi_{\mathbb{R}^+}$, related to the domain $\mathbb{R}^+$ given by

$$\mathbb{R}^+ = \{\alpha \in \mathbb{R};\ \alpha > 0\}\,. \tag{1.68}$$

From Eqs. (1.54), (1.59), (1.66) and (1.67) it follows that, when $u(x)$ is real, Eq. (1.53) can be rewritten as (cf. Eq. (1.26))

$$2\,\mathrm{Re}\left[\int_{\alpha\in\mathbb{R}} \exp(-j2\pi\alpha x)\chi_{\mathbb{R}^+}(\alpha)U(\alpha)\mathrm{d}\alpha\right] = u(x)\,. \tag{1.69}$$

It is noted that in the numerical discretization a symmetrical domain around $\alpha = 0$ has to be chosen in order to compute Eq. (1.69) numerically. This requires space allocation of zero values for negative values of $\alpha$.

*Step functions in the $x$-domain*

The step function $w(x)$ is defined on the domain $\mathbb{D} = \{x \in \mathbb{R}; 0 < x_0 < x\}$ and given by means of the characteristic function $\chi_{\mathbb{D}}$ (see Fig. 1.5) according to

$$w(x) = \chi_{\mathbb{D}}(x)u(x)\,. \tag{1.70}$$

The even part $w^e$ and odd part $w^o$ are given by

$$w^e(x) = \frac{1}{2}\chi_{\mathbb{D}}(x)u(x) + \frac{1}{2}\chi_{\mathbb{D}}(-x)u(-x) \tag{1.71}$$

and

$$w^o(x) = \frac{1}{2}\chi_{\mathbb{D}}(x)u(x) - \frac{1}{2}\chi_{\mathbb{D}}(-x)u(-x)\,. \tag{1.72}$$

Since

$$w(x) = 2\chi_{\mathbb{R}^+}(x)w^e(x) = 2\chi_{\mathbb{R}^+}(x)w^o(x)\,, \tag{1.73}$$

we deduce from Eqs. (1.63) and (1.65) that $w(x)$ either follows from the even part or from the odd part of the Fourier-transform counterpart in the $\alpha$-domain. Hence

$$2\chi_{\mathbb{R}^+}(x)\int_{\alpha\in\mathbb{R}} \exp(-j2\pi\alpha x)W^e(\alpha)\mathrm{d}\alpha = w(x)\,, \tag{1.74}$$

or

$$2\chi_{\mathbb{R}^+}(x)\int_{\alpha\in\mathbb{R}} \exp(-j2\pi\alpha x)W^o(\alpha)\mathrm{d}\alpha = w(x)\,, \tag{1.75}$$

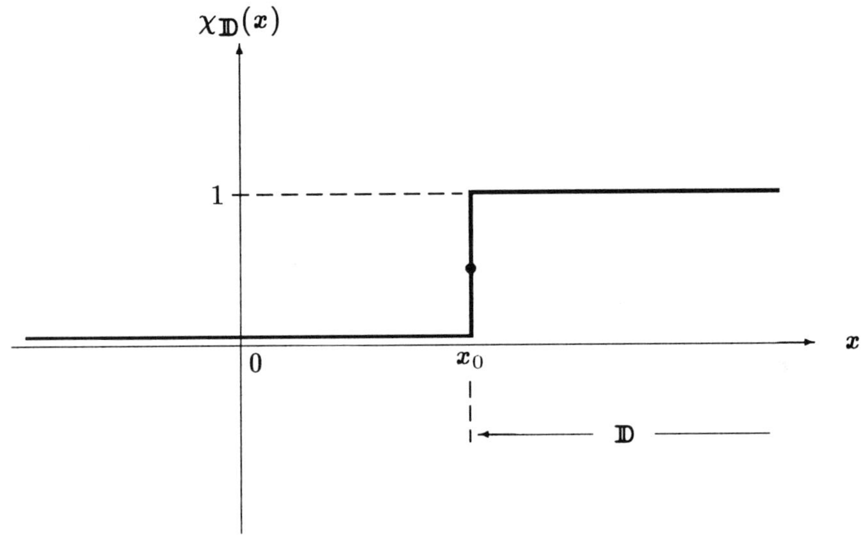


Figure 1.5. The domain $\mathbb{D}$ and its characteristic function $\chi_{\mathbb{D}}$.

in which $W^e(\alpha)$ and $W^o(\alpha)$ are the even and odd part of $W(\alpha)$, respectively. Note that for this class of functions, we can reconstruct the function $w(x)$ either from Eq. (1.74) or from Eq. (1.75).

*Real step functions in the $x$-domain*

When $w(x)$ is real, the results from Eqs. (1.66) and (1.67) can be combined with those from Eqs. (1.74) and (1.75) yielding

$$4\chi_{\mathbb{R}^+}(x)\mathrm{Re}\left[\int_{\alpha\in\mathbb{R}} \exp(-j2\pi\alpha x)\chi_{\mathbb{R}^+}(\alpha)W^e(\alpha)\mathrm{d}\alpha\right] = w(x)\,, \qquad (1.76)$$

or

$$4\chi_{\mathbb{R}^+}(x)\mathrm{Re}\left[\int_{\alpha\in\mathbb{R}} \exp(-j2\pi\alpha x)\chi_{\mathbb{R}^+}(\alpha)W^o(\alpha)\mathrm{d}\alpha\right] = w(x)\,. \qquad (1.77)$$

It is noted that in the numerical discretization a symmetrical domain around $\alpha = 0$ has to be chosen in order to compute either Eq. (1.76) or (1.77) numerically. This requires space allocation of zero values for negative values of $\alpha$.

*Discretization*

In seismic processing methods, integral transformations of the type of Eqs. (1.52) and (1.53) are carried out by means of numerical techniques. We only know the pertinent function values $u(x)$ in discrete points in a finite interval

$$\mathbb{D} = \{x \in \mathbb{R};\ x_{min}^{(u)} \leq x \leq x_{max}^{(u)}\} . \tag{1.78}$$

To study the consequences of the discretization, we depart from the continuous Fourier transform given by Eq. (1.52) and consider a discretization in $\alpha$ with sample interval $\Delta\alpha$

$$U_m = \int_{x\in\mathbb{R}} \exp(j2\pi m\Delta\alpha x)u(x)\mathrm{d}x , \tag{1.79}$$

where $U_m$ is defined as

$$U_m = U(m\Delta\alpha) . \tag{1.80}$$

Next, we write the infinite integral on the right-hand side of Eq. (1.79) as an infinite sum of integrals over finite intervals with length $\frac{1}{\Delta\alpha}$,

$$U_m = \sum_{n=-\infty}^{\infty} \int_{\frac{2n-1}{2\Delta\alpha}}^{\frac{2n+1}{2\Delta\alpha}} \exp(j2\pi m\Delta\alpha x)u(x)\mathrm{d}x . \tag{1.81}$$

Note that the center positions of the finite intervals are chosen such that the point of symmetry $x = 0$ of the original transformation, Eq. (1.79), is contained in the contribution for $n = 0$. This is necessary to guarantee that the symmetry properties are maintained in the discretized version. Changing the variable of integration, we rewrite Eq. (1.81) as

$$U_m = \sum_{n=-\infty}^{\infty} \int_{\frac{-1}{2\Delta\alpha}}^{\frac{1}{2\Delta\alpha}} \exp(j2\pi m\Delta\alpha x)u(x + \frac{n}{\Delta\alpha})\mathrm{d}x . \tag{1.82}$$

We interchange the order of summation and integration, leading to

$$U_m = \int_{\frac{-1}{2\Delta\alpha}}^{\frac{1}{2\Delta\alpha}} \exp(j2\pi m\Delta\alpha x)u^{per}(x)\mathrm{d}x , \tag{1.83}$$

where $u^{per}$ is a periodic function with period $\frac{1}{\Delta\alpha}$, given by

$$u^{per}(x) = \sum_{n=-\infty}^{\infty} u(x + \frac{n}{\Delta\alpha}) . \tag{1.84}$$

In view of the symmetry requirements of the Fourier transform we define a symmetric domain around $x = 0$. To this end we extend the supporting domain $\mathbb{D}$ to the extended domain

$$\mathbb{D}_E = \{x \in \mathbb{R};\ |x| \leq X^E\}\,, \tag{1.85}$$

where

$$X^E = \max\left(|x^{(u)}_{min}|, |x^{(u)}_{max}|\right)\,. \tag{1.86}$$

This is accomplished by extending the definition of the function $u(x)$ as

$$u^E(x) = \begin{cases} u(x)\,, & x \in \mathbb{D}\,, \\ 0\,, & x \notin \mathbb{D}\,. \end{cases} \tag{1.87}$$

It is convenient to take

$$\frac{1}{2\Delta\alpha} = X^E\,. \tag{1.88}$$

Taking this value for $\Delta\alpha$ we observe that $u^{per}(x) = u^E(x)$ for $|x| < \frac{1}{2\Delta\alpha}$. Hence, we rewrite Eq. (1.83) as

$$U^E_m = \int_{-\frac{1}{2\Delta\alpha}}^{\frac{1}{2\Delta\alpha}} \exp(j2\pi m\Delta\alpha x)u^E(x)\mathrm{d}x\,. \tag{1.89}$$

The integral on the right-hand side of Eq. (1.89) is computed by employing the trapezoidal integration rule (RALSTON and RABINOWITZ, 1978, p. 120) of a periodic function with $N$ sample intervals of length $\Delta x$. This procedure leads to the discrete Fourier transform $U^E_m$ of the extended discrete function $u^E_n$,

$$U^E_m = \Delta x \sum_{n=-\frac{1}{2}N+1}^{\frac{1}{2}N} \exp(j2\pi mn\Delta\alpha\Delta x)u^E_n\,, \tag{1.90}$$

with

$$u^E_n = u^E(n\Delta x)\,, \tag{1.91}$$

and

$$\Delta x = \frac{1}{N\Delta\alpha} = \frac{2X^E}{N}\,. \tag{1.92}$$

Using Eq. (1.92) in Eq. (1.90), we arrive at the final result

$$U^E_m = DF_N\{u^E_n\}(m) = \Delta x \sum_{n=-\frac{1}{2}N+1}^{\frac{1}{2}N} \exp(j2\pi\frac{mn}{N})u^E_n\,,$$

$$m = -\tfrac{1}{2}N+1, \cdots, \tfrac{1}{2}N\,. \tag{1.93}$$

In numerical computations, the values of $U_m^E$ are taken as approximate results of the Fourier transform of the localized function defined on $\mathbb{D}$. The inverse discrete Fourier transformation leads to

$$DF_N^{-1}\{U_m^E\}(n) = \Delta\alpha \sum_{m=-\frac{1}{2}N+1}^{\frac{1}{2}N} \exp(-j2\pi\frac{mn}{N})U_m^E = u_n^E \,,$$

$$n = -\tfrac{1}{2}N + 1, \cdots, \tfrac{1}{2}N \,. \tag{1.94}$$

The original function is then recovered as

$$u(x) = u_n^E, \quad x = n\Delta x, \quad x \in \mathbb{D} \,. \tag{1.95}$$

Note that our definitions of the discrete Fourier transforms are different from the standard ones, as far as normalization and numbering are concerned (OPPENHEIM *et al.*, 1983, p. 297). However, we prefer the situation where the normalization and numbering of the discrete transform is closely related to the definition and symmetry properties of the continuous transform.

*Discrete convolution*

The continuous convolution of the functions $u(x)$ and $v(x)$ is given by the expression (cf. Eqs. (1.19) and (1.37))

$$C_x\{u,v\}(x) = \int_{x'\in\mathbb{R}} u(x-x')v(x')\mathrm{d}x' \,. \tag{1.96}$$

We assume that the supporting domains of $u(x)$ and $v(x)$ are given by

$$\mathbb{D}_u = \{x \in \mathbb{R};\; x_{min}^{(u)} \le x \le x_{max}^{(u)}\} \tag{1.97}$$

and

$$\mathbb{D}_v = \{x \in \mathbb{R};\; x_{min}^{(v)} \le x \le x_{max}^{(v)}\} \,, \tag{1.98}$$

respectively. By inspection it follows that the supporting domain of the function $C_x\{u,v\}(x)$ is given by

$$\mathbb{D}_C = \{x \in \mathbb{R};\; x_{min}^{(u)} + x_{min}^{(v)} \le x \le x_{max}^{(u)} + x_{max}^{(v)}\} \,. \tag{1.99}$$

Outside this domain the convolution $C_x\{u,v\}(x) = 0$. In view of the symmetry properties of the Fourier transform we introduce an extended, symmetric domain common for the functions $u(x)$, $v(x)$ and $C_x\{u,v\}(x)$, i.e.,

$$\mathbb{D}_E = \{x \in \mathbb{R};\; |x| \le X^E\} \,, \tag{1.100}$$

where

$$X^E = \max\left(|x_{min}^{(u)} + x_{min}^{(v)}|, |x_{max}^{(u)} + x_{max}^{(v)}|, |x_{min}^{(u)}|, |x_{max}^{(u)}|, |x_{min}^{(v)}|, |x_{max}^{(v)}|\right) . \tag{1.101}$$

This is accomplished by extending the definitions of the functions $u(x)$ and $v(x)$ as

$$u^E(x) = \begin{cases} u(x)\,, & x \in \mathbb{D}_u\,, \\ 0\,, & x \notin \mathbb{D}_u\,, \end{cases} \tag{1.102}$$

and

$$v^E(x) = \begin{cases} v(x)\,, & x \in \mathbb{D}_v\,, \\ 0\,, & x \notin \mathbb{D}_v\,. \end{cases} \tag{1.103}$$

Observing that $u^E(x - x')v^E(x')$ vanishes outside the extended domain $\mathbb{D}_E$, Eq. (1.96) is replaced by

$$C_x\{u^E, v^E\}(x) = \int_{x' \in \mathbb{D}_E} u^E(x - x')v^E(x')\mathrm{d}x' . \tag{1.104}$$

The integral on the right-hand side of Eq. (1.104) is computed by employing the trapezoidal integration rule with $N$ sample intervals of length $\Delta x$. We arrive at the discrete convolution

$$C_n = \Delta x \sum_{i=-\frac{1}{2}N+1}^{\frac{1}{2}N} u_{n-i}^E v_i^E\,, \tag{1.105}$$

where

$$C_n = C_x\{u^E, v^E\}(n\Delta x)\,, \tag{1.106}$$

$$u_n^E = u^E(n\Delta x)\,, \tag{1.107}$$

$$v_n^E = v^E(n\Delta x)\,, \tag{1.108}$$

in which

$$\Delta x = \frac{2X^E}{N} . \tag{1.109}$$

Using the definitions of the discrete Fourier transforms we observe that

$$DF_N\{C_n\}(m) = DF_N\{u_n^E\}(m)\; DF_N\{v_n^E\}(m) . \tag{1.110}$$

The final result is obtained as

$$C_n = DF_N^{-1}\left\{DF_N\{u_n^E\}(m)\; DF_N\{v_n^E\}(m)\right\}(n)\,, \tag{1.111}$$

for $n = -\frac{1}{2}N + 1, \cdots, \frac{1}{2}N$. In numerical computations, the values of $C_n$ are taken as approximate results of $C_x\{u,v\}(x)$, $x = n\Delta x$ and $x \in \mathbb{D}_E$.

*Discrete correlation*

The continuous correlation of the functions $u(x)$ and $v(x)$ is given by the expression (cf. Eqs. (1.21) and (1.39))

$$C'_x\{u,v\}(x) = \int_{x' \in \mathbb{R}} u(x+x')v(x')\mathrm{d}x' \,. \tag{1.112}$$

We assume that the supporting domains of $u(x)$ and $v(x)$ are given by

$$\mathbb{D}_u = \{x \in \mathbb{R};\ x^{(u)}_{min} \leq x \leq x^{(u)}_{max}\} \tag{1.113}$$

and

$$\mathbb{D}_v = \{x \in \mathbb{R};\ x^{(v)}_{min} \leq x \leq x^{(v)}_{max}\}\,, \tag{1.114}$$

respectively. By inspection it follows that the supporting domain of the function $C'_x\{u,v\}(x)$ is given by

$$\mathbb{D}_{C'} = \{x \in \mathbb{R};\ x^{(u)}_{min} - x^{(v)}_{max} \leq x \leq x^{(u)}_{max} - x^{(v)}_{min}\}\,. \tag{1.115}$$

Outside this domain the correlation $C'_x\{u,v\}(x) = 0$. In view of the symmetry properties of the Fourier transform we introduce an extended, symmetric domain common for the functions $u(x)$, $v(x)$ and $C'_x\{u,v\}(x)$, i.e.,

$$\mathbb{D}_E = \{x \in \mathbb{R};\ |x| \leq X^E\}\,, \tag{1.116}$$

where

$$X^E = \max\left(|x^{(u)}_{min} - x^{(v)}_{max}|, |x^{(u)}_{max} - x^{(v)}_{min}|, |x^{(u)}_{min}|, |x^{(u)}_{max}|, |x^{(v)}_{min}|, |x^{(v)}_{max}|\right)\,. \tag{1.117}$$

This is accomplished by extending the definitions of the functions $u(x)$ and $v(x)$ as

$$u^E(x) = \begin{cases} u(x)\,, & x \in \mathbb{D}_u\,, \\ 0\,, & x \notin \mathbb{D}_u\,, \end{cases} \tag{1.118}$$

and

$$v^E(x) = \begin{cases} v(x)\,, & x \in \mathbb{D}_v\,, \\ 0\,, & x \notin \mathbb{D}_v\,. \end{cases} \tag{1.119}$$

Observing that $u^E(x')v^E(x'-x)$ vanishes outside the extended domain $\mathbb{D}_E$, Eq. (1.112) is replaced by

$$C'_x\{u^E, v^E\}(x) = \int_{x' \in \mathbb{D}_E} u^E(x')v^E(x'-x)\mathrm{d}x' . \tag{1.120}$$

The integral on the right-hand side of Eq. (1.120) is computed by employing the trapezoidal integration rule with $N$ sample intervals of length $\Delta x$. We then arrive at the discrete correlation

$$C'_n = \Delta x \sum_{i=-\frac{1}{2}N+1}^{\frac{1}{2}N} u^E_{n+i} v^E_i , \tag{1.121}$$

where

$$C'_n = C'_x\{u^E, v^E\}(n\Delta x) , \tag{1.122}$$

$$u^E_n = u^E(n\Delta x) , \tag{1.123}$$

$$v^E_n = v^E(n\Delta x) , \tag{1.124}$$

in which

$$\Delta x = \frac{2X^E}{N} . \tag{1.125}$$

Using the definitions of the discrete Fourier transforms we observe that

$$DF_N\{C'_n\}(m) = DF_N\{u^E_n\}(m)\, DF_N\{v^E_{-n}\}(m) . \tag{1.126}$$

The final result is obtained as

$$C'_n = DF_N^{-1}\left\{DF_N\{u^E_n\}(m)\, DF_N\{v^E_{-n}\}(m)\right\}(n) , \tag{1.127}$$

for $n = -\frac{1}{2}N+1, \cdots, \frac{1}{2}N$. In numerical computations, the values of $C'_n$ are taken as approximate results of $C'_x\{u, v\}(x)$, $x = n\Delta x$ and $x \in \mathbb{D}_E$.

# Chapter 2

# Iterative Solution of Integral Equations

Many problems in seismic wave motion are formulated through integral equations. In the present chapter the iterative solution of an integral equation is discussed. To have a measure for the accuracy attained, we select the global root-mean-square error in the equality sign of the integral equation that has to be satisfied by the exact solution. For a given sequence of expansion functions used to represent the unknown wavefield quantities, the minimization of the relevant error leads to a particular method of moments. For configurations of realistic size and degree of complexity, this leads to the numerical solution of a large system of linear algebraic equations which is intractable for a direct numerical implementation. In this chapter we develop some iterative techniques, where the intermediate step of the solution of a large system of equations is superfluous. In all the iterative techniques we take the integrated square error with respect to the original operator equation as a measure of deviation of the approximate solution from the exact one. Variational techniques are employed to arrive at a minimum error. Several schemes are presented to control the successive decline of the error. A symmetrization and preconditioning procedure of the integral equation is also discussed. In the special case that the operator is of the convolution type, Fourier transformations can be used advantageously to compute the operator; further, we derive an approximate inverse operator that can be used as an efficient preconditioner.

## 2.1. The integral equation

In this section we consider the integral equations that arise from the application of the acoustic reciprocity theorem, see Chapter 5, both in the frequency domain and the time domain. All of these are of the general form

$$\int_{x' \in \mathbb{D}} L(x, x')\, u(x')\, \mathrm{d}x' = f(x), \quad \text{when } x \in \mathbb{D}\,. \tag{2.1}$$

In this equation, $u$ is the unknown wavefield quantity in the relevant spatial or space-time domain, $f$ is a known wavefield, and $L$ is the kernel of the integral equation. In general, $x$ and $x'$ stand for the relevant coordinate variables (for example, the Cartesian coordinates $\{x_1, x_2, x_3\}$ in three-dimensional space in the frequency-domain formulation, and $\{x_1, x_2, x_3, t\}$ in the corresponding time-domain formulation); $u$ and $f$ are vector valued, $L$ yields the proper matrix relationship, and $\mathbb{D}$ is the (space or space-time) domain for which Eq. (2.1) holds. $\mathbb{D}$ is a subspace of $\mathbb{R}^p$, where $p$ is the dimension of the space under consideration. We further assume that the integral equation has a unique solution, i.e., $u(x) = 0$ for all $x \in \mathbb{D}$, if and only if $f(x) = 0$ for all $x \in \mathbb{D}$.

In almost all situations encountered in practice, the integral equation of Eq. (2.1) can only be solved approximately with the aid of numerical techniques. How good an approximate solution is, can be quantified only after one has chosen a particular quantitative *error*. To discuss this aspect, we introduce an operator formalism together with an inner product of two functions defined on $\mathbb{D}$ (with the associated norm). To write Eq. (2.1) in an operator form, the (bounded) linear operator $L$ acting on a function $u \in \mathbb{D}$ is introduced by

$$Lu = L\{u\}(x) = \int_{x' \in \mathbb{D}} L(x, x')\, u(x')\, \mathrm{d}x'\,. \tag{2.2}$$

Note that the right-hand side of Eq. (2.2) is defined for all $x \in \mathbb{R}^p$ (see Section 2.7). Equation (2.1) is equivalent to the operator equation

$$Lu = f, \quad \text{when } x \in \mathbb{D}\,. \tag{2.3}$$

Further, the inner product of two integrable functions $u$ and $v$ defined on $\mathbb{D}$ is taken as

$$\langle u, v \rangle = \int_{x \in \mathbb{D}} u(x)\, v^{\star}(x)\, \mathrm{d}x\,, \tag{2.4}$$

where the star denotes complex conjugate. The norm of a function $u$ is, in accordance with Eq. (2.4), defined as real positive quantity

$$\|u\| = \langle u, u \rangle^{\frac{1}{2}} . \tag{2.5}$$

The (Hermitean) adjoint operator $L^*$ of $L$ is defined as that one for which

$$\langle Lu, v \rangle = \langle u, L^* v \rangle , \tag{2.6}$$

for all functions $u$ and $v$ defined on $\mathbb{D}$. If $L^* = L$, the operator is selfadjoint. It is noted that in most seismic problems the operator is, however, not selfadjoint. Combining Eq. (2.2) with Eq. (2.6) it follows that

$$L^* v = L^*\{v\}(x) = \int_{x' \in \mathbb{D}} L^*(x', x)\, v(x')\, \mathrm{d}x' , \tag{2.7}$$

where $L^*(x', x)$ denotes the complex conjugate of the transpose of the matrix kernel $L(x', x)$. Note that in the functional dependencies of the matrix kernel of Eq. (2.7) the coordinates $x$ and $x'$ have the reverse order of the ones in Eq. (2.2).

For any function $u^{app}$ differing from the exact solution $u$ of Eq. (2.3) we define the residual as

$$r = f - Lu^{app} , \tag{2.8}$$

and the global root-mean-square error in the satisfaction of the equality sign in Eq. (2.3) as

$$\mathrm{ERR} = \langle r, r \rangle^{\frac{1}{2}} = \|r\| , \tag{2.9}$$

being the norm of $r$. Note that $\mathrm{ERR} \geq 0$ and that $\mathrm{ERR} = 0$ if and only if $u^{app} = u$. In seismic problems, $f$ is related to the source wavefield in the domain $\mathbb{D}$; therefore, in numerical implementations, we normalize the root-mean-square error according to

$$\widehat{\mathrm{ERR}} = \frac{\|r\|}{\|f\|} , \tag{2.10}$$

with the properties $\widehat{\mathrm{ERR}} = 0$ if $u^{app} = u$ and $\widehat{\mathrm{ERR}} = 1$ if $u^{app} = 0$. The error defined in Eqs. (2.9) and (2.10) is used as a measure for the accuracy attained in all the various iterative schemes to be dealt with.

## 2.2. Direct minimization of the error

In this section we first discuss a direct (i.e., non-iterative) approximation to the solution of the operator equation (2.3). To construct an approximate solution, the unknown function $u$ is expanded in terms of a given, appropriately chosen, sequence of linearly independent expansion functions $\{\phi_n;\ n = 1, \cdots, N\}$ that are defined on $\mathbb{D}$ and belong to the same vector space as to which $u$ belongs. Let, for some $N \geq 1$,

$$u_N = \sum_{n=1}^{N} \alpha_n^{(N)} \phi_n \ , \tag{2.11}$$

and

$$r_N = f - Lu_N \ . \tag{2.12}$$

Then the problem is to determine, for given $N$, the sequence of expansion coefficients $\{\alpha_n^{(N)};\ n = 1, \cdots, N\}$ such that $\langle r_N, r_N \rangle$ is minimized. The relevant values of $\{\alpha_n^{(N)}\}$ are denoted as the optimum values $\{\alpha_n^{opt}\}$. Assuming that the optimum exists, let

$$\alpha_n^{(N)} = \alpha_n^{opt} + \delta\alpha_n \quad \text{for } n = 1, \cdots, N \ , \tag{2.13}$$

where $\delta\alpha_n$ is arbitrary. Let, further,

$$r_N^{opt} = f - Lu_N^{opt} = f - \sum_{n=1}^{N} \alpha_n^{opt} L\phi_n \ , \tag{2.14}$$

then

$$\begin{aligned} \langle r_N, r_N \rangle = \langle r_N^{opt}, r_N^{opt} \rangle - 2\,\mathrm{Re}\left[ \sum_{m=1}^{N} \delta\alpha_m^{\star} \langle r_N^{opt}, L\phi_m \rangle \right] \\ + \langle \sum_{n=1}^{N} \delta\alpha_n L\phi_n, \sum_{m=1}^{N} \delta\alpha_m L\phi_m \rangle \ . \end{aligned} \tag{2.15}$$

Now, the last term on the right-hand side of this equation is always positive if $\{\delta\alpha_1, \cdots, \delta\alpha_N\} \neq \{0, \cdots, 0\}$. Hence, if

$$\langle r_N^{opt}, L\phi_m \rangle = 0 \quad \text{for } m = 1, \cdots, N \ , \tag{2.16}$$

we have constructed the situation that for $\{\delta\alpha_1, \cdots, \delta\alpha_N\} = \{0, \cdots, 0\}$ the quantity $\langle r_N^{opt}, r_N^{opt} \rangle$ is the absolute minimum of $\langle r_N, r_N \rangle$. Substitution of Eq. (2.14) in Eq. (2.16) yields the system of linear algebraic equations

$$\sum_{n=1}^{N} \alpha_n^{opt} \langle L\phi_n, L\phi_m \rangle = \langle f, L\phi_m \rangle \quad \text{for } m = 1, \cdots, N \,, \tag{2.17}$$

to be solved for $\{\alpha_n^{opt}\}$. From Eqs. (2.16) and (2.14) it also follows that

$$\langle r_N^{opt}, Lu_N^{opt} \rangle = 0 \,. \tag{2.18}$$

The resulting value of the error is

$$\mathrm{ERR}_N = \langle r_N^{opt}, r_N^{opt} \rangle^{\frac{1}{2}} = \langle r_N^{opt}, f \rangle^{\frac{1}{2}} \,, \tag{2.19}$$

where Eqs. (2.14) and (2.18) has been used. If this value does not meet the accuracy requirements set on the solution of the operator equation, it can be reduced either by selecting a more appropriate sequence of expansion functions (which is difficult to realize in practice) or by increasing $N$. Note that Eq. (2.17) would also result from the application of the method of moments (HARRINGTON, 1968) or Galerkin's method (KANTOROVICH and KRYLOV, 1964, p. 151), provided that the sequence of testing functions is chosen as $\{(L\phi_m)^*;\ m = 1, \cdots, N\}$ when the sequence of expansion functions is $\{\phi_n;\ n = 1, \cdots, N\}$, which choice gives the best result in the integrated-square-error sense. For problems of realistic size and complexity the value $N$ soon becomes so large that the storage requirements exceed the capacity of even present-day large computer systems. The problem of excessive computation time and computer storage requirements for a direct numerical solution (e.g., by Gaussian elimination, e.g. see RALSTON and RABINOWITZ, 1978, p. 415) of the system of equations can be circumvented by using suitable iterative techniques. Another argument in favor of solving the pertinent system of equations iteratively is the evident fact that we should not solve the system of equations to a higher degree of accuracy than is needed.

## 2.3. Recursive minimization of the error

In this section we develop a recursive method for calculating the approximate solution to the operator equation (2.3). In a direct procedure for solving such an equation approximately a (finite) sequence of expansion functions is somehow selected beforehand, and the sequence of expansion coefficients is solved from a system of linear algebraic equations that follows from - in our case - minimizing the norm of the error in the residual. In an iterative procedure the elements of the sequence of expansion functions are recursively generated from the operator equation to be solved, one in each iteration step. To achieve this, the successive residuals in the iteration process are at one's disposal. The sequence of expansion coefficients grows with the number of iterations. Since at the $N^{\text{th}}$ step, $N$ presumably linearly independent expansion functions have been generated, $N$ expansion coefficients are available to represent the $N^{\text{th}}$ approximation to the solution of the operator equation. Here, too, the minimization of the norm of the residual at the $N^{\text{th}}$ step will be employed to generate the system of linear algebraic equations that the sequence of expansion coefficients must satisfy.

Let $u_N$ denote the $N^{\text{th}}$ approximation to the solution of the operator equation

$$Lu = f, \quad \text{for } \boldsymbol{x} \in \mathbb{D}\,, \tag{2.20}$$

and let $\{\phi_n;\ n = 1, \cdots, N\}$ be the recursively generated sequence of expansion functions. Then, we take

$$\begin{aligned} u_0 &= 0\,, \\ u_N &= u_{N-1} + u_N^{cor} \quad \text{for } N = 1, \cdots\,, \end{aligned} \tag{2.21}$$

where $u_N^{cor}$ is the correction to $u_{N-1}$ to arrive at $u_N$. The correction is now expressed as

$$u_N^{cor} = \sum_{n=1}^{N} \alpha_n^{(N)} \phi_n \quad \text{for } N = 1, \cdots\,, \tag{2.22}$$

where $\{\alpha_n^{(N)};\ n = 1, \cdots, N\}$ is the sequence of expansion coefficients of $u_N^{cor}$. The residuals are found as

$$\begin{aligned} r_0 &= f\,, \\ r_N &= f - Lu_N \quad \text{for } N = 1, \cdots\,. \end{aligned} \tag{2.23}$$

From Eqs. (2.21) and (2.23) it follows that

$$r_N = r_{N-1} - Lu_N^{cor}\,. \tag{2.24}$$

Substituting Eq. (2.22) in Eq. (2.24), we arrive at

$$r_N = r_{N-1} - \sum_{n=1}^{N} \alpha_n^{(N)} L\phi_n \quad \text{for } N = 1, \cdots . \tag{2.25}$$

Taking into account that at the $N^{\text{th}}$ step $r_{N-1}$ is known, minimization of the norm of $r_N$ leads to a system of linear algebraic equations in the $N$ expansion coefficients $\{\alpha_n^{(N)};\ n = 1, \cdots, N\}$. On account of Eqs. (2.14) and (2.16) this system of equations follows from

$$\langle r_N, L\phi_m \rangle = 0 \quad \text{for } m = 1, \cdots, N . \tag{2.26}$$

Substitution of Eqs. (2.24) and (2.22) in these equations leads to

$$\sum_{n=1}^{N} \alpha_n^{(N)} \langle L\phi_n, L\phi_m \rangle = \langle r_{N-1}, L\phi_m \rangle \quad \text{for } m = 1, \cdots, N . \tag{2.27}$$

Now, on account of Eq. (2.26) only the right-hand side of Eq. (2.27) for $m = N$ may differ from zero. For non-zero values of the coefficients $\{\alpha_n^{(N)};\ n = 1, \cdots, N\}$, this should be the case and hence

$$\langle r_{N-1}, L\phi_N \rangle \neq 0 , \tag{2.28}$$

which is denoted as the *improvement condition.* If Eq. (2.28) is satisfied, the coefficients $\{\alpha_n^{(N)};\ n = 1, \cdots, N\}$ can be solved from Eq. (2.27).

First of all it is observed that the vanishing of the right-hand sides in Eq. (2.27) for $m = 1, \cdots, N - 1$ entails the property that all $\alpha_n^{(N)}$ for $n = 1, \cdots, N - 1$ are proportional to $\alpha_N^{(N)}$. In view of this, we introduce the sequence of functions $\{\psi_N;\ N = 1, \cdots\}$ as

$$\psi_N = \frac{u_N^{cor}}{\alpha_N^{(N)}} \quad \text{for } N = 1, \cdots , \tag{2.29}$$

or (cf. Eq. (2.22))

$$\begin{aligned} \psi_1 &= \phi_1 , \\ \psi_N &= \phi_N + \sum_{n=1}^{N-1} \frac{\alpha_n^{(N)}}{\alpha_N^{(N)}} \phi_n \quad \text{for } N = 2, \cdots . \end{aligned} \tag{2.30}$$

Evidently, Eq. (2.30) expresses $\psi_N$ as a linear combination of $\{\phi_n;\ n = 1,\cdots,N\}$. Since the reverse is also true, $\phi_N$ can be expressed as a linear combination of $\{\psi_n;\ n = 1,\cdots,N\}$. In view of this, Eq. (2.30) can be rewritten as

$$\begin{aligned} \psi_1 &= \phi_1 , \\ \psi_N &= \phi_N + \sum_{n=1}^{N-1} \beta_n^{(N)} \psi_n \quad \text{for } N = 2,\cdots . \end{aligned} \tag{2.31}$$

Owing to the orthogonality properties of $\{L\psi_N\}$ to be discussed below, the coefficients $\{\beta_n^{(N)};\ n = 1,\cdots,N-1\}$ can readily be determined. Since any $\psi_N$ is a linear combination of the expansion functions $\{\phi_n;\ n = 1,\cdots,N\}$, Eq. (2.26) leads to

$$\langle r_N, L\psi_m\rangle = 0 \quad \text{for } m = 1,\cdots,N . \tag{2.32}$$

Using Eqs. (2.24) and (2.29) in Eq. (2.32), we arrive at the orthogonality relation

$$\langle L\psi_N, L\psi_m\rangle = 0 \quad \text{for } m = 1,\cdots,N-1 . \tag{2.33}$$

Since Eq. (2.33) holds for any $N = 2,\cdots$, the sequence of $\{\psi_n\}$ satisfies the orthogonality relationship

$$\langle L\psi_n, L\psi_m\rangle = 0 , \quad m \neq n . \tag{2.34}$$

Note that our procedure of Eqs. (2.30) - (2.34) is equivalent with a Gram-Schmidt orthogonalization procedure (see e.g., KANTOROVICH and KRYLOV, 1964, pp. 45-47) of the sequence $L\phi_m$ into an orthogonal sequence $L\psi_m$. Combining Eq. (2.34) with Eq. (2.31) we obtain

$$\beta_n^{(N)} = -\frac{\langle L\phi_N, L\psi_n\rangle}{\|L\psi_n\|^2} \quad \text{for } n = 1,\cdots,N-1 . \tag{2.35}$$

Through substitution of Eq. (2.35) in Eq. (2.31), the sequence $\{\psi_N;\ N = 1,\cdots\}$ has been constructed.

The value of $u_N^{cor}$ finally follows from (cf. Eq. (2.29))

$$u_N^{cor} = \alpha_N^{(N)} \psi_N \quad \text{for } N = 1,\cdots , \tag{2.36}$$

which leads to

$$\langle Lu_N^{cor}, L\psi_N\rangle = \alpha_N^{(N)} \langle L\psi_N, L\psi_N\rangle . \tag{2.37}$$

However (cf. Eq. (2.24)),

$$\langle Lu_N^{cor}, L\psi_N\rangle = \langle r_{N-1} - r_N, L\psi_N\rangle . \tag{2.38}$$

Application of the orthogonality relations of Eq. (2.26), with $m = N$, leads to the result

$$\langle Lu_N^{cor}, L\psi_N\rangle = \langle r_{N-1}, L\psi_N\rangle = \langle r_{N-1}, L\phi_N\rangle . \tag{2.39}$$

Combining Eq. (2.39) with Eq. (2.37), we arrive at

$$\alpha_N^{(N)} = \frac{\langle r_{N-1}, L\phi_N\rangle}{||L\psi_N||^2} . \tag{2.40}$$

With this, the determination of $u_N^{cor}$ has been completed and the iterative scheme based on error minimization has been defined.

More specifically we consider the case that the function $\phi_N$ that is generated at the $N^{\text{th}}$ step of iteration is linearly related to the residual $r_{N-1}$ at the previous step. Then,

$$\phi_N = Tr_{N-1} \text{ for } N = 1, \cdots , \tag{2.41}$$

where $T$ is a unique (bounded) linear operator on $\mathbb{D}$. Then the following computational scheme is arrived at. In this scheme, the operator $LTu$ denotes the repeated operation $L\{Tu\}(x)$.

*Computational scheme for an arbitrary operator T*

The iterative scheme starts with the initial values

$$u_0 = 0, \quad r_0 = f, \quad \text{ERR}_0 = ||f|| . \tag{2.42}$$

Next, the scheme puts

$$\begin{aligned}
\psi_1 &= Tr_0 , \\
B_1 &= ||L\psi_1||^2 , \\
\alpha_1^{(1)} &= \frac{\langle r_0, LTr_0\rangle}{B_1} , \\
u_1 &= u_0 + \alpha_1^{(1)}\psi_1 , \\
r_1 &= r_0 - \alpha_1^{(1)} L\psi_1 , \\
\text{ERR}_1 &= ||r_1|| ,
\end{aligned} \tag{2.43}$$

and computes successively for $N = 2, \cdots$,

$$
\begin{aligned}
\beta_n^{(N)} &= -\frac{\langle LTr_{N-1}, L\psi_n \rangle}{B_n} \quad \text{for } n = 1, \cdots, N-1 \,, \\
\psi_N &= Tr_{N-1} + \sum_{n=1}^{N-1} \beta_n^{(N)} \psi_n \,, \\
B_N &= \|L\psi_N\|^2 \,, \\
\alpha_N^{(N)} &= \frac{\langle r_{N-1}, LTr_{N-1} \rangle}{B_N} \,, \\
u_N &= u_{N-1} + \alpha_N^{(N)} \psi_N \,, \\
r_N &= f - Lu_N = r_{N-1} - \alpha_N^{(N)} L\psi_N \,, \\
\text{ERR}_N &= \|r_N\| \,.
\end{aligned}
\tag{2.44}
$$

The important orthogonality relations implicitly used in the iterative scheme are

$$
\begin{aligned}
\langle L\psi_n, L\psi_m \rangle &= 0 \ \text{for } m \neq n \,, \\
\langle r_N, L\psi_m \rangle &= 0 \ \text{for } m = 1, \cdots, N \,, \\
\langle r_N, LTr_m \rangle &= 0 \ \text{for } m = 0, \cdots, N-1 \,.
\end{aligned}
\tag{2.45}
$$

In this scheme, for each $N = 1, \cdots$, the values of $\psi_N$, $L\psi_N$ and $B_N$ are stored. This means that at the $N^{\text{th}}$ step of iteration, we need computer storage for the updated values $u_N$ and $r_N$, as well as some background storage for the values of $\psi_m$, $L\psi_m$ and $B_m$, for $m = 1, \cdots, N$. The computation time and computer storage required for each step of iteration increase with an increasing number of iterations.

## 2.4. Selfadjoint operator $LT$

In this section we now investigate the consequences to the scheme of the previous section in case the operator $LT$ is selfadjoint. For such operators the property

$$\langle u, LTv \rangle = \langle LTu, v \rangle \tag{2.46}$$

holds. Then, the last orthogonality relation of (2.45) can be written as

$$\langle r_n, LTr_m\rangle = \langle LTr_n, r_m\rangle = 0 \text{ for } m \neq n\,. \tag{2.47}$$

The quantity

$$\langle r_N, LTr_N\rangle = \langle LTr_N, r_N\rangle = \langle r_N, LTr_N\rangle^{\star} \tag{2.48}$$

is real valued. Therefore the expansion coefficient (cf. Eqs. (2.40) and (2.41)),

$$\alpha_N^{(N)} = \frac{\langle r_{N-1}, LTr_{N-1}\rangle}{\|L\psi_N\|^2} \tag{2.49}$$

is also a real quantity. From this equation we directly observe that

$$\|L\psi_n\|^2 = \frac{\langle r_{n-1}, LTr_{n-1}\rangle}{\alpha_n^{(n)}} \quad \text{for } n = 1,\cdots,N\,. \tag{2.50}$$

Further, from Eqs. (2.24) and (2.29) we have

$$L\psi_n = -\frac{r_n - r_{n-1}}{\alpha_n^{(n)}} \quad \text{for } n = 1,\cdots,N\,. \tag{2.51}$$

Using Eqs. (2.50), (2.51) and (2.41) in Eq. (2.35), the expression for $\beta_n^{(N)}$ becomes

$$\beta_n^{(N)} = \frac{\langle LTr_{N-1}, r_n\rangle - \langle LTr_{N-1}, r_{n-1}\rangle}{\langle r_{n-1}, LTr_{n-1}\rangle} \quad \text{for } n = 1,\cdots,N-1\,. \tag{2.52}$$

Taking into account the orthogonality relations of Eq. (2.47), we arrive at

$$\beta_n^{(N)} = \begin{cases} 0 & \text{for } n = 1,\cdots,N-2\,, \\[2ex] \dfrac{\langle r_{N-1}, LTr_{N-1}\rangle}{\langle r_{N-2}, LTr_{N-2}\rangle} & \text{for } n = N-1\,. \end{cases} \tag{2.53}$$

Hence, only $\beta_{N-1}^{(N)}$ differs from zero and has to be determined.

*Computational scheme for a selfadjoint operator LT*

The iterative scheme starts with the initial values

$$u_0 = 0, \quad r_0 = f, \quad \mathrm{ERR}_0 = \|f\|\,. \tag{2.54}$$

Next, the scheme puts

$$\begin{aligned}
A_1 &= \langle r_0, LTr_0\rangle\,, \\
\psi_1 &= Tr_0\,, \\
B_1 &= \|L\psi_1\|^2\,, \\
u_1 &= u_0 + \frac{A_1}{B_1}\psi_1\,, \\
r_1 &= r_0 - \frac{A_1}{B_1}L\psi_1\,, \\
\mathrm{ERR}_1 &= \|r_1\|\,,
\end{aligned} \tag{2.55}$$

and computes successively for $N = 2, \cdots$,

$$\begin{aligned}
A_N &= \langle r_{N-1}, LTr_{N-1}\rangle\,, \\
\psi_N &= Tr_{N-1} + \frac{A_N}{A_{N-1}}\psi_{N-1}\,, \\
B_N &= \|L\psi_N\|^2\,, \\
u_N &= u_{N-1} + \frac{A_N}{B_N}\psi_N\,, \\
r_N &= f - Lu_N = r_{N-1} - \frac{A_N}{B_N}L\psi_N\,, \\
\mathrm{ERR}_N &= \|r_N\|\,.
\end{aligned} \tag{2.56}$$

In this scheme, we need computer storage for the updated values $u_N$, $\psi_N$, $A_N$ and $r_N$. The computation time and computer storage required for each step of iteration remain the same for all iterations $N = 2, \cdots$. This scheme is equivalent to one of the conjugate-gradient schemes in the literature (Golub and O'Leary, 1989).

The algorithm of the present conjugate-gradient scheme can be further simplified as follows. If we introduce the function $w_N$ as

$$w_N = \frac{\psi_N}{A_N} \tag{2.57}$$

and use it in the recursion relation $\psi_N = Tr_{N-1} + \frac{A_N}{A_{N-1}}\psi_{N-1}$ of Eq. (2.56), we find

$$w_N = w_{N-1} + \frac{1}{A_N}Tr_{N-1}, \quad N = 2, \cdots . \tag{2.58}$$

Using the expression of $B_N$ of Eq. (2.56), we can change the iteration scheme as

$$\begin{aligned} u_N &= u_{N-1} + \frac{w_N}{\|Lw_N\|^2}, \\ r_N &= r_{N-1} - \frac{Lw_N}{\|Lw_N\|^2}, \quad N = 2, \cdots . \end{aligned} \tag{2.59}$$

In summary, the following computational scheme is arrived at.

*Simplified computational scheme for a selfadjoint operator LT*

The iterative scheme starts with the initial values

$$u_0 = 0, \ \ r_0 = f, \ \ \mathrm{ERR}_0 = \|f\| . \tag{2.60}$$

Next, the scheme puts

$$\begin{aligned} A_1 &= \langle r_0, LTr_0 \rangle , \\ w_1 &= \frac{1}{A_1} Tr_0 , \\ C_1 &= \|Lw_1\|^2 , \\ u_1 &= u_0 + \frac{1}{C_1} w_1 , \\ r_1 &= r_0 - \frac{1}{C_1} Lw_1 , \\ \mathrm{ERR}_1 &= \|r_1\| , \end{aligned} \tag{2.61}$$

and computes successively for $N = 2, \cdots$ ,

$$\begin{aligned} A_N &= \langle r_{N-1}, LTr_{N-1} \rangle , \\ w_N &= w_{N-1} + \frac{1}{A_N} Tr_{N-1} , \\ C_N &= \|Lw_N\|^2 , \\ u_N &= u_{N-1} + \frac{1}{C_N} w_N , \\ r_N &= f - Lu_N = r_{N-1} - \frac{1}{C_N} Lw_N , \\ \mathrm{ERR}_N &= \|r_N\| . \end{aligned} \tag{2.62}$$

In this scheme, we need computer storage for the updated values $u_N$, $w_N$ and $r_N$. The computation time and computer storage required for each step of iteration remain the same for all iterations $N = 2, \cdots$ .

## 2.5. The Neumann expansion

In this section we discuss the application of the Neumann expansion to the operator equation

$$Lu = f \quad \text{for } x \in \mathbb{D} . \tag{2.63}$$

To this end the operator is rewritten as

$$L = I - K , \tag{2.64}$$

where $I$ is the identity operator. Substitution of Eq. (2.64) in Eq. (2.63) leads to

$$u = Ku + f \quad \text{for } x \in \mathbb{D} . \tag{2.65}$$

Operator equations of the type of Eq. (2.65) naturally arise from integral equations of the second kind. The iterative scheme defined by

$$\begin{aligned} u_0 &= 0 , \\ u_N &= f + Ku_{N-1} \quad \text{for } N = 1, \cdots , \end{aligned} \tag{2.66}$$

is known as the Neumann iterative solution. We define the repeated operator

$$K^N u = K\{K^{N-1}u\}(x) \quad \text{for } N = 1, \cdots , \tag{2.67}$$

and $K^0 u = u(x)$. Then, the repeated application of Eq. (2.66) leads to

$$u_N = \sum_{n=0}^{N-1} K^n f \quad \text{for } N = 1, \cdots , \tag{2.68}$$

and hence

$$r_N = f - Lu_N = f - u_N + Ku_N = K^N f , \tag{2.69}$$

which yields

$$ERR_N = \|K^N f\| . \tag{2.70}$$

Convergence of the Neumann expansion is therefore established if $||K^N f|| \to 0$ as $N \to \infty$. Introducing the norm of the operator $K$ as

$$||K|| = \sup_{f \neq 0} \frac{||Kf||}{||f||}, \tag{2.71}$$

we may write $||Kf|| \leq ||K|| \, ||f||$, and using Eq. (2.67) we find

$$||K^N f|| \leq ||K|| \, ||K^{N-1} f|| . \tag{2.72}$$

Repeated application of this inequality leads to

$$||K^N f|| \leq ||K||^N \, ||f|| , \tag{2.73}$$

hence, by combining Eqs. (2.70) and (2.73), it follows that the convergence of the Neumann expansion is guaranteed for any admissible $f$, provided that the operator norm

$$||K|| < 1 . \tag{2.74}$$

In practical applications we do not know the value of this norm and the rate of convergence can only be tested numerically.

In order to relate the Neumann expansion to our general iterative scheme, we note that the second relation of Eq. (2.66) is equivalent to

$$u_N = u_{N-1} + r_{N-1} \quad \text{for } N = 1, \cdots, \tag{2.75}$$

where

$$\begin{aligned} r_0 &= f , \\ r_N &= f - L u_N \quad \text{for } N = 1, \cdots . \end{aligned} \tag{2.76}$$

Comparing Eq. (2.75) with Eq. (2.21), the Neumann expansion can be characterized through

$$u_N^{cor} = r_{N-1} \quad \text{for } N = 1, \cdots, \tag{2.77}$$

i.e., the correction function $u_N^{cor}$ is taken to be the residual of the previous step. It is noted that the Neumann iterative solution only converges for a very restrictive class of integral equations (cf. Eq. (2.71)). However, variational techniques derived in this chapter remove this restriction (VAN DEN BERG, 1991; KLEINMAN and VAN DEN BERG, 1991).

## 2.6. Special choices of the operator $T$

In this section we now investigate the consequences of some particular choices of the operator $T$.

*Residuals as expansion functions ($T = I$)*

Firstly, the Neumann expansion suggests the residual of the previous step to be taken as a particular choice for the expansion function $\phi_N$ in the recursive minimization scheme of Section 2.3, i.e.,

$$\phi_N = r_{N-1} \quad \text{for } n = 1, \cdots . \tag{2.78}$$

This is equivalent to putting the operator $T$ introduced in Section 2.3 equal to the identity operator, i.e.,

$$T = I . \tag{2.79}$$

The improvement condition of Eq. (2.28) is then replaced by

$$\langle r_{N-1}, L r_{N-1} \rangle \neq 0 . \tag{2.80}$$

Whether or not this condition is satisfied, depends on the particular form of the operator $L$ under consideration. The relevant iteration scheme now follows by replacing the operator $T$ by the identity operator $I$ in Eqs. (2.42) - (2.44) for non-selfadjoint operators $L$, and in either Eqs. (2.54) - (2.56) or Eqs. (2.60) - (2.62) for selfadjoint, not necessarily positive, operators $L$. The scheme of Eqs. (2.54) - (2.56) differs slightly from the standard conjugate-gradient schemes for positive operators ($\langle Lu, u \rangle = \langle u, Lu \rangle > 0$ for all $u \neq 0$ on $\mathbb{D}$) that are given in the literature (GOLUB and VAN LOAN, 1983, section 10.2), where the quantity $\langle Lu, u \rangle - \langle f, u \rangle - \langle u, f \rangle$ is minimized.

*Preconditioning ($T = P$)*

We first observe that, if $T$ is chosen equal to the inverse $L^{-1}$ of $L$, then $\phi_1 = \psi_1 = L^{-1} f$ and the recursive scheme of Section 2.3 will terminate in the first iteration; we then have arrived at the exact solution. For this reasoning we take $T$ equal to a suitably chosen preconditioning operator $P$, where the operator $LP$ resembles the identity operator more than $L$ itself does. Thus, the method depends on the availability of an approximate inverse to the operator $L$. Accordingly, we take

$$\phi_N = Pr_{N-1} \tag{2.81}$$

or

$$T = P, \tag{2.82}$$

The relevant iteration scheme now follows by replacing the operator $T$ by the preconditioning operator $P$ in Eqs. (2.42) - (2.44) for non-selfadjoint operators $LP$, and in either Eqs. (2.54) - (2.56) or Eqs. (2.60) - (2.62) for selfadjoint operators $LP$.

*Symmetrization ($T = L^*$)*

When the operator $L$ is not selfadjoint, the previous scheme for $T = I$ leads to a recursive minimization scheme, in which the computer storage of the expansion functions required for each step of iteration increases with an increasing number of iterations. However, when we take

$$\phi_N = L^* r_{N-1} \tag{2.83}$$

or

$$T = L^*, \tag{2.84}$$

the operator

$$LT = LL^* = L^* L \tag{2.85}$$

is selfadjoint and we can now use the simple iteration scheme of either Eqs. (2.54) - (2.56) or Eqs. (2.60) - (2.62) for selfadjoint operators $LT$ with $T$ replaced by $L^*$. Since

$$\langle r_{N-1}, L\phi_N \rangle = \langle L^* r_{N-1}, \phi_N \rangle = \langle \phi_N, \phi_N \rangle = \langle L^* r_{N-1}, L^* r_{N-1} \rangle \neq 0, \tag{2.86}$$

the improvement condition (cf. Eq. (2.28)) is automatically satisfied and the orthogonality relation of Eq. (2.47) simplifies to

$$\langle L^* r_{m-1}, L^* r_{n-1} \rangle = \langle \phi_m, \phi_n \rangle = 0 \text{ for } m \neq n. \tag{2.87}$$

Hence, the expansion functions generated according to Eq. (2.83) form an orthogonal sequence.

The scheme of Eqs. (2.54) - (2.56), with $T$ replaced by $L^*$ and the expression for $A_N$ replaced by

$$A_N = ||L^* r_{N-1}||^2 \text{ for } N = 1, \cdots , \tag{2.88}$$

is known as the conjugate-gradient scheme for a non-selfadjoint operator $L$ (VAN DEN BERG, 1984).

*Preconditioning and symmetrization ($T = PP^*L^*$)*

When the operator $LP$ is not selfadjoint, the scheme for $T = P$ leads to a recursive minimization scheme, in which the computer storage of the expansion functions required for each step of iteration increases with an increasing number of iterations. However, when we take

$$\phi_N = PP^*L^* r_{N-1} \tag{2.89}$$

or

$$T = PP^*L^* , \tag{2.90}$$

the operator

$$LT = LPP^*L^* = (LP)(LP)^* \tag{2.91}$$

is selfadjoint and we can now use the simple iteration scheme of either Eqs. (2.54) - (2.56) or Eqs. (2.60) - (2.62) for selfadjoint operators $LT$ with $T$ replaced by $PP^*L^*$ and the expression for $A_N$ replaced by

$$A_N = ||P^*L^* r_{N-1}||^2 \text{ for } N = 1, \cdots , \tag{2.92}$$

is a conjugate-gradient scheme for a preconditioned non-selfadjoint operator $L$. Note that the error criterion applies to the original operator equation. This differs from standard preconditioned conjugate-gradient schemes, where the error is minimized in the range of the preconditioned operator equation.

In the next section, we consider the case that the integral operator $L$ has a convolution kernel. Then, for the evaluation of the operator expressions, an advantageous combination of the conjugate-gradient algorithm and the Fourier transform can be employed. In addition, for this special case, an efficient preconditioning operator can be devised to improve the rate of convergence.

## 2.7. Operators of convolution type

In this section we consider an operator equation obtained in an analysis in the Laplace-transform domain. For some particular value of the Laplace-transform parameter, we write this equation in the operator notation. In the special case that in the operator equation

$$L\hat{u} = \hat{f}, \quad \boldsymbol{x} \in \mathbb{D}\,, \tag{2.93}$$

the operator has a convolution kernel with respect to the spatial variables, we take advantage of this structure. Although both $L\hat{u}$ and $\hat{f}$ assume values on $\mathbb{D}$, the operator has a natural extension to the complete space $\mathbb{R}^p$. $\mathbb{D}$ is considered as a proper subspace of $\mathbb{R}^p$. It is assumed that the operator has a convolution structure of the form

$$L\hat{u} = L\chi_{\mathbb{D}}\hat{u} = \int_{\boldsymbol{x}'\in\mathbb{R}^p} \hat{g}(\boldsymbol{x}-\boldsymbol{x}')\chi_{\mathbb{D}}(\boldsymbol{x}')\hat{u}(\boldsymbol{x}')\mathrm{d}\boldsymbol{x}', \quad \boldsymbol{x} \in \mathbb{R}^p\,, \tag{2.94}$$

in which we have assumed that $\hat{g}(\boldsymbol{x})$ is defined for all $\boldsymbol{x} \in \mathbb{R}^p$. Here, the characteristic function (see Fig. 1.3)

$$\chi_{\mathbb{D}} = \{1, \frac{1}{2}, 0\} \quad \text{when } \boldsymbol{x} \in \{\mathbb{D}, \partial\mathbb{D}, \mathbb{D}'\} \tag{2.95}$$

is introduced. In order to take advantage of the convolution structure of the operator, we define the $p$-dimensional Fourier transform by the operator $F$ and the inverse by $F^{-1}$, i.e.,

$$\tilde{u} = F\{\hat{u}\} = \int_{\boldsymbol{x}\in\mathbb{R}^p} \exp(js\boldsymbol{\alpha}\cdot\boldsymbol{x})\hat{u}(\boldsymbol{x})\mathrm{d}\boldsymbol{x}, \quad s\boldsymbol{\alpha} \in \mathbb{R}^p\,, \tag{2.96}$$

$$F^{-1}\{\tilde{u}\} = \frac{1}{(2\pi)^p}\int_{s\boldsymbol{\alpha}\in\mathbb{R}^p} \exp(-js\boldsymbol{\alpha}\cdot\boldsymbol{x})\tilde{u}(js\boldsymbol{\alpha})\mathrm{d}\boldsymbol{\alpha} = \hat{u}, \quad \boldsymbol{x} \in \mathbb{R}^p\,, \tag{2.97}$$

where $\boldsymbol{\alpha}\cdot\boldsymbol{x} = \sum_{q=1}^{p} \alpha_q x_q$. Then, the Fourier transform of the convolution of Eq. (2.94) can be written as

$$F\{L\chi_{\mathbb{D}}\hat{u}\} = \tilde{g}\, F\{\chi_{\mathbb{D}}\hat{u}\}\,, \tag{2.98}$$

where

$$\tilde{g} = F\{\hat{g}\}\,. \tag{2.99}$$

In our iterative schemes, the different operator expressions can now be computed with the help of the Fourier transformations.

*Computation of the operator*

When the operator of Eq. (2.93) is of the convolution type, the operator expression $L\chi_{\mathbb{D}}\hat{u}$ can be computed with the use of Eq. (2.98) as

$$L\chi_{\mathbb{D}}\hat{u} = F^{-1}\{\tilde{g}\,F\{\chi_{\mathbb{D}}\hat{u}\}\} \,. \tag{2.100}$$

It is noted that in the actual numerical computations this operator expression can be computed by using the discrete versions of the forward and inverse Fourier transformations.

*Computation of the preconditioning operator*

A preconditioning operator is now constructed as follows. Let us write Eq. (2.98) as

$$F\{\chi_{\mathbb{D}}\hat{u}\} = \tilde{g}^{-1}F\{L\chi_{\mathbb{D}}\hat{u}\} \,. \tag{2.101}$$

Using Eqs. (2.93) and (2.94) we may write

$$\begin{aligned} F\{\chi_{\mathbb{D}}\hat{u}\} &= \tilde{g}^{-1}F\{\chi_{\mathbb{D}}L\chi_{\mathbb{D}}\hat{u}\} + \tilde{g}^{-1}F\{\chi_{\mathbb{D}'}L\chi_{\mathbb{D}}\hat{u}\} \\ &= \tilde{g}^{-1}F\{\chi_{\mathbb{D}}\hat{f}\} + \tilde{g}^{-1}F\{\chi_{\mathbb{D}'}L\chi_{\mathbb{D}}\hat{u}\} \,, \end{aligned} \tag{2.102}$$

where $\chi_{\mathbb{D}'} = 1 - \chi_{\mathbb{D}}$. The inverse Fourier transformation of Eq. (2.102) yields

$$\hat{u} = F^{-1}\Big\{\tilde{g}^{-1}F\{\chi_{\mathbb{D}}\hat{f}\}\Big\} + F^{-1}\Big\{\tilde{g}^{-1}F\{\chi_{\mathbb{D}'}L\chi_{\mathbb{D}}\hat{u}\}\Big\}, \quad x \in \mathbb{D} \,. \tag{2.103}$$

This is an integral equation of the second kind and in operator notation it may be written as

$$\hat{u} = P\chi_{\mathbb{D}}\hat{f} + Q\chi_{\mathbb{D}}\hat{u}, \quad x \in \mathbb{D}. \tag{2.104}$$

Now $P$ is taken to be an approximate inverse of $L$ and $Q\chi_{\mathbb{D}}\hat{u}$ is the error in solving Eq. (2.93). The latter error in constructing the approximate inverse is due to the non-zero values of $Lu$ on $\mathbb{D}'$. In the remainder the operator $P$ will be employed as a preconditioner and its extension to $\mathbb{R}^p$ is given by

$$P\chi_{\mathbb{D}}\hat{v} = F^{-1}\Big\{\tilde{g}^{-1}F\{\chi_{\mathbb{D}}\hat{v}\}\Big\}, \quad x \in \mathbb{R}^p \,. \tag{2.105}$$

It is noted that in the numerical work the operator expression $P\chi_{\mathbb{D}}\hat{v}$ can also be computed by using the discrete versions of the forward and inverse Fourier transformations.

*Computation of the adjoint operators*

To complete the discussion we remark that the adjoint operators necessary in our iterative scheme are given by

$$L^* \hat{v} = \int_{x' \in \mathbb{D}} \hat{g}^*(x' - x)\hat{v}(x')\mathrm{d}x', \quad x \in \mathbb{D}\,, \tag{2.106}$$

or in terms of the Fourier transforms, which are the forms actually employed in the computations,

$$L^* \chi_{\mathbb{D}} \hat{v} = F^{-1}\{\tilde{g}^* F\{\chi_{\mathbb{D}} \hat{v}\}\} \tag{2.107}$$

and

$$P^* \chi_{\mathbb{D}} \hat{u} = F^{-1}\Big\{(\tilde{g}^*)^{-1} F\{\chi_{\mathbb{D}} \hat{u}\}\Big\}\,, \tag{2.108}$$

where $\tilde{g}$ is given in Eq. (2.99) and $\tilde{g}^*$ is the complex conjugate of the transpose. Again, it is noted that in actual numerical computations we use the discrete versions of the forward and inverse Fourier transformations.

# Chapter 3

# Basic Equations in Acoustics

In the present chapter we discuss the acoustic wave equations in their most simple form. These equations form the basis of our treatment of seismic wave problems. The fundamentals can be found in the classical treatises by LORD RAYLEIGH (1894), LAMB (1925), LOVE (1944), RSCHEVKIN (1963), MORSE and INGARD (1968), SKUDRZYK (1971), HANISH (1981) and in the series edited by MASON (1964 - $\cdots$). The application of acoustic waves in seismic processing has been discussed by BERKHOUT (1987). We present the partial differential equations for the acoustic pressure, the particle velocity, the mass flow density and the volume change. Next, we discuss the constitutive relations that define how the mass flow density and the volume change are related to the acoustic pressure and the particle velocity. The relational parameters are the mass density and the compressibility. At discontinuities, the partial differential equations are supplemented with boundary conditions.

## 3.1. The acoustic wave equations

The acoustic wave equations are representative for the action of mechanical forces and the influence of inertia during the acoustic wave motion as well as of the deformation that take place during this wave motion. The acoustic wave motion is a dynamical state of matter that is superimposed on a static equilibrium state. In this respect, we shall only retain the first-order terms to describe the acoustic wave motion. Further, as can be done in practically all terrestrial applications, the influence of gravity is neglected (see BÅTH and BERKHOUT, 1984, p. 400).

*The equation of motion*

Applying Newton's law of motion to a representative elementary domain of a fluid, we arrive at the local equation of motion which we write as

$$\partial_k p + \dot{\Phi}_k = f_k \, . \tag{3.1}$$

In this equation,

$p$ = acoustic pressure (Pa),

$\dot{\Phi}_k$ = mass-flow density rate ($kg/m^2s^2$),

$f_k$ = volume source density of volume force ($N/m^3$).

The acoustic pressure is representative for the contact forces between adjacent elements of the fluid, the mass-flow density rate for the inertia properties of a portion of the fluid, and the volume source density of volume force for the action of volume forces. In our applications forces of this kind will be employed to represent the action of acoustic sources of the "dipole" type.

*The deformation equation*

By following an elementary portion of the fluid for a short while on its course and observing the changes in dimensions and shape of this portion, the following deformation equation is obtained:

$$\partial_k v_k - \dot{\Theta}^i = q \, . \tag{3.2}$$

In this equation,

$v_k$ = particle velocity (m/s),

$\dot{\Theta}^i$ = induced part of the cubic dilatation rate ($s^{-1}$),

$q$ = volume source density of injection rate ($s^{-1}$).

The particle velocity is the spatially averaged drift velocity of the particles constituting the fluid, the cubic dilatation rate is representative for the relative changes in volume of a fluid portion per time unit. The induced part of the latter is related to the action of the pressure via an equation of state; the volume density of injection rate is the remaining part, which is representative for the action of acoustic sources of the "monopole" type.

*The constitutive relations*

The constitutive relations express the mass flow density rate $\dot{\Phi}_k$ and the induced cubic dilatation rate $\dot{\Theta}^i$ in terms of the particle velocity $v_k$ and the acoustic pressure $p$. These relations are established by a physical experiment. In view of the assumed passivity of the fluid, we require that for any type of fluid we have $\{\dot{\Phi}_k, \dot{\Theta}^i\} \to 0$ as $\{v_k, p\} \to 0$.

When the values of $\{\dot{\Phi}_k, \dot{\Theta}^i\}$ are linearly related to the values of $\{v_k, p\}$, we denote the fluid as *linear*. If this is not the case, the fluid is denoted as non-linear. Since the constitutive relations apply to a particular piece of matter, linearity in behavior is to be understood in the so-called Lagrangian sense, i.e., time rates are observed by a co-moving observer.

When the operators that express the values of $\{\dot{\Phi}_k, \dot{\Theta}^i\}$ in terms of the values of $\{v_k, p\}$ are time invariant, the fluid is denoted as *time invariant.*

When the constitutive relations express the values of $\{\dot{\Phi}_k, \dot{\Theta}^i\}$ at some instant in terms of the values of $\{v_k, p\}$ at the same instant only, the fluid is denoted as *instantaneously reacting.* When, on the other hand, the values of $\{\dot{\Phi}_k, \dot{\Theta}^i\}$ are expressed in terms of the values of $\{v_k, p\}$ at all previous instants, the fluid is said to show relaxation; the property that only the past is involved in relaxation phenomena, is known as the principle of causality.

When the values of $\{\dot{\Phi}_k, \dot{\Theta}^i\}$ at some position are related to the values of $\{v_k, p\}$ at the same position only, the fluid is denoted as *locally reacting.*

If at a point in space the constitutive operators are orientation invariant, the fluid is denoted as *isotropic* at that point. If this property does not apply, the fluid is denoted as anisotropic.

In a domain in space where the constitutive operators are shift invariant, the fluid is denoted as *homogeneous*; in a domain in space where the shift invariance does not apply, the fluid is denoted as *inhomogeneous.*

For a wide class of fluids, the mass flow density rate $\dot{\Phi}_k$ only depends on the particle velocity $v_k$ (and not on the acoustic pressure), and the induced cubic dilatation rate $\dot{\Theta}^i$ only depends on the acoustic pressure $p$ (and not on the particle velocity). Further, in this book, we shall only deal with fluids that, in addition, are linear, time invariant, instantaneously reacting, locally reacting and isotropic in its acoustic behavior, while inhomogeneous fluids will be admitted. We then have

$$\dot{\Phi}_k(\boldsymbol{x},t) = \rho(\boldsymbol{x}) D_t v_k(\boldsymbol{x},t) , \tag{3.3}$$

and

$$\dot{\Theta}^i(\boldsymbol{x},t) = -\kappa(\boldsymbol{x}) D_t p(\boldsymbol{x},t) , \tag{3.4}$$

where

$\rho =$ volume density of mass (kg/m$^3$),

$\kappa =$ compressibility (Pa$^{-1}$),

are the constitutive coefficients and

$$D_t = \partial_t + v_k \partial_k \tag{3.5}$$

is the time derivative that an observer experiences when co-moving with the fluid (with speed $v_k$). In a domain where the constitutive coefficients introduced here change with position, the fluid is inhomogeneous; in a domain where they are constant, the fluid is homogeneous.

### 3.1.1. Low-velocity approximation

The system of equations that consists of the equation of motion, the deformation equation and the constitutive relations discussed in the previous section, is non-linear in the particle velocity due to the occurrence of the latter in the operator $D_t$ of Eq. (3.5). Fortunately, in seismic practice, the quantities associated with the acoustic wavefield are small-amplitude variations on the equilibrium state of the earth. Then, results of a sufficient accuracy are obtained by solving the linearized equations. Therefore, we shall employ Eqs. (3.3) - (3.4) in their low-velocity approximation, i.e., we replace $D_t$ as defined by Eq. (3.5) by

$$D_t = \partial_t . \tag{3.6}$$

The basic acoustic wave equations are then given by

$$\partial_k p + \rho \partial_t v_k = f_k \, , \tag{3.7}$$

$$\partial_k v_k + \kappa \partial_t p = q \, . \tag{3.8}$$

These equations are the theoretical fundaments of our seismic model.

### 3.1.2. Acoustic boundary conditions

In those domains in a fluid where the constitutive parameters change continuously with position, the acoustic pressure and the particle velocity are continuously differentiable functions of position and satisfy the differential equations (3.7) and (3.8). Now, in practice, it often occurs that fluids with different material parameters are in contact along interfaces. Across such an interface, the constitutive parameters show a jump discontinuity. From the equation of motion and the deformation rate equation it then follows that at least some components of the particle velocity, and possibly the pressure, show a jump discontinuity across the interface. Consequently, the pressure and/or the particle velocity are no longer continuously differentiable, and Eqs. (3.7) and (3.8) cease to hold. To interrelate the acoustic wavefield quantities at either side of the interface, a certain set of boundary conditions is needed. To arrive at these boundary conditions, we shall assume that along the interface the two fluids remain in touch, but do not mix. Then, the component of the particle velocity normal to the interface should be continuous across the interface. Otherwise, the fluid at one side of the interface would either move away from the fluid at the other side of the interface or mix with it. The components of the particle velocity parallel to the interface may be different at each side of the interface, since along the interface the two fluids may slide with respect to each other. To arrive at the conditions to be laid upon the acoustic pressure, we proceed as follows.

Let S denote the interface and assume that S has everywhere a unique tangent plane. Let, further, $\nu_k$ denote the unit vector along the normal to S such that upon traversing S in the direction of $\nu_k$, we pass from the domain $\mathbb{D}_2$ to the domain $\mathbb{D}_1$, $\mathbb{D}_1$ and $\mathbb{D}_2$ being located at either side of S (Fig. 3.1). Suppose, now that some (or all) acoustic wave quantities jump across S. In

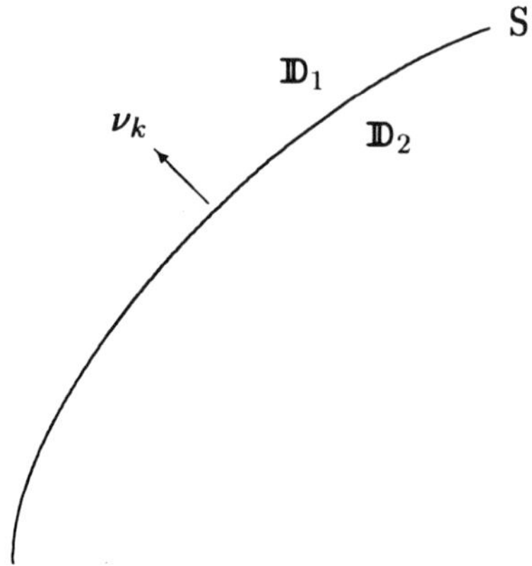


Figure 3.1. Interface between two media with different acoustic properties.

the direction parallel to S, all acoustic wave quantities still vary in a continuously differentiable manner, and hence the partial derivatives parallel to S give no problem in Eqs. (3.1) and (3.2). The partial derivatives perpendicular to S, on the contrary, meet functions that show a jump discontinuity across S; these give rise to surface Dirac distributions (surface impulse functions) located on S. Distributions of this kind would, however, physically be representative of surface sources located on S. In the absence of such surface sources, the absence of surface impulse functions in the partial derivatives perpendicular to S should be enforced. The latter is done by requiring that these normal derivatives only meet functions that are continuous across S. This procedure leads to

$$p \text{ is continuous across S} \tag{3.9}$$

and

$$\nu_k v_k \text{ is continuous across S} . \tag{3.10}$$

Equations (3.9) and (3.10) are the boundary conditions at a sourcefree interface between two different fluids. Equation (3.10) has already been conjectured on physical grounds, but is here shown to be consistent with the deformation rate equation (Eq. (3.2)).

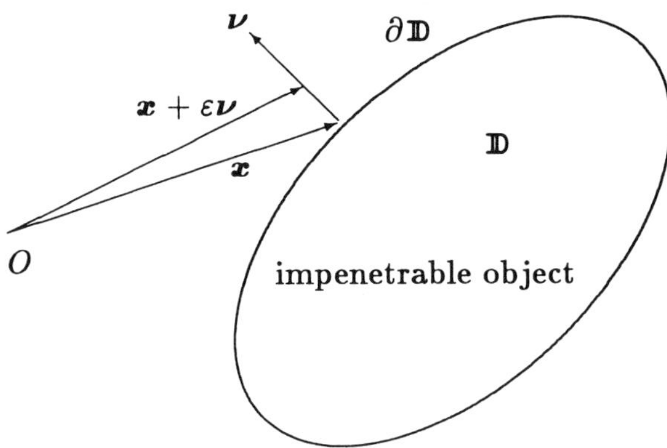


Figure 3.2. Limiting procedure approaching the boundary of an impenetrable object.

*Impenetrable object (void)*

A subdomain $\mathbb{D}$ of the fluid is denoted as a *void* if in it the acoustic pressure is negligibly small, while the continuity of the acoustic pressure across the boundary surface $\partial\mathbb{D}$ of the void is maintained. Consequently, the boundary condition upon approaching the boundary surface $\partial\mathbb{D}$ of a void via its exterior is given by (Fig. 3.2)

$$\lim_{\varepsilon \downarrow 0} p(\boldsymbol{x} + \varepsilon\boldsymbol{\nu}, t) = 0 \quad \text{for any } \boldsymbol{x} \in \partial\mathbb{D} \, , \tag{3.11}$$

where $\boldsymbol{\nu}$ is the unit vector along the normal to $\partial\mathbb{D}$ pointing away from $\mathbb{D}$. We are not free to prescribe the normal component of the particle velocity in this case. In fact, the normal component of the particle velocity will, in general, have a non-zero value at $\partial\mathbb{D}$, while it is not defined in $\mathbb{D}$. A subdomain of the fluid can in practice be considered as a void if the acoustic pressure in it is negligibly small compared to the pressure in other parts of the fluid. For example, an air bubble in sea water can in most cases be regarded as a void.

*Impenetrable object (perfectly rigid)*

A material body, occupying a domain $\mathbb{D}$ in the fluid, is denoted as a *perfectly rigid* object if it cannot be deformed and if its surface is impenetrable to the surrounding fluid. If in addition the object is held immovable by external means, we have (Fig. 3.2)

$$\lim_{\varepsilon \downarrow 0} \nu_k v_k(\boldsymbol{x} + \varepsilon \boldsymbol{\nu}, t) = 0 \text{ for any } \boldsymbol{x} \in \partial \mathbb{D} . \tag{3.12}$$

If Eq. (3.12) holds, we are not free to prescribe the acoustic pressure on $\partial \mathbb{D}$. In fact, the acoustic pressure will, in general, have a non-zero value at $\partial \mathbb{D}$, while it is not defined in $\mathbb{D}$.

## 3.2. The acoustic equations in the Laplace-transform domain

In a large number of cases met in practice, we are interested in the behavior of acoustic wavefields in linear and time-invariant configurations. Mathematically, one can take advantage of this situation by carrying out a Laplace transformation with respect to time and considering the equations governing the acoustic wavefield in the corresponding Laplace-transform domain or $s$-domain. In the $s$-domain relations, the time coordinate has been eliminated, and a wavefield problem in space remains in which the transform parameter $s$ occurs. Causality of the wavefield is taken into account by taking $\mathrm{Re}(s) > 0$, and requiring that all causal wavefield quantities are analytic functions of $s$ in the right half $0 < \mathrm{Re}(s) < \infty$ of the complex $s$-plane. In a number of wavefield problems, the transform parameter $s$ is profitably chosen to be real and positive. Further, by considering the limiting case $s = j\omega$, where $j$ is the imaginary unit and $\omega$ is real and positive, the complex steady-state representation of sinusoidally in time oscillating wavefields of angular (or circular) frequency $\omega$ follows, the complex representation having the complex time factor $\exp(j\omega t)$. For arbitrary complex values of $s$ in the domain of analyticity, all $s$-domain wavefield quantities are Laplace transforms of real-valued functions of the time coordinate $t$. As a consequence, the $s$-domain wavefield quantities are real valued for real and positive values of $s$. On account of Schwarz's principle of reflection, the relevant functions then take complex conjugate values for conjugate points in the complex $s$-plane (TITCHMARSH, 1939, p. 155).

*The s-domain acoustic wave equations*

We subject the acoustic wavefield equations (3.7) and (3.8) to a Laplace transformation over the interval $\mathrm{T}=\{t\in\mathbb{R};t>t_0\}$, (see Section 1.2.1). For completeness, we allow a non-vanishing acoustic wavefield to be present at $t=t_0$, although in the majority of cases we are interested in the causal wavefield generated by sources that are switched on at the instant $t=t_0$, in which case the initial values of the acoustic wavefield are taken to be zero. Since (cf. Eq. (1.15))

$$\int_{t_0}^{\infty}\exp(-st)\partial_t v_k(\boldsymbol{x},t)\mathrm{d}t=-v_k(\boldsymbol{x},t_0)\exp(-st_0)+s\hat{v}_k(\boldsymbol{x},s) \tag{3.13}$$

and

$$\int_{t_0}^{\infty}\exp(-st)\partial_t p(\boldsymbol{x},t)\mathrm{d}t=-p(\boldsymbol{x},t_0)\exp(-st_0)+s\hat{p}(\boldsymbol{x},s)\,, \tag{3.14}$$

we arrive at

$$\partial_k\hat{p}+s\rho\hat{v}_k=\hat{f}_k+\rho v_k(\boldsymbol{x},t_0)\exp(-st_0)\,, \tag{3.15}$$

$$\partial_k\hat{v}_k+s\kappa\hat{p}=\hat{q}+\kappa p(\boldsymbol{x},t_0)\exp(-st_0)\,. \tag{3.16}$$

From Eqs. (3.15) and (3.16) it follows that, in the $s$-domain, one can take into account the influence of a non-vanishing initial acoustic wavefield by properly incorporating it in the $s$-domain volume densities of external volume force and external volume injection rate. In the remainder of our analysis, it will be tacitly understood that non-zero initial acoustic wavefield values have been accounted for in this manner. Our acoustic wave equations in the s-domain have then the final form

$$\partial_k\hat{p}+s\rho\hat{v}_k=\hat{f}_k\,, \tag{3.17}$$

$$\partial_k\hat{v}_k+s\kappa\hat{p}=\hat{q}\,. \tag{3.18}$$

After transforming back to the time domain, the reconstructed acoustic wavefield values are zero in the interval $t\in\mathrm{T}'$, where $\mathrm{T}'=\{t\in\mathbb{R};t<t_0\}$, and equal to the actual wavefield values when $t\in\mathrm{T}$. In addition, many of the Laplace inversion algorithms yield half the wavefield values at the instant $t\in\partial\mathrm{T}$, where $\partial\mathrm{T}=\{t\in\mathbb{R};t=t_0\}$. Notationally, this can be expressed by employing the characteristic function $\chi_{\mathrm{T}}=\chi_{\mathrm{T}}(t)$ of the set

T defined by Eq. (1.13). With this notation, we have for any space-time function $u = u(\boldsymbol{x}, t)$

$$\text{Inverse Laplace transform of } \hat{u}(\boldsymbol{x}, s) = \chi_{\mathrm{T}}(t)\, u(\boldsymbol{x}, t). \tag{3.19}$$

*The steady-state analysis*

The behavior of acoustic waves of interest is often characterized by the results of a steady-state analysis. In such an analysis, all acoustic wave quantities are taken to depend sinusoidally on time with a common angular frequency, $\omega$ say. Each purely real quantity can then be associated with a complex counterpart and a common time factor $\exp(j\omega t)$. In doing so, the original quantities in the time domain are found from the complex counterparts as

$$\{p(\boldsymbol{x}, t), v_k(\boldsymbol{x}, t), q(\boldsymbol{x}, t), f_k(\boldsymbol{x}, t)\}$$
$$= \mathrm{Re}\left[\{\hat{p}(\boldsymbol{x}, j\omega), \hat{v}_k(\boldsymbol{x}, j\omega), \hat{q}(\boldsymbol{x}, j\omega), \hat{f}_k(\boldsymbol{x}, j\omega)\}\exp(j\omega t)\right] . \tag{3.20}$$

Substitution of these complex representations in the basic equations in the time domain yields, except for the common time factor $\exp(j\omega t)$, a set of basic equations

$$\partial_k \hat{p} + j\omega\rho\hat{v}_k = \hat{f}_k \,, \tag{3.21}$$

$$\partial_k \hat{v}_k + j\omega\kappa\hat{p} = \hat{q} \,. \tag{3.22}$$

These equations are identical to the one of the Laplace-transform domain, viz. Eqs. (3.17) - (3.18) with $s = j\omega$. Hence, we interpret the steady-state analysis as a limiting case of the time Laplace-transform analysis in which $s \to j\omega$ via $\mathrm{Re}(s) > 0$. In this way we have ensured that the causality remains satisfied.

### 3.2.1. Boundary conditions in the Laplace-transform domain

The boundary conditions that have been discussed in Section 3.1 apply, in the linearized low-velocity approximation, to time-invariant boundaries. As a consequence of this, these boundary conditions can directly be transferred to the $s$-domain. At the interface S of two different fluids we therefore have (cf. Eqs. (3.9) and (3.10))

$$\hat{p} \text{ is continuous across S} \tag{3.23}$$

and

$$\nu_k \hat{v}_k \text{ is continuous across S} , \tag{3.24}$$

while from Eq. (3.11) we have

$$\lim_{\varepsilon \downarrow 0} \hat{p}(\boldsymbol{x} + \varepsilon \boldsymbol{\nu}, s) = 0 \quad \text{for } \boldsymbol{x} \in \text{boundary of a void} , \tag{3.25}$$

and from Eq. (3.12) we have

$$\lim_{\varepsilon \downarrow 0} \nu_k \hat{v}_k(\boldsymbol{x} + \varepsilon \boldsymbol{\nu}, s) = 0 \tag{3.26}$$

for $\boldsymbol{x} \in$ boundary of an immovable, perfectly rigid object.

This concludes the discussion of the general framework of the acoustic wave equations and the acoustic boundary conditions. Before turning our attention to the acoustic reciprocity theorems, we first discuss in the next chapter the acoustic wavefield in an unbounded, homogeneous medium; in particular, the acoustic wavefield from a point source (the Green's function) is derived.

# Chapter 4

# Radiation in an Unbounded, Homogeneous Medium

In this chapter we calculate the acoustic wavefield that is causally related to the action of sources of bounded extent in an unbounded homogeneous medium. Mathematically, one can take advantage of this spatial invariance of the configuration by carrying out a spatial Fourier transformation. The latter is applied to the wavefield quantities that satisfy the $s$-domain equations discussed in Chapter 3. Then, an algebraic problem remains, in which the spatial Fourier-transform parameter $js\alpha$ and the Laplace-transform parameter $s$ occur. The acoustic scalar and vector potentials are introduced. Inversion to the spatial domain and to the time domain yields the desired representations for the wavefield quantities. Specifically, the acoustic wavefield of a point source is calculated.

## 4.1. Source representations in the spectral domain

The $s$-domain particle velocity and acoustic pressure in a homogeneous fluid with constant volume density of mass $\rho$ and constant compressibility $\kappa$ satisfy the $s$-domain acoustic wavefield equations (cf. Eqs. (3.17) and (3.18))

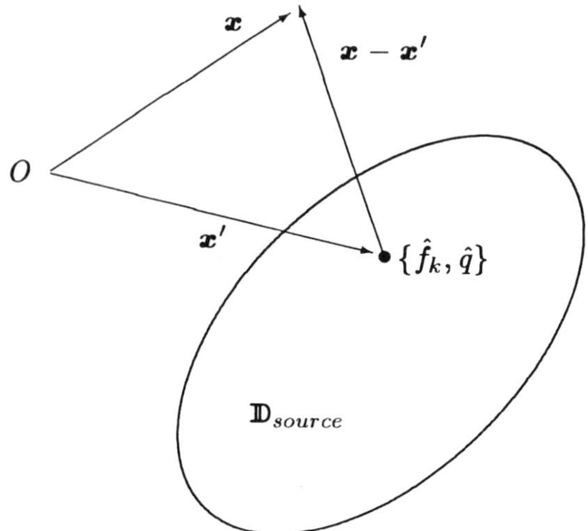


Figure 4.1. Source domain in an unbounded homogeneous embedding.

$$\partial_k \hat{p} + s\rho \hat{v}_k = \hat{f}_k \, , \tag{4.1}$$

$$\partial_k \hat{v}_k + s\kappa \hat{p} = \hat{q} \, . \tag{4.2}$$

We assume that $\hat{f}_k$ and $\hat{q}$ only differ from zero in some bounded subdomain $\mathbb{D}_{source}$ of $\mathbb{R}^3$ (Fig. 4.1). To solve Eqs. (4.1) and (4.2) we subject these equations to a three-dimensional Fourier transformation (cf. Section 1.2.2) over the entire three-dimensional configuration space $\mathbb{R}^3$. We further assume that $\hat{v}_k$ and $\hat{p}$ show, for $\mathrm{Re}(s) > 0$, an exponential decay as $|\boldsymbol{x}| \to \infty$. Then, the spatial derivative $\partial_k$ may be replaced by a multiplication with $-js\alpha_k$ in the spectral domain. With this, Eqs. (4.1) and (4.2) transform into

$$- js\alpha_k \tilde{p} + s\rho \tilde{v}_k = \tilde{f}_k \, , \tag{4.3}$$

$$- js\alpha_k \tilde{v}_k + s\kappa \tilde{p} = \tilde{q} \, , \tag{4.4}$$

where

$$\{\tilde{p}, \tilde{v}_k\}(js\boldsymbol{\alpha}, s) = \int_{\boldsymbol{x}\in\mathbb{R}^3} \exp(js\alpha_q x_q)\{\hat{p}, \hat{v}_k\}(\boldsymbol{x}, s)\mathrm{dV} \, , \tag{4.5}$$

$$\{\tilde{q}, \tilde{f}_k\}(js\boldsymbol{\alpha}, s) = \int_{\boldsymbol{x}\in\mathbb{D}_{source}} \exp(js\alpha_q x_q)\{\hat{q}, \hat{f}_k\}(\boldsymbol{x}, s)\mathrm{dV} , \tag{4.6}$$

and $\boldsymbol{\alpha}$ is the angular-slowness vector.

To solve $\{\tilde{p}, \tilde{v}_k\}$ from Eqs. (4.3) and (4.4) we multiply Eq. (4.4) by $s\rho$ and substitute Eq. (4.3) into Eq. (4.4). The result is

$$s^2(\alpha_q\alpha_q + \kappa\rho)\tilde{p} = s\rho\tilde{q} + js\alpha_k\tilde{f}_k . \tag{4.7}$$

From Eq. (4.7) the expression for the acoustic pressure in the angular-slowness-vector space is obtained as

$$\tilde{p} = \tilde{G}(s\rho\tilde{q} + js\alpha_k\tilde{f}_k) , \tag{4.8}$$

in which

$$\tilde{G} = \frac{1}{s^2(\alpha_q\alpha_q + \kappa\rho)} . \tag{4.9}$$

From Eqs. (4.3) and (4.8) the expression for the particle velocity in the angular-slowness-vector space follows as

$$\tilde{v}_l = \frac{1}{s\rho}\tilde{f}_l + js\alpha_l\tilde{G}\tilde{q} + \frac{1}{s\rho}js\alpha_l js\alpha_k\tilde{G}\tilde{f}_k . \tag{4.10}$$

To elucidate the structure of the right-hand sides of Eqs. (4.8) and (4.10), we introduce the angular-slowness-vector-domain acoustic scalar potential

$$\tilde{\Psi} = \tilde{G}\tilde{q} \tag{4.11}$$

and the angular-slowness-vector-domain acoustic vector potential

$$\tilde{W}_k = \tilde{G}\tilde{f}_k . \tag{4.12}$$

In terms of these, Eq. (4.8) can be rewritten as

$$\tilde{p} = s\rho\tilde{\Psi} + js\alpha_k\tilde{W}_k \tag{4.13}$$

and Eq. (4.10) as

$$\tilde{v}_l = \frac{1}{s\rho}\tilde{f}_l + js\alpha_l\tilde{\Psi} + \frac{1}{s\rho}js\alpha_l js\alpha_k\tilde{W}_k . \tag{4.14}$$

Equations (4.13) and (4.14) are the desired source representations in the angular-slowness-vector domain or spectral domain.

## 4.2. Source representations in the $s$-domain

The inverse spatial Fourier transformation of Eqs. (4.13) and (4.14) will be carried out; this will result into the $s$-domain expressions for the acoustic pressure and the particle velocity of the acoustic wavefield radiated by the sources. Using the rule that $-js\alpha_k$ corresponds to $\partial_k$, we obtain the $s$-domain source representations

$$\hat{p} = s\rho\hat{\Psi} - \partial_k \hat{W}_k \tag{4.15}$$

and

$$\hat{v}_l = \frac{1}{s\rho}\hat{f}_l - \partial_l \hat{\Psi} + \frac{1}{s\rho}\partial_l \partial_k \hat{W}_k \,. \tag{4.16}$$

The expressions for the $s$-domain acoustic scalar and vector potentials $\hat{\Psi}$ and $\hat{W}_k$ are obtained by carrying out the inverse spatial Fourier transformation of Eqs. (4.11) and (4.12), respectively. Since the product of two functions in the angular-slowness-vector space corresponds to the convolution of these functions in the spatial domain (cf. Section 1.2.2), we obtain

$$\hat{\Psi}(\boldsymbol{x}, s) = \int_{\boldsymbol{x}' \in \mathbb{D}_{source}} \hat{G}(\boldsymbol{x} - \boldsymbol{x}', s)\hat{q}(\boldsymbol{x}', s)\mathrm{dV}, \quad \boldsymbol{x} \in \mathbb{R}^3 , \tag{4.17}$$

and

$$\hat{W}_k(\boldsymbol{x}, s) = \int_{\boldsymbol{x}' \in \mathbb{D}_{source}} \hat{G}(\boldsymbol{x} - \boldsymbol{x}', s)\hat{f}_k(\boldsymbol{x}', s)\mathrm{dV}, \quad \boldsymbol{x} \in \mathbb{R}^3 , \tag{4.18}$$

in which the Green's function

$$\hat{G}(\boldsymbol{x}, s) = \frac{1}{(2\pi)^3} \int_{s\boldsymbol{\alpha} \in \mathbb{R}^3} \exp(-js\alpha_q x_q)\tilde{G}(js\boldsymbol{\alpha}, s)\mathrm{dV} \,. \tag{4.19}$$

Note that in the right-hand sides of Eqs. (4.17) and (4.18) we have taken care to distribute the arguments over the functions such that the integration is carried out over the fixed source domain $\mathbb{D}_{source}$.

Equations (4.15) - (4.16) constitute the solution to the $s$-domain acoustic radiation problem in an unbounded homogeneous medium. Owing to the simple way in which the Laplace-transform parameter $s$ occurs, the inversion back to the time domain can easily be carried out. Before going to the time domain, we first consider the Green's function and the far-field expressions.

*The $s$-domain expression of the Green's function*

In this subsection we discuss the spatial Fourier integral representation of the Green's function (cf. Eq. (4.19)), in which the spectral representation of the Green's function is given by (cf. Eq. (4.9))

$$\tilde{G}(js\boldsymbol{\alpha}, s) = \frac{1}{s^2\alpha_q\alpha_q + \hat{\gamma}^2}, \tag{4.20}$$

with

$$\hat{\gamma} = \frac{s}{c}, \tag{4.21}$$

where the acoustic wave speed $c$ is given by

$$c = (\kappa\rho)^{-\frac{1}{2}}. \tag{4.22}$$

It is easily verified that $\tilde{G}$ is the three-dimensional Fourier transform, over the entire configuration space $\mathbb{R}^3$, of the function $\hat{G} = \hat{G}(\boldsymbol{x}, s)$ that satisfies the three-dimensional modified Helmholtz equation

$$(\partial_q\partial_q - \hat{\gamma}^2)\,\hat{G} = -\delta(\boldsymbol{x}), \tag{4.23}$$

where $\delta(\boldsymbol{x})$ is the three-dimensional Dirac distribution (impulse function) operative at $\boldsymbol{x} = 0$. The simplest way to evaluate the right-hand side of Eq. (4.19), where $\tilde{G}(js\boldsymbol{\alpha}, s)$ is given by Eq. (4.20), is to introduce spherical coordinates in the $s\boldsymbol{\alpha}$-domain with origin at $s\alpha_k = 0$ and the direction $\boldsymbol{x}$ as polar axis. Let $\hat{\beta} = |s\boldsymbol{\alpha}|$ and $\theta$ the polar angle between $s\boldsymbol{\alpha}$ and $\boldsymbol{x}$. Then the range of integration is $0 \le \hat{\beta} < \infty$, $0 \le \theta \le \pi$, $0 \le \phi < 2\pi$, where $\phi$ is the azimuth angle in the plane perpendicular to $\boldsymbol{x}$. In the integral we use $s\alpha_q x_q = \hat{\beta}|\boldsymbol{x}|\cos(\theta)$, $s^2\alpha_q\alpha_q = \hat{\beta}^2$ and $\mathrm{d}V = \hat{\beta}^2\sin(\theta)\mathrm{d}\hat{\beta}\mathrm{d}\theta\mathrm{d}\phi$. In the resulting integral we first carry out the integration with respect to $\phi$; this merely amounts to a multiplication by a factor of $2\pi$. Next we carry out the integration with respect to $\theta$, which is elementary. After this we have

$$\hat{G}(\boldsymbol{x}, s) = \frac{1}{4\pi^2 j|\boldsymbol{x}|}\int_0^\infty \frac{\exp(j\hat{\beta}|\boldsymbol{x}|) - \exp(-j\hat{\beta}|\boldsymbol{x}|)}{\hat{\beta}^2 + \hat{\gamma}^2}\hat{\beta}\mathrm{d}\hat{\beta}. \tag{4.24}$$

This can be written as

$$\hat{G}(\boldsymbol{x}, s) = \frac{1}{4\pi^2 j|\boldsymbol{x}|}\int_{-\infty}^\infty \frac{\exp(j\hat{\beta}|\boldsymbol{x}|)}{\hat{\beta}^2 + \hat{\gamma}^2}\hat{\beta}\mathrm{d}\hat{\beta}. \tag{4.25}$$

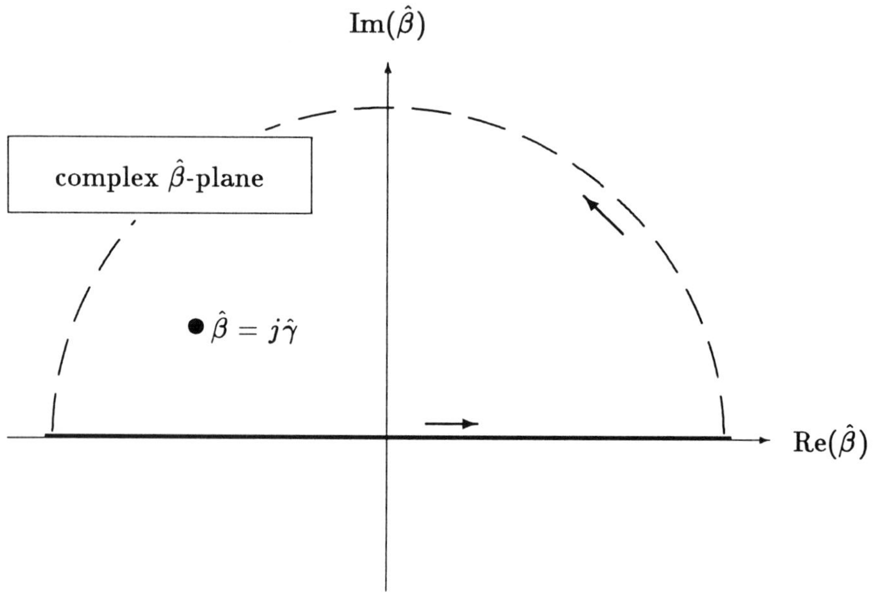


Figure 4.2. Contour integration in the complex $\hat{\beta}$-plane.

The integral at the right-hand side of Eq. (4.25) is evaluated by continuing the integrand analytically into the complex $\hat{\beta}$-plane, supplementing the path of integration by a semi-circle situated in the upper half-plane $0 \leq \mathrm{Im}(\hat{\beta}) < \infty$ and of infinitely large radius, and applying the theorem of residues (Fig. 4.2). On account of Jordan's lemma the contribution from the semi-circle at infinity vanishes. Further, the only singularity of the integrand in the upper half of the complex $\hat{\beta}$-plane is the simple pole $\hat{\beta} = j\hat{\gamma}$. Taking into account the residue of this pole, we finally arrive at the Green's function in the $s$-domain

$$\hat{G}(\boldsymbol{x}, s) = \frac{\exp(-\hat{\gamma}|\boldsymbol{x}|)}{4\pi|\boldsymbol{x}|} . \tag{4.26}$$

This is the well-known expression for the spherical wave due to a point source in a homogeneous medium.

# 4.3. Far-field radiation characteristics in the $s$-domain

In many applications one is often particularly interested in the behavior of the radiated wavefield at large distances from the radiating structures. To investigate this behavior, we consider the leading term in the expansion of the right-hand sides of Eqs. (4.15) - (4.16) as $|\boldsymbol{x}| \to \infty$; this term is denoted as the far-field approximation of the relevant acoustic wavefield. The region in which the far-field approximation sufficiently accurately represents the wavefield values is denoted as the far-field region.

To arrive at the far-field representations we observe that (see Fig. 4.3)

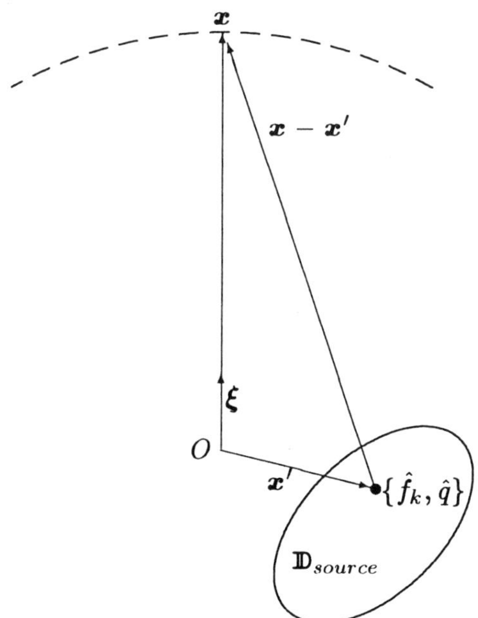


Figure 4.3. Configuration for the far-field radiation-characteristics.

$$
\begin{aligned}
|\boldsymbol{x} - \boldsymbol{x}'| &= \left[(x_q - x'_q)(x_q - x'_q)\right]^{\frac{1}{2}} \\
&= |\boldsymbol{x}| \left[1 - \frac{2x_q x'_q}{|\boldsymbol{x}|^2} + \frac{|\boldsymbol{x}'|^2}{|\boldsymbol{x}|^2}\right]^{\frac{1}{2}} ,
\end{aligned}
\tag{4.27}
$$

and hence

$$
|\boldsymbol{x} - \boldsymbol{x}'| = |\boldsymbol{x}| - \xi_q x'_q + \mathrm{Order}(|\boldsymbol{x}|^{-1}) \text{ as } |\boldsymbol{x}| \to \infty , \tag{4.28}
$$

where

$$
\xi_k = \frac{x_k}{|\boldsymbol{x}|} \tag{4.29}
$$

is the unit vector in the direction of observation. For the derivatives of $|\boldsymbol{x} - \boldsymbol{x}'|$ we further have

$$
\partial_k |\boldsymbol{x} - \boldsymbol{x}'| = \frac{x_k - x'_k}{|\boldsymbol{x} - \boldsymbol{x}'|} , \tag{4.30}
$$

and hence

$$
\partial_k |\boldsymbol{x} - \boldsymbol{x}'| = \xi_k + \mathrm{Order}(|\boldsymbol{x}|^{-1}) \text{ as } |\boldsymbol{x}| \to \infty . \tag{4.31}
$$

Using these results, the Green's function of Eq. (4.26) can, in the far-field region, be approximated by

$$
\hat{G}(\boldsymbol{x} - \boldsymbol{x}', s) = \hat{G}(\boldsymbol{x}, s)\exp(\hat{\gamma}\xi_q x'_q) + \mathrm{Order}(|\boldsymbol{x}|^{-2}) \text{ as } |\boldsymbol{x}| \to \infty , \tag{4.32}
$$

and its spatial derivatives by

$$
\partial_k \hat{G}(\boldsymbol{x} - \boldsymbol{x}', s) = -\hat{\gamma}\xi_k \hat{G}(\boldsymbol{x}, s)\exp(\hat{\gamma}\xi_q x'_q) + \mathrm{Order}(|\boldsymbol{x}|^{-2}) \text{ as } |\boldsymbol{x}| \to \infty . \tag{4.33}
$$

Using these results in the expressions of Eqs. (4.17) and (4.18) for the potentials we obtain their far-field approximations

$$
\{\hat{\Psi}, \hat{W}_k\}(\boldsymbol{x}, s) = \{\hat{\Psi}^{\infty}, \hat{W}_k^{\infty}\}(\boldsymbol{\xi}, s)\hat{G}(\boldsymbol{x}, s) + \mathrm{Order}(|\boldsymbol{x}|^{-2}) \text{ as } |\boldsymbol{x}| \to \infty , \tag{4.34}
$$

in which

$$
\{\hat{\Psi}^{\infty}, \hat{W}_k^{\infty}\}(\boldsymbol{\xi}, s) = \int_{\boldsymbol{x}' \in \mathbb{D}_{source}} \exp(\hat{\gamma}\xi_q x'_q)\{\hat{q}, \hat{f}_k\}(\boldsymbol{x}', s)\mathrm{dV} . \tag{4.35}
$$

Using them in the expressions of Eqs. (4.15) and (4.16) for the acoustic pressure and particle velocity, we obtain the far-field approximations

$$
\{\hat{p}, \hat{v}_k\}(\boldsymbol{x}, s) = \{\hat{p}^{\infty}, \hat{v}_k^{\infty}\}(\boldsymbol{\xi}, s)\hat{G}(\boldsymbol{x}, s) + \mathrm{Order}(|\boldsymbol{x}|^{-2}) \text{ as } |\boldsymbol{x}| \to \infty , \tag{4.36}
$$

in which

$$\hat{p}^{\infty} = s\rho\hat{\Psi}^{\infty} + \hat{\gamma}\xi_k\hat{W}_k^{\infty} , \tag{4.37}$$

$$\hat{v}_l^{\infty} = \hat{\gamma}\xi_l\hat{\Psi}^{\infty} + \frac{\hat{\gamma}^2}{s\rho}\xi_l\xi_k\hat{W}_k^{\infty} , \tag{4.38}$$

where we have used the fact that $\{\hat{q}, \hat{f}_k\}$ vanishes outside the source domain. As Eq. (4.36) shows, the acoustic pressure and particle velocity have, in the far-field region, the structure of a spherical wave that expands radially from the origin of the coordinate system (which is also denoted as the phase center of the far-field approximation), the latter being chosen in the neighborhood of the radiating sources, with an amplitude that depends on the direction of observation and that decreases inversely proportional to the distance from the chosen origin. The amplitude radiation characteristics $\{\hat{p}^{\infty}, \hat{v}_k^{\infty}\}$ depend only on the direction of observation $\boldsymbol{\xi}$, and on $s$.

Upon inspecting the dependence of $\hat{\Psi}^{\infty}$ and $\hat{W}_k^{\infty}$ on $\boldsymbol{\xi}$, a comparison of Eq. (4.35) with Eq. (4.6) shows that

$$\{\hat{\Psi}^{\infty}, \hat{W}_k^{\infty}\}(\boldsymbol{\xi}, s) = \{\tilde{q}, \tilde{f}_k\}(\hat{\gamma}\boldsymbol{\xi}, s) . \tag{4.39}$$

Consequently, in the far-field region only the spatial Fourier transforms of the source distribution at the subset of angular-wave-vector values $js\boldsymbol{\alpha} = \hat{\gamma}\boldsymbol{\xi}$ are "seen". Since $s\boldsymbol{\alpha}$ is real, the observation is only true if $s$ is imaginary valued. In the latter case we introduce the real slowness vector $\boldsymbol{p} = j\boldsymbol{\alpha}$ (see Eq. (1.42)). Then, Eq. (4.39) is rewritten as

$$\{\hat{\Psi}^{\infty}, \hat{W}_k^{\infty}\}(\boldsymbol{\xi}, j\omega) = \{\tilde{q}, \tilde{f}_k\}(j\omega\boldsymbol{p}, j\omega), \quad \boldsymbol{p} = \frac{\boldsymbol{\xi}}{c} . \tag{4.40}$$

## 4.4. Source representations in the time domain

Owing to the simple structure of Eqs. (4.15) - (4.16) in which the Laplace-transform parameter $s$ occur, the inversion of the acoustic wavefield quantities back to the time domain can now easily be carried out. Evidently, the time-domain equivalents of Eqs. (4.15) -(4.16) contain the time-differentiated and the time-integrated forms of the acoustic scalar and vector potentials. For the latter, we employ the notation

$$\{\mathcal{I}_t\Psi, \mathcal{I}_t W_k\}(\boldsymbol{x},t) = \{\int_{t_0}^{t} \Psi(\boldsymbol{x},t')\mathrm{d}t', \int_{t_0}^{t} W_k(\boldsymbol{x},t')\mathrm{d}t'\}\,. \tag{4.41}$$

Using Eq. (1.23), we obtain the time-domain equivalents of Eqs. (4.15) - (4.16) as

$$\rho\partial_t\Psi - \partial_k W_k = \chi_{\mathrm{T}}(t)p(\boldsymbol{x},t)\,, \tag{4.42}$$

$$\frac{1}{\rho}\mathcal{I}_t f_l - \partial_l\Psi + \frac{1}{\rho}\partial_l\partial_k\mathcal{I}_t W_k = \chi_{\mathrm{T}}(t)v_l(\boldsymbol{x},t)\,. \tag{4.43}$$

The time-domain equivalent of the Green's function is easily obtained from Eq. (4.26) as

$$G(\boldsymbol{x},t) = \frac{\delta(t-\frac{|\boldsymbol{x}|}{c})}{4\pi|\boldsymbol{x}|}\,. \tag{4.44}$$

Therefore, the time-domain equivalents of the scalar and vector potentials of Eqs. (4.17) and (4.18) are obtained as

$$\Psi(\boldsymbol{x},t) = \int_{\boldsymbol{x}'\in\mathbb{D}_{source}} \frac{q(\boldsymbol{x}',t-\frac{|\boldsymbol{x}-\boldsymbol{x}'|}{c})}{4\pi|\boldsymbol{x}-\boldsymbol{x}'|}\mathrm{dV}\,, \tag{4.45}$$

and

$$W_k(\boldsymbol{x},t) = \int_{\boldsymbol{x}'\in\mathbb{D}_{source}} \frac{f_k(\boldsymbol{x}',t-\frac{|\boldsymbol{x}-\boldsymbol{x}'|}{c})}{4\pi|\boldsymbol{x}-\boldsymbol{x}'|}\mathrm{dV}\,. \tag{4.46}$$

Expressions (4.45) and (4.46) are known as retarded potentials: the time argument in the integrands is delayed by the travel time $\frac{|\boldsymbol{x}-\boldsymbol{x}'|}{c}$ that the acoustic wave needs to traverse the distance $|\boldsymbol{x}-\boldsymbol{x}'|$ from the source point $\boldsymbol{x}'$ to the observation point $\boldsymbol{x}$ with the speed $c$.

## 4.5. Far-field characteristics in the time domain

In this section we consider the far-field radiation characteristics that result from the time-domain source-type representations of Eqs. (4.42) - (4.43) as $|\boldsymbol{x}| \to \infty$. Using Eq. (4.28) we obtain for the retarded time argument occurring in Eq. (4.45) and (4.46) the expression

$$t - \frac{|\boldsymbol{x} - \boldsymbol{x}'|}{c} = t - \frac{|\boldsymbol{x}|}{c} + \frac{\xi_q x'_q}{c} + \text{Order}(|\boldsymbol{x}|^{-1}) \text{ as } |\boldsymbol{x}| \to \infty . \tag{4.47}$$

With this the far-field representations for the vector potentials are obtained as

$$\{\Psi, W_k\}(\boldsymbol{x}, t) = \frac{\{\Psi^\infty, W_k^\infty\}(\boldsymbol{\xi}, t - \frac{|\boldsymbol{x}|}{c})}{4\pi|\boldsymbol{x}|} + \text{Order}(|\boldsymbol{x}|^{-2}) \text{ as } |\boldsymbol{x}| \to \infty, \tag{4.48}$$

in which

$$\{\Psi^\infty, W_k^\infty\}(\boldsymbol{\xi}, t) = \int_{\boldsymbol{x}' \in \mathbb{D}_{source}} \{q, f_k\}(\boldsymbol{x}', t + \frac{\xi_q x'_q}{c}) \mathrm{dV} . \tag{4.49}$$

We further obtain for the spatial derivatives

$$\partial_k\{\Psi, W_k\}(\boldsymbol{x}, t) = -\frac{\xi_k}{c} \frac{\{\partial_t \Psi^\infty, \partial_t W_k^\infty\}(\boldsymbol{\xi}, t - \frac{|\boldsymbol{x}|}{c})}{4\pi|\boldsymbol{x}|} + \text{Order}(|\boldsymbol{x}|^{-2})$$

$$\text{as } |\boldsymbol{x}| \to \infty , \tag{4.50}$$

where derivatives with respect to $x_k$ have been interchanged for a derivative with respect to time. Using Eqs. (4.48) - (4.50) in the expressions of Eqs. (4.42) - (4.43), we arrive at

$$\{p, v_k\}(\boldsymbol{x}, t) = \frac{\{p^\infty, v_k^\infty\}(\boldsymbol{\xi}, t - \frac{|\boldsymbol{x}|}{c})}{4\pi|\boldsymbol{x}|} + \text{Order}(|\boldsymbol{x}|^{-2}) \text{ as } |\boldsymbol{x}| \to \infty , \tag{4.51}$$

in which

$$p^\infty = \rho \partial_t \Psi^\infty + \frac{\xi_k}{c} \partial_t W_k^\infty , \tag{4.52}$$

$$v_l^\infty = \frac{\xi_l}{c} \partial_t \Psi^\infty + \kappa \xi_l \xi_k \partial_t W_k^\infty . \tag{4.53}$$

As Eq. (4.51) shows, the acoustic pressure and particle velocity have, in the far-field region, the structure of a spherical wave that expands radially from the origin of the coordinate system, the latter being chosen in the neighborhood of the radiating sources, with an amplitude that depends on the direction of observation and that decreases inversely proportional to the distance from the chosen origin. The amplitude radiation characteristics $\{p^\infty, v_k^\infty\}$ depend only on the direction of observation $\boldsymbol{\xi}$, and on the pulse shapes of the source distributions. Note that in the right-hand sides of Eqs. (4.52) and (4.53) via Eq. (4.50) only the time-differentiated pulse shapes of the volume source densities occur.

Upon inspecting the dependence of $\Psi^\infty$ and $W_k^\infty$ on $\boldsymbol{\xi}$, a comparison of Eq. (4.49) with Eq. (1.45) shows that

$$\{\Psi^\infty, W_k^\infty\}(\boldsymbol{\xi}, t) = \{\check{q}, \check{f}_k\}(\frac{\boldsymbol{\xi}}{c}, t) , \tag{4.54}$$

where $\check{q}$ and $\check{f}_k$ denote the three-dimensional Radon transform of $q$ and $f_k$, respectively. Consequently, in the far-field region only the Radon transform of source distributions at the subset of Radon-transform parameters $\boldsymbol{p} = \frac{\boldsymbol{\xi}}{c}$ are "seen".

## 4.6. The Cagniard-de Hoop method

An alternative method to arrive at the time-domain representation of the Green's function is the Cagniard-de Hoop method (DE HOOP, 1960). To illustrate the essence of this technique in obtaining time-domain versions from the corresponding spectral representations, we start with the expression of the Green's function given in Eq. (4.20):

$$\tilde{G}(js\boldsymbol{\alpha}, s) = \frac{1}{s^2\alpha_q\alpha_q + \hat{\gamma}^2} . \tag{4.55}$$

As a first step, we perform an inverse Fourier transform with respect to real parameter $s\alpha_3$ and denote the results as $\overline{G}(js\alpha_1, js\alpha_2, x_3, s)$, according to the definition of the Fourier transform of Eq. (1.46). The two-dimensional Fourier transform of the Green's function becomes

$$\overline{G}(js\alpha_1, js\alpha_2, x_3, s) = \frac{1}{2\pi}\int_{(s\alpha_3)\in\mathbb{R}} \exp(-js\alpha_3 x_3)\frac{1}{(s\alpha_3)^2 + s^2\Gamma^2}\mathrm{d}(s\alpha_3) , \tag{4.56}$$

where

$$\Gamma = \left(\frac{1}{c^2} + \alpha_1^2 + \alpha_2^2\right)^{\frac{1}{2}} , \quad \mathrm{Re}(\Gamma) > 0 . \tag{4.57}$$

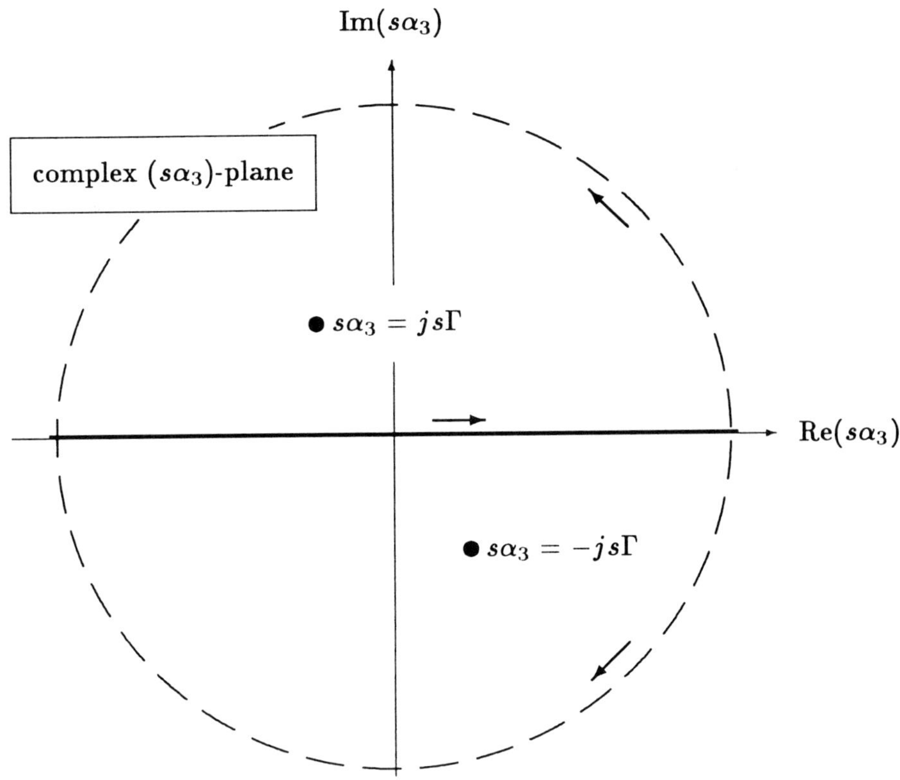


Figure 4.4. Contour integration in the complex ($s\alpha_3$)-plane.

When $x_3 \leq 0$, the integral at the right-hand side of Eq. (4.56) is evaluated by continuing the integrand analytically into the complex ($s\alpha_3$)-plane, supplementing the path of integration by a semi-circle situated in the upper half-plane $0 \leq \text{Im}(s\alpha_3) < \infty$ and of infinitely large radius, and applying the theorem of residues (see Fig. 4.4). On account of Jordan's lemma the contribution from the semi-circle at infinity vanishes. Further, the only singularity of the integrand in the upper half of the complex ($s\alpha_3$)-plane is the simple pole at $s\alpha_3 = js\Gamma$. Taking into account the residue of this pole, we arrive at

$$\overline{G}(js\alpha_1, js\alpha_2, x_3, s) = \frac{\exp(s\Gamma x_3)}{2s\Gamma}, \quad x_3 \leq 0\,. \tag{4.58}$$

When $x_3 \geq 0$, the integral at the right-hand side of Eq. (4.56) is evaluated by continuing the integrand analytically into the complex $(s\alpha_3)$-plane, supplementing the path of integration by a semi-circle situated in the lower half-plane $-\infty < \mathrm{Im}(s\alpha_3) \leq 0$ and of infinitely large radius, and applying the theorem of residues (see Fig. 4.4). On account of Jordan's lemma the contribution from the semi-circle at infinity vanishes. Further, the only singularity of the integrand in the upper half of the complex $(s\alpha_3)$-plane is the simple pole at $s\alpha_3 = -js\Gamma$. Taking into account the residue of this pole, we arrive at

$$\overline{G}(js\alpha_1, js\alpha_2, x_3, s) = \frac{\exp(-s\Gamma x_3)}{2s\Gamma}, \quad x_3 \geq 0 \, . \tag{4.59}$$

With the aid of Fourier's inversion theorem we obtain the following expression for $\hat{G}(\boldsymbol{x}, s)$:

$$\hat{G}(\boldsymbol{x}, s) = \frac{1}{(2\pi)^2} \int_{(s\alpha_1, s\alpha_2) \in \mathbb{R}^2} \frac{\exp(-js\alpha_1 x_1 - js\alpha_2 x_2 - s\Gamma|x_3|)}{2s\Gamma} \mathrm{dA} \, . \tag{4.60}$$

This is the well-known plane-wave representation of the Green's function for complex values of $s$, $\mathrm{Re}(s) > 0$.

As a next step, we shall try to cast the integral on the right-hand side of Eq. (4.60) in such a form that $G(\boldsymbol{x}, t)$ can be found by inspection. To that end, we first restrict the Laplace-transform parameter to real values ($s > 0$). The slowness vector $(\alpha_1, \alpha_2)$ is taken real as well. Then, the general plane-wave representation of Eq. (4.60) is written as

$$\hat{G}(\boldsymbol{x}, s) = \frac{s}{(2\pi)^2} \int_{(\alpha_1, \alpha_2) \in \mathbb{R}^2} \frac{\exp(-js\alpha_1 x_1 - js\alpha_2 x_2 - s\Gamma|x_3|)}{2\Gamma} \mathrm{dA} \, . \tag{4.61}$$

It will be advantageous to deform the integration surface such that the expression in the exponential function in the right-hand side of Eq. (4.61) becomes real. Therefore, we introduce the polar coordinates in the $(x_1, x_2)$-plane through

$$x_1 = r\cos(\phi), \quad x_2 = r\sin(\phi), \quad r = (x_1^2 + x_2^2)^{\frac{1}{2}}, \tag{4.62}$$

where $0 \leq r < \infty$ and $0 \leq \phi < 2\pi$. Next we replace the variables of integration $\alpha_1$ and $\alpha_2$ in Eq. (4.61) by $p$ and $q$ through

$$\begin{aligned} \alpha_1 &= -jp\cos(\phi) - q\sin(\phi), \\ \alpha_2 &= -jp\sin(\phi) + q\cos(\phi), \end{aligned} \tag{4.63}$$

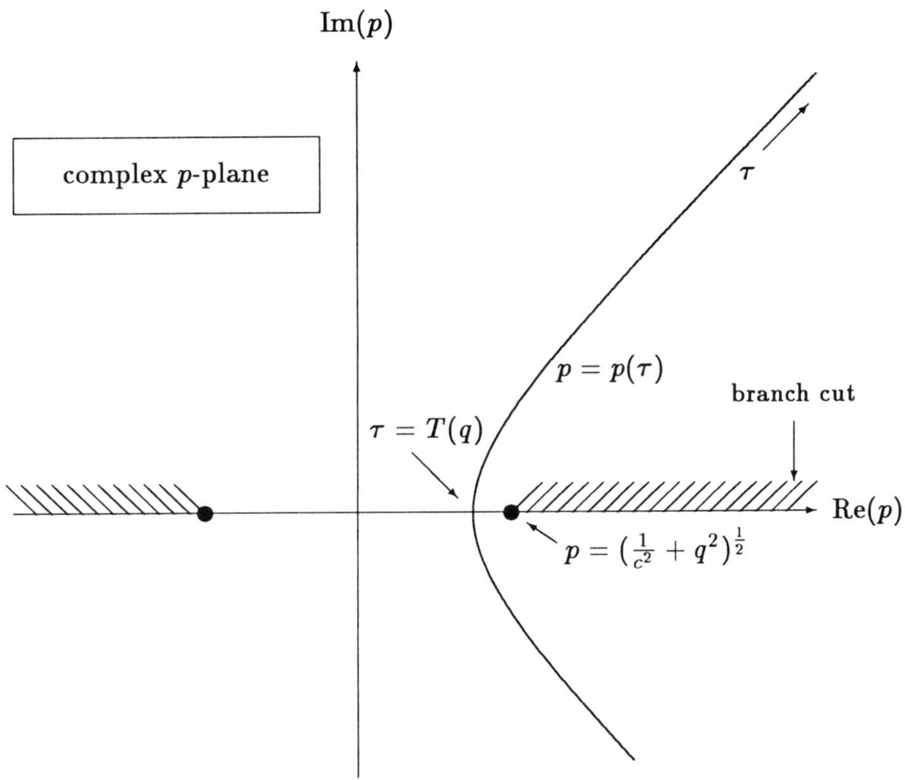


Figure 4.5. Complex $p$-plane with Cagniard-de Hoop path $p(\tau)$.

which entails $\alpha_1^2 + \alpha_2^2 = q^2 - p^2$, $\mathrm{d}\alpha_1 \mathrm{d}\alpha_2 = -j\mathrm{d}p\mathrm{d}q$ and $j\alpha_1 x_1 + j\alpha_2 x_2 = pr$. Hence, Eq. (4.61) is replaced by

$$\hat{G}(\boldsymbol{x}, s) = \frac{s}{(2\pi)^2} \int_{-\infty}^{\infty} \mathrm{d}q \int_{-j\infty}^{j\infty} \frac{\exp[-s(pr + \Gamma|x_3|)]}{2j\Gamma} \mathrm{d}p \,, \tag{4.64}$$

where

$$\Gamma = \left(\frac{1}{c^2} + q^2 - p^2\right)^{\frac{1}{2}} . \tag{4.65}$$

The next step towards the solution of the transient problem is to perform the integration in the complex $p$-plane along such a path that the right-hand side of Eq. (4.64) can be recognized as the Laplace transform of a certain function of time. We therefore extend the integrand of Eq. (4.64) in the complex $p$-plane. Since we want to keep the integrand single-valued, we introduce a branch cut for the square-root expression $\Gamma$ along $\{(\frac{1}{c^2}+q^2)^{\frac{1}{2}} < |\mathrm{Re}(p)| < \infty, \mathrm{Im}(p)=0\}$ and keep $\mathrm{Re}(\Gamma) \geq 0$ in the entire cut $p$-plane. On virtue of Cauchy's theorem and Jordan's lemma, the integration path along the imaginary $p$-axis is now deformed into a path along

$$pr + \Gamma|x_3| = \tau \,, \tag{4.66}$$

where $\tau$ is real and positive. If $\frac{r}{c} < \tau < \infty$, Eq. (4.66) represents the branch of a hyperbola. The parametric representation of this Cagniard-de Hoop path

$$p = \frac{r\tau + j|x_3|\{\tau^2 - [T(q)]^2\}^{\frac{1}{2}}}{|\boldsymbol{x}|^2} \,, \tag{4.67}$$

with

$$T(q) = |\boldsymbol{x}|\left(\frac{1}{c^2}+q^2\right)^{\frac{1}{2}} \tag{4.68}$$

and

$$|\boldsymbol{x}| = (r^2 + x_3^2)^{\frac{1}{2}} \,, \tag{4.69}$$

denotes the part of the hyperbola in the upper half of the $p$-plane, while $p = p^\star$ (the star denotes complex conjugation) represents the lower half (see Fig. 4.5). Using Eq. (4.67) together with Eq. (4.66), the Jacobian of this transformation is found as

$$\frac{\partial p}{\partial \tau} = \frac{r + j\tau|x_3|\{\tau^2 - [T(q)]^2\}^{-\frac{1}{2}}}{|\boldsymbol{x}|^2} = \frac{j\Gamma}{\{\tau^2 - [T(q)]^2\}^{\frac{1}{2}}} \,. \tag{4.70}$$

After this deformation and the use of Schwarz's principle of reflection, we are able to rewrite Eq. (4.64) as

$$\hat{G}(\boldsymbol{x}, s) = \frac{s}{4\pi^2}\int_{-\infty}^{\infty} \mathrm{d}q \int_{T(q)}^{\infty} \frac{\exp(-s\tau)}{\{\tau^2 - [T(q)]^2\}^{\frac{1}{2}}}\mathrm{d}\tau \,. \tag{4.71}$$

Next we interchange the order of integration leading to

$$\hat{G}(\boldsymbol{x}, s) = \frac{s}{4\pi^2|\boldsymbol{x}|}\int_{T_0}^{\infty} \exp(-s\tau)\mathrm{d}\tau \int_{-Q(\tau)}^{Q(\tau)} \frac{1}{\{[Q(\tau)]^2 - q^2\}^{\frac{1}{2}}}\mathrm{d}q \,, \tag{4.72}$$

where

$$T_0 = T(0) = \frac{|\boldsymbol{x}|}{c} \tag{4.73}$$

and

$$Q(\tau) = \frac{[\tau^2 - (\frac{|\boldsymbol{x}|}{c})^2]^{\frac{1}{2}}}{|\boldsymbol{x}|} . \tag{4.74}$$

Introducing the new variable of integration $q = Q(\tau)\sin(\psi)$, we arrive at

$$\hat{G}(\boldsymbol{x}, s) = \frac{s}{4\pi^2|\boldsymbol{x}|} \int_{T_0}^{\infty} \exp(-s\tau)\mathrm{d}\tau \int_{-\frac{1}{2}\pi}^{\frac{1}{2}\pi} \mathrm{d}\psi , \tag{4.75}$$

or

$$\hat{G}(\boldsymbol{x}, s) = \frac{\exp(-sT_0)}{4\pi|\boldsymbol{x}|} , \tag{4.76}$$

which corresponds to Eq. (4.26) with $\hat{\gamma}|\boldsymbol{x}| = sT_0$. The time-domain equivalent of the Green's function is (cf. Eq. (4.44))

$$G(\boldsymbol{x}, t) = \frac{\delta(t - T_0)}{4\pi|\boldsymbol{x}|} . \tag{4.77}$$

This final result concludes the discussion of the Cagniard-de Hoop method, (see also AKI and RICHARDS, 1980, section 6.5). We have shown the essential steps in the latter technique. We remark that this method can be extended to the case of an arbitrary planar layered medium (VAN DER HIJDEN, 1987). In the latter case, the Cagniard-de Hoop path has to be constructed for every generalized-ray constituent.

## 4.7. The acoustic wavefield of point sources

In this section we calculate the acoustic pressure and particle velocity of the wavefield emitted by two point sources of different type, viz. the acoustic monopole transducer and the acoustic dipole transducer.

*The monopole transducer*

An acoustic monopole transducer is a point source of volume injection, i.e., a source of volume injection whose maximum diameter is negligibly small with respect to the distance from the point of observation to the position of the source and negligibly small with respect to the spatial extent of the emitted wave. We start our analysis in the $s$-domain. Let the monopole source be located at $\boldsymbol{x}^S$, then the source densities are given by

$$\hat{q}(\boldsymbol{x}, s) = \hat{q}^S(s)\delta(\boldsymbol{x} - \boldsymbol{x}^S)\,, \tag{4.78}$$

$$\hat{f}_k^S(\boldsymbol{x}, s) = 0\,, \tag{4.79}$$

where $\hat{q}^S(s)$ is the $s$-domain time rate of volume injection. Then the scalar potential $\hat{\Psi}$ and the vector potential $\hat{W}_k$ become (see Eqs. (4.17) and (4.18))

$$\hat{\Psi}(\boldsymbol{x}, s) = \hat{q}^S(s)\,\hat{G}(\boldsymbol{x} - \boldsymbol{x}^S, s)\,, \tag{4.80}$$

$$\hat{W}_k(\boldsymbol{x}, s) = 0\,. \tag{4.81}$$

The acoustic pressure and the particle velocity of the wavefield emitted by the monopole transducer is then given (see Eqs. (4.15) and (4.16)) as

$$\hat{p}(\boldsymbol{x}, s) = s\rho\hat{q}^S(s)\,\hat{G}(\boldsymbol{x} - \boldsymbol{x}^S, s) \tag{4.82}$$

and

$$\hat{v}_l(\boldsymbol{x}, s) = -\hat{q}^S(s)\,\partial_l\hat{G}(\boldsymbol{x} - \boldsymbol{x}^S, s)\,, \tag{4.83}$$

where

$$\hat{G}(\boldsymbol{x} - \boldsymbol{x}^S, s) = \frac{\exp(-\frac{s}{c}|\boldsymbol{x} - \boldsymbol{x}^S|)}{4\pi|\boldsymbol{x} - \boldsymbol{x}^S|}\,. \tag{4.84}$$

From these $s$-domain quantities, the time-domain equivalents can directly be obtained, viz.,

$$\rho\partial_t\frac{q^S(t - \frac{|\boldsymbol{x}-\boldsymbol{x}^S|}{c})}{4\pi|\boldsymbol{x} - \boldsymbol{x}^S|} = \chi_{\mathrm{T}}(t)p(\boldsymbol{x}, t)\,, \tag{4.85}$$

$$-\,\partial_l\frac{q^S(t - \frac{|\boldsymbol{x}-\boldsymbol{x}^S|}{c})}{4\pi|\boldsymbol{x} - \boldsymbol{x}^S|} = \chi_{\mathrm{T}}(t)v_l(\boldsymbol{x}, t)\,, \tag{4.86}$$

in which $q^S(t)$ is the time rate of volume injection.

*The dipole transducer*

An acoustic dipole transducer is a point source of volume force, i.e., a source of volume force whose maximum diameter is negligibly small with respect to the distance from the point of observation to the position of the source and negligibly small with respect to the spatial extent of the emitted wave. We start our analysis in the $s$-domain. Let the dipole source be located at $\boldsymbol{x}^S$, then the source densities are given by

$$\hat{q}^S(\boldsymbol{x}, s) = 0\,, \tag{4.87}$$

$$\hat{f}_k(\boldsymbol{x}, s) = \hat{f}_k^S(s)\delta(\boldsymbol{x} - \boldsymbol{x}^S)\,, \tag{4.88}$$

where $\hat{f}_k^S(s)$ is the $s$-domain time rate of volume force. Then the scalar potential $\hat{\Psi}$ and the vector potential $\hat{W}_q$ become (see Eqs. (4.17) and (4.18))

$$\hat{\Psi}(\boldsymbol{x}, s) = 0\,, \tag{4.89}$$

$$\hat{W}_k(\boldsymbol{x}, s) = \hat{f}_k^S(s)\,\hat{G}(\boldsymbol{x} - \boldsymbol{x}^S, s)\,. \tag{4.90}$$

The acoustic pressure and the particle velocity of the wavefield emitted by the dipole transducer is then given (see Eqs. (4.15) and (4.16)) as

$$\hat{p}(\boldsymbol{x}, s) = -\hat{f}_k^S(s)\partial_k\hat{G}(\boldsymbol{x} - \boldsymbol{x}^S, s) \tag{4.91}$$

and

$$\hat{v}_l(\boldsymbol{x}, s) = \frac{1}{s\rho}\hat{f}_l^S(s)\delta(\boldsymbol{x} - \boldsymbol{x}^S) + \frac{1}{s\rho}\hat{f}_k^S(s)\partial_k\partial_l\hat{G}(\boldsymbol{x} - \boldsymbol{x}^S, s)\,. \tag{4.92}$$

From these $s$-domain quantities, the time-domain equivalents can directly be obtained, viz.,

$$-\,\partial_k\frac{f_k^S(t - \frac{|\boldsymbol{x}-\boldsymbol{x}^S|}{c})}{4\pi|\boldsymbol{x} - \boldsymbol{x}^S|} = \chi_{\mathrm{T}}(t)p(\boldsymbol{x}, t)\,, \tag{4.93}$$

$$\frac{1}{\rho}\mathcal{S}_t f_l^S(t)\,\delta(\boldsymbol{x} - \boldsymbol{x}^S) + \frac{1}{\rho}\partial_l\partial_k\frac{\mathcal{S}_t f_k^S(t - \frac{|\boldsymbol{x}-\boldsymbol{x}^S|}{c})}{4\pi|\boldsymbol{x} - \boldsymbol{x}^S|} = \chi_{\mathrm{T}}(t)v_l(\boldsymbol{x}, t)\,, \tag{4.94}$$

in which $f_k^S(t)$ is the time rate of volume force.

The concept of a point source is of importance in our analysis of the acoustic wavefield of the Green's state, to be discussed in Chapter 7.

# Chapter 5

# Reciprocity Theorems

In this chapter we discuss the reciprocity theorems. These theorems constitute the fundament of the seismic wave theory. In the reciprocity theorems we consider a time-invariant, bounded, domain $\mathbb{D}$ in space in which two non-identical acoustic states can occur. The two states will be distinguished by the superscripts $A$ and $B$, respectively. Neither the source distributions of the acoustic wavefields in the two states, nor the fluids present in the two states need to be the same. The boundary surface of $\mathbb{D}$ is denoted by $\partial\mathbb{D}$; the normal vector $\nu_k$ on $\partial\mathbb{D}$ is directed away from $\mathbb{D}$. The complement of $\mathbb{D} \cup \partial\mathbb{D}$ in $\mathbb{R}^3$ is denoted by $\mathbb{D}'$ (see Fig. 5.1). We characterize the acoustic properties of the fluids by the volume density of mass $\rho = \rho(\boldsymbol{x})$ and the compressibility $\kappa = \kappa(\boldsymbol{x})$.

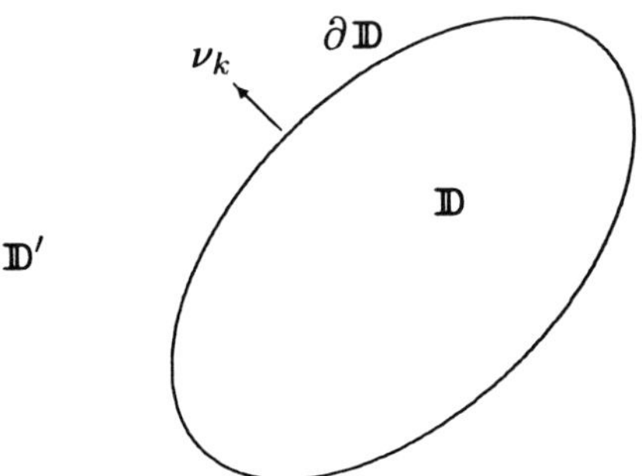


Figure 5.1. Configuration for the application of the reciprocity theorem.

## 5.1. The $s$-domain field reciprocity theorem

We start with the basic acoustic wavefield equations in the $s$-domain, as discussed in Chapter 3, cf. Eqs. (3.17) and (3.18). State $A$ is characterized by the acoustic wavefield $\{\hat{p}^A, \hat{v}_k^A\}$, the constitutive parameters $\{\rho^A, \kappa^A\}$ and the source distributions $\{\hat{q}^A, \hat{f}_k^A\}$. Similarly, State $B$ is characterized by the acoustic wavefield $\{\hat{p}^B, \hat{v}_k^B\}$, the constitutive parameters $\{\rho^B, \kappa^B\}$ and the source distributions $\{\hat{q}^B, \hat{f}_k^B\}$ (cf. Table 5.1). The acoustic wavefield equations pertaining to State $A$ are then

$$\partial_k \hat{p}^A + s\rho^A \hat{v}_k^A = \hat{f}_k^A , \tag{5.1}$$

$$\partial_k \hat{v}_k^A + s\kappa^A \hat{p}^A = \hat{q}^A . \tag{5.2}$$

Similarly, the acoustic wavefield equations pertaining to State $B$ are

$$\partial_k \hat{p}^B + s\rho^B \hat{v}_k^B = \hat{f}_k^B , \tag{5.3}$$

$$\partial_k \hat{v}_k^B + s\kappa^B \hat{p}^B = \hat{q}^B . \tag{5.4}$$

Table 5.1. States in the field reciprocity theorem

| | State $A$ | State $B$ |
|---|---|---|
| Field state | $\{\hat{p}^A, \hat{v}_k^A\}(\boldsymbol{x}, s)$ | $\{\hat{p}^B, \hat{v}_k^B\}(\boldsymbol{x}, s)$ |
| Material state | $\{\rho^A, \kappa^A\}(\boldsymbol{x})$ | $\{\rho^B, \kappa^B\}(\boldsymbol{x})$ |
| Source state | $\{\hat{q}^A, \hat{f}_k^A\}(\boldsymbol{x}, s)$ | $\{\hat{q}^B, \hat{f}_k^B\}(\boldsymbol{x}, s)$ |
| Domain $\mathbb{D}$ (see Fig. 5.1) | | |

If, in the domain $\mathbb{D}$, surfaces of discontinuity in acoustic properties are present, Eqs. (5.1) - (5.4) are supplemented by boundary conditions of the type discussed in Section 3.2, both in State $A$ and in State $B$. In the field reciprocity relation, the interaction quantity between the two states is

$$\partial_k(\hat{p}^A\hat{v}_k^B - \hat{p}^B\hat{v}_k^A) = \hat{v}_k^B\partial_k\hat{p}^A + \hat{p}^A\partial_k\hat{v}_k^B - \hat{v}_k^A\partial_k\hat{p}^B - \hat{p}^B\partial_k\hat{v}_k^A \,. \qquad (5.5)$$

Upon multiplying Eq. (5.1) by $\hat{v}_k^B$, Eq. (5.2) by $\hat{p}^B$, Eq. (5.3) by $\hat{v}_k^A$ and Eq. (5.4) by $\hat{p}^A$, and using the results in Eq. (5.5), we arrive at

$$\begin{aligned}\partial_k(\hat{p}^A\hat{v}_k^B - \hat{p}^B\hat{v}_k^A) &= s(\rho^B-\rho^A)\hat{v}_k^A\hat{v}_k^B - s(\kappa^B-\kappa^A)\hat{p}^A\hat{p}^B \\ &\quad + \hat{f}_k^A\hat{v}_k^B + \hat{q}^B\hat{p}^A - \hat{f}_k^B\hat{v}_k^A - \hat{q}^A\hat{p}^B \,. \qquad (5.6)\end{aligned}$$

Equation (5.6) is the local form of Rayleigh's reciprocity theorem.

Integration of Eq. (5.6) over the domain $\mathbb{D}$ with boundary $\partial\mathbb{D}$ and the use of Gauss' integral theorem in the resulting left-hand side lead to

$$\begin{aligned}\int_{\boldsymbol{x}\in\partial\mathbb{D}}(\hat{p}^A\hat{v}_k^B &- \hat{p}^B\hat{v}_k^A)\nu_k\mathrm{dA} \\ &= \int_{\boldsymbol{x}\in\mathbb{D}}[s(\rho^B-\rho^A)\hat{v}_k^A\hat{v}_k^B - s(\kappa^B-\kappa^A)\hat{p}^A\hat{p}^B]\mathrm{dV} \\ &\quad + \int_{\boldsymbol{x}\in\mathbb{D}}(\hat{f}_k^A\hat{v}_k^B + \hat{q}^B\hat{p}^A - \hat{f}_k^B\hat{v}_k^A - \hat{q}^A\hat{p}^B)\mathrm{dV} \,. \qquad (5.7)\end{aligned}$$

Equation (5.7) is Rayleigh's reciprocity theorem in its global form for the domain $\mathbb{D}$ (LORD RAYLEIGH, 1894; Dover 1945, vol. II, pp. 145-147; he denotes it as Helmholtz's theorem). First of all, it is remarked that the first and the second term on the right-hand side of Eq. (5.6), as well as the first integral on the right-hand side of Eq. (5.7), vanish in case the fluids in the two states are chosen such that $\rho^A = \rho^B$ and $\kappa^A = \kappa^B$. In particular, the relevant relations hold for one and the same fluid. Under these conditions, the interaction between the two states is only related to the source distributions in the two states. If, in addition, these source distributions vanish in some domain, the relevant interactions (local or global) are zero in that domain.

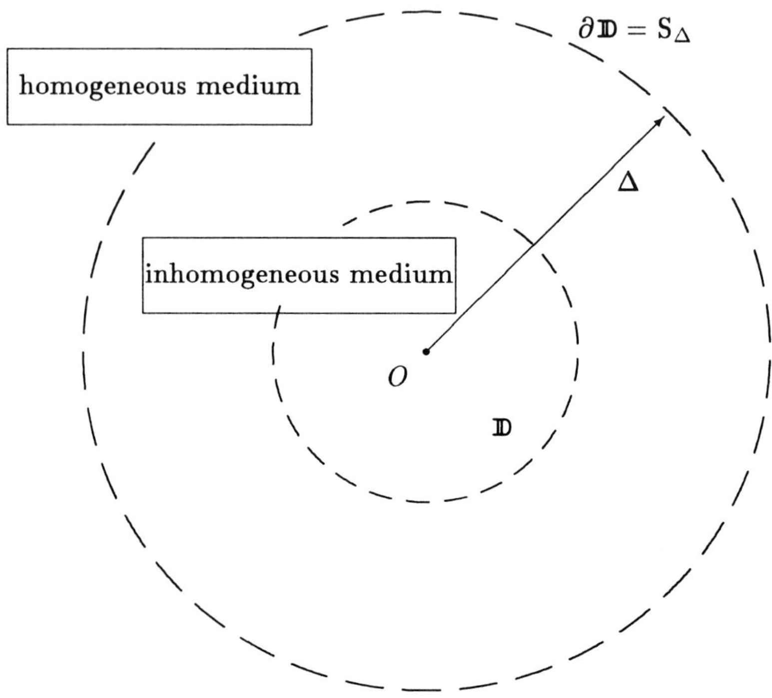


Figure 5.2. Unbounded configuration for the application of the acoustic reciprocity theorem.

*The limiting case of an unbounded domain*

In quite a number of cases it is desirable to apply the reciprocity theorem of Eq. (5.7) to an unbounded domain. These cases will always be handled as the limiting one that occurs if $\mathbb{D}$ is taken to be the domain interior to a sphere $S_\Delta$ of radius $\Delta$ and center at the origin of the chosen coordinate system, and the limit $\Delta \to \infty$ is considered (see Fig. 5.2). Obviously, then we must evaluate the left-hand side of Eq. (5.7) on $S_\Delta$. To this end, we shall always assume that outside some sphere of bounded radius, and center at the origin of the chosen coordinate system, the fluid is homogeneous with position independent mass density and compressibility. Taking the acoustic wavefield in both states to be causally related to the action of their sources, the latter being non-vanishing in some bounded subdomain of space only, we then can, for sufficiently large values of $\Delta$, use on $S_\Delta$ the far-field approximation (cf. Section 4.3) of the radiated wavefields, both in State $A$ and State $B$. Using these relations and taking $\rho^A = \rho^B$ and $\kappa^A = \kappa^B$ in this exterior domain, it follows that, as $|\boldsymbol{x}| \to \infty$, the terms in the expansion of $\hat{p}\hat{v}_k$ of Order $(\Delta^{-2})$ in the integral over $S_\Delta$ cancel. The next term in the expansion of $\hat{p}\hat{v}_k$ as $|\boldsymbol{x}| \to \infty$ is of Order $(\Delta^{-3})$. Therefore,

$$\hat{p}^A \hat{v}_k^B - \hat{p}^B \hat{v}_k^A = \text{Order}\,(\Delta^{-3}) \text{ as } \Delta \to \infty\,. \tag{5.8}$$

Consequently (note that the area of $\mathrm{S}_\Delta$ is $4\pi\Delta^2$), we obtain the causality condition

$$\int_{\boldsymbol{x}\in \mathrm{S}_\Delta} (\hat{p}^A \hat{v}_k^B - \hat{p}^B \hat{v}_k^A)\nu_k \mathrm{dA} = \text{Order}\,(\Delta^{-1}) \text{ as } \Delta \to \infty\,, \tag{5.9}$$

and the left-hand side of Eq. (5.7) vanishes in the limit $\Delta \to \infty$. Further, the integration domain $\mathbb{D}$ of the integrals in the right-hand side of Eq. (5.7) becomes the domain $\mathbb{R}^3$.

## 5.2. The time-domain reciprocity theorem of convolution type

In the time-domain reciprocity theorem of the convolution type, the two states are now defined in the space-time domain. State $A$ is characterized by the acoustic wavefield $\{p^A, v_k^A\}$, the constitutive parameters $\{\rho^A, \kappa^A\}$ and the source distributions $\{q^A, f_k^A\}$. Similarly, State $B$ is characterized by the acoustic wavefield $\{p^B, v_k^B\}$, the constitutive parameters $\{\rho^B, \kappa^B\}$ and the source distributions $\{q^B, f_k^B\}$ (cf. Table 5.2). By applying the standard rules for inversion from the $s$-domain to the time domain, the local form of the time-domain reciprocity is directly obtained from Rayleigh's reciprocity theorem of Eq. (5.6). Denoting $C_t\{\cdot,\cdot\} = C_t\{\cdot,\cdot\}(\boldsymbol{x},t)$ as the temporal convolution (cf. Eq. (1.19)) we arrive at

$$\begin{aligned}
&\partial_k[C_t\{p^A, v_k^B\} - C_t\{v_k^A, p^B\}] \\
&= (\rho^B - \rho^A)\partial_t C_t\{v_k^A, v_k^B\} - (\kappa^B - \kappa^A)\partial_t C_t\{p^A, p^B\} \\
&\quad + C_t\{f_k^A, v_k^B\} + C_t\{p^A, q^B\} - C_t\{v_k^A, f_k^B\} - C_t\{q^A, p^B\}\,.
\end{aligned} \tag{5.10}$$

Table 5.2. States in the reciprocity theorem of convolution type

| | State $A$ | State $B$ |
|---|---|---|
| Field state | $\{p^A, v_k^A\}(\boldsymbol{x}, t)$ | $\{p^B, v_k^B\}(\boldsymbol{x}, t)$ |
| Material state | $\{\rho^A, \kappa^A\}(\boldsymbol{x})$ | $\{\rho^B, \kappa^B\}(\boldsymbol{x})$ |
| Source state | $\{q^A, f_k^A\}(\boldsymbol{x}, t)$ | $\{q^B, f_k^B\}(\boldsymbol{x}, t)$ |
| Domain $\mathbb{D}$ (see Fig. 5.1) | | |

Integration of Eq. (5.10) over the domain $\mathbb{D}$ with boundary $\partial\mathbb{D}$ and the use of Gauss' integral theorem in the resulting left-hand side lead to

$$\begin{aligned}&\int_{\boldsymbol{x}\in\partial\mathbb{D}}[C_t\{p^A, v_k^B\} - C_t\{v_k^A, p^B\}]\nu_k \mathrm{dA} \\ &\quad = \int_{\boldsymbol{x}\in\mathbb{D}}[(\rho^B - \rho^A)\partial_t C_t\{v_k^A, v_k^B\} - (\kappa^B - \kappa^A)\partial_t C_t\{p^A, p^B\}]\mathrm{dV} \\ &\qquad + \int_{\boldsymbol{x}\in\mathbb{D}}[C_t\{f_k^A, v_k^B\} + C_t\{p^A, q^B\} - C_t\{v_k^A, f_k^B\} - C_t\{q^A, p^B\}]\mathrm{dV}\,. \end{aligned} \tag{5.11}$$

Equation (5.11) is the global form of the time-domain reciprocity theorem of the convolution type for the domain $\mathbb{D}$. First of all, it is remarked that the first and the second term on the right-hand side of Eq. (5.10), as well as the first integral on the right-hand side of Eq. (5.11), vanish in case the fluids in the two states are chosen such that $\rho^A = \rho^B$ and $\kappa^A = \kappa^B$. In particular, the relevant relations hold for one and the same fluid. Under these conditions, the interaction between the two states is only related to the source distributions in the two states. If, in addition, these source distributions vanish in some domain, the relevant interactions (local or global) are zero in that domain.

## 5.3. The $s$-domain power reciprocity theorem

In the $s$-domain power reciprocity theorem we again consider two states $A$ and $B$ in a bounded domain (Fig. 5.1.). State $A$ is characterized by the acoustic wavefield $\{\hat{p}^A, \hat{v}_k^A\} = \{\hat{p}^A, \hat{v}_k^A\}(\boldsymbol{x}, s)$, the constitutive parameters $\{\rho^A, \kappa^A\}$ and the source distributions $\{\hat{q}^A, \hat{f}_k^A\} = \{\hat{q}^A, \hat{f}_k^A\}(\boldsymbol{x}, s)$. The acoustic wavefield equations pertaining to State $A$ are then

$$\partial_k \hat{p}^A + s\rho^A \hat{v}_k^A = \hat{f}_k^A , \tag{5.12}$$

$$\partial_k \hat{v}_k^A + s\kappa^A \hat{p}^A = \hat{q}^A . \tag{5.13}$$

Now, State $B$ is characterized by the anti-causal acoustic wavefield $\{\hat{p}^B, \hat{v}_k^B\} = \{\hat{p}^B, \hat{v}_k^B\}(\boldsymbol{x}, -s)$, the constitutive parameters $\{\rho^B, \kappa^B\}$ and the source distributions $\{\hat{q}^B, \hat{f}_k^B\} = \{\hat{q}^B, \hat{f}_k^B\}(\boldsymbol{x}, -s)$ (cf. Table 5.3). State $B$ is the anti-causal counterpart of the causal wavefield $\{\hat{p}^B, \hat{v}_k^B\}(\boldsymbol{x}, s)$ by simply replacing $s$ by $-s$. The acoustic wavefield equations pertaining to the anti-causal State $B$ are

$$\partial_k \hat{p}^B - s\rho^B \hat{v}_k^B = \hat{f}_k^B , \tag{5.14}$$

$$\partial_k \hat{v}_k^B - s\kappa^B \hat{p}^B = \hat{q}^B . \tag{5.15}$$

Table 5.3. States in the power reciprocity theorem

| | State $A$ | State $B$ |
|---|---|---|
| Field state | $\{\hat{p}^A, \hat{v}_k^A\}(\boldsymbol{x}, s)$ | $\{\hat{p}^B, \hat{v}_k^B\}(\boldsymbol{x}, -s)$ |
| Material state | $\{\rho^A, \kappa^A\}(\boldsymbol{x})$ | $\{\rho^B, \kappa^B\}(\boldsymbol{x})$ |
| Source state | $\{\hat{q}^A, \hat{f}_k^A\}(\boldsymbol{x}, s)$ | $\{\hat{q}^B, \hat{f}_k^B\}(\boldsymbol{x}, -s)$ |
| Domain $\mathbb{D}$ (see Fig. 5.1) | | |

If, in the domain $\mathbb{D}$, surfaces of discontinuity in acoustic properties are present, Eqs. (5.12) - (5.15) are supplemented by boundary conditions of the type discussed in Section 3.2, both in State $A$ and in State $B$. The interaction quantity between the two states is

$$\partial_k(\hat{p}^A\hat{v}_k^B + \hat{p}^B\hat{v}_k^A) = \hat{v}_k^B\partial_k\hat{p}^A + \hat{p}^A\partial_k\hat{v}_k^B + \hat{v}_k^A\partial_k\hat{p}^B + \hat{p}^B\partial_k\hat{v}_k^A . \quad (5.16)$$

Upon multiplying Eq. (5.12) by $\hat{v}_k^B$, Eq. (5.13) by $\hat{p}^B$, Eq. (5.14) by $\hat{v}_k^A$ and Eq. (5.15) by $\hat{p}^A$, and using the results in Eq. (5.16), we arrive at

$$\begin{aligned}\partial_k(\hat{p}^A\hat{v}_k^B + \hat{p}^B\hat{v}_k^A) &= s(\rho^B - \rho^A)\hat{v}_k^A\hat{v}_k^B + s(\kappa^B - \kappa^A)\hat{p}^A\hat{p}^B \\ &\quad + \hat{f}_k^A\hat{v}_k^B + \hat{q}^B\hat{p}^A + \hat{f}_k^B\hat{v}_k^A + \hat{q}^A\hat{p}^B . \quad (5.17)\end{aligned}$$

Equation (5.17) is the local form of the power reciprocity theorem.

Integration of Eq. (5.17) over the domain $\mathbb{D}$ with boundary $\partial\mathbb{D}$ and the use of Gauss' integral theorem in the resulting left-hand side lead to

$$\begin{aligned}\int_{\boldsymbol{x}\in\partial\mathbb{D}}(\hat{p}^A\hat{v}_k^B + \hat{p}^B\hat{v}_k^A)\nu_k\mathrm{dA} \\ &= \int_{\boldsymbol{x}\in\mathbb{D}}[s(\rho^B - \rho^A)\hat{v}_k^A\hat{v}_k^B + s(\kappa^B - \kappa^A)\hat{p}^A\hat{p}^B]\mathrm{dV} \\ &\quad + \int_{\boldsymbol{x}\in\mathbb{D}}(\hat{f}_k^A\hat{v}_k^B + \hat{q}^B\hat{p}^A + \hat{f}_k^B\hat{v}_k^A + \hat{q}^A\hat{p}^B)\mathrm{dV} . \quad (5.18)\end{aligned}$$

Equation (5.18) is the power reciprocity theorem in its global form for the domain $\mathbb{D}$. First of all, it is remarked that the first and the second term on the right-hand side of Eq. (5.17), as well as the first integral on the right-hand side of Eq. (5.18), vanish in case the fluids in the two states are chosen such that $\rho^A = \rho^B$ and $\kappa^A = \kappa^B$. In particular, the relevant relations hold for one and the same fluid. Under these conditions, the interaction between the two states is only related to the source distributions in the two states. If, in addition, these source distributions vanish in some domain, the relevant interactions (local or global) are zero in that domain.

*Power conservation in the frequency domain*

As we have seen in Section 3.2, the steady-state analysis is arrived at by letting $s \to j\omega$. Noting that

$$\{\hat{p}^{B*}(\boldsymbol{x}, j\omega), \hat{v}_k^{B*}(\boldsymbol{x}, j\omega\} = \{\hat{p}^B(\boldsymbol{x}, -j\omega), \hat{v}_k^B(\boldsymbol{x}, -j\omega)\} , \quad (5.19)$$

the local form of the power reciprocity theorem of Eq. (5.17) is then rewritten as

$$\partial_k(\hat{p}^A\hat{v}_k^{B\star} + \hat{p}^{B\star}\hat{v}_k^A) = j\omega(\rho^B - \rho^A)\hat{v}_k^A\hat{v}_k^{B\star} + j\omega(\kappa^B - \kappa^A)\hat{p}^A\hat{p}^{B\star}$$
$$+ \hat{f}_k^A\hat{v}_k^{B\star} + \hat{q}^{B\star}\hat{p}^A + \hat{f}_k^{B\star}\hat{v}_k^A + \hat{q}^A\hat{p}^{B\star} . \qquad (5.20)$$

Taking $\{\hat{p}^A, \hat{v}_k^A\} = \{\hat{p}^B, \hat{v}_k^B\} = \{\hat{p}, \hat{v}_k\}(\boldsymbol{x}, j\omega)$, $\rho^A = \rho^B = \rho(\boldsymbol{x})$ and $\kappa^A = \kappa^B = \kappa(\boldsymbol{x})$, we arrive at the local form of the power conservation theorem in the frequency domain as

$$\partial_k(\hat{p}\hat{v}_k^\star + \hat{p}^\star\hat{v}_k) = \hat{f}_k\hat{v}_k^\star + \hat{q}^\star\hat{p} + \hat{f}_k^\star\hat{v}_k + \hat{q}\hat{p}^\star . \qquad (5.21)$$

Integration of Eq. (5.21) over the domain $\mathbb{D}$ with boundary $\partial\mathbb{D}$ and the use of Gauss' integral theorem in the resulting left-hand side lead to

$$\int_{\boldsymbol{x}\in\partial\mathbb{D}} (\hat{p}\hat{v}_k^\star + \hat{p}^\star\hat{v}_k)\nu_k \mathrm{dA}$$
$$= \int_{\boldsymbol{x}\in\mathbb{D}} (\hat{f}_k\hat{v}_k^\star + \hat{q}^\star\hat{p} + \hat{f}_k^\star\hat{v}_k + \hat{q}\hat{p}^\star)\mathrm{dV} . \qquad (5.22)$$

Equation (5.22) is the power conservation theorem in the frequency domain in its global form for the domain $\mathbb{D}$.

## 5.4. The time-domain reciprocity theorem of correlation type

In the time-domain reciprocity theorem of the correlation type, the two states are now defined in the space-time domain. State $A$ is characterized by the acoustic wavefield $\{p^A, v_k^A\}$, the constitutive parameters $\{\rho^A, \kappa^A\}$ and the source distributions $\{q^A, f_k^A\}$. Similarly, State $B$ is characterized by the acoustic wavefield $\{p^B, v_k^B\}$, the constitutive parameters $\{\rho^B, \kappa^B\}$ and the source distributions $\{q^B, f_k^B\}$ (cf. Table 5.4). By applying the standard rules for inversion from the $s$-domain to the time domain, the local form of the time-domain reciprocity is directly obtained from the power reciprocity theorem of Eq. (5.17).

Table 5.4. States in the reciprocity theorem of correlation type

| | State $A$ | State $B$ |
|---|---|---|
| Field state | $\{p^A, v_k^A\}(\boldsymbol{x}, t)$ | $\{p^B, v_k^B\}(\boldsymbol{x}, t)$ |
| Material state | $\{\rho^A, \kappa^A\}(\boldsymbol{x})$ | $\{\rho^B, \kappa^B\}(\boldsymbol{x})$ |
| Source state | $\{q^A, f_k^A\}(\boldsymbol{x}, t)$ | $\{q^B, f_k^B\}(\boldsymbol{x}, t)$ |
| Domain $\mathbb{D}$ (see Fig. 5.1) | | |

In order to arrive at valid correlation results, we first have to assume that the Laplace-transform parameter is imaginary ($s \to j\omega$). Denoting $C_t'\{\cdot,\cdot\} = C_t'\{\cdot,\cdot\}(\boldsymbol{x}, t)$ as the temporal correlation (cf. Eq. (1.21)) we arrive at

$$\partial_k[C_t'\{p^A, v_k^B\} + C_t'\{v_k^A, p^B\}]$$

$$= (\rho^B - \rho^A)\partial_t C_t'\{v_k^A, v_k^B\} + (\kappa^B - \kappa^A)\partial_t C_t'\{p^A, p^B\}$$

$$+ C_t'\{f_k^A, v_k^B\} + C_t'\{p^A, q^B\} + C_t'\{v_k^A, f_k^B\} + C_t'\{q^A, p^B\}\,. \tag{5.23}$$

Note that this result can directly be derived from the acoustic wavefield equations in the time domain (DE HOOP, 1988).

Integration of Eq. (5.23) over the domain $\mathbb{D}$ with boundary $\partial\mathbb{D}$ and the use of Gauss' integral theorem in the resulting left-hand side lead to

$$\int_{\boldsymbol{x}\in\partial\mathbb{D}} [C_t'\{p^A, v_k^B\} + C_t'\{v_k^A, p^B\}]\nu_k \mathrm{dA}$$

$$= \int_{\boldsymbol{x}\in\mathbb{D}} [(\rho^B - \rho^A)\partial_t C'_t\{v_k^A, v_k^B\} + (\kappa^B - \kappa^A)\partial_t C'_t\{p^A, p^B\}]\mathrm{dV}$$

$$+ \int_{\boldsymbol{x}\in\mathbb{D}} [C'_t\{f_k^A, v_k^B\} + C'_t\{p^A, q^B\} + C'_t\{v_k^A, f_k^B\} + C'_t\{q^A, p^B\}]\mathrm{dV} \,. \tag{5.24}$$

Equation (5.24) is the global form of the time-domain reciprocity theorem of the correlation type for the domain $\mathbb{D}$. First of all, it is remarked that the first and the second term on the right-hand side of Eq. (5.23), as well as the first integral on the right-hand side of Eq. (5.24), vanish in case the fluids in the two states are chosen such that $\rho^A = \rho^B$ and $\kappa^A = \kappa^B$. In particular, the relevant relations hold for one and the same fluid. Under these conditions, the interaction between the two states is only related to the source distributions in the two states. If, in addition, these source distributions vanish in some domain, the relevant interactions (local or global) are zero in that domain.

*Power conservation in the time domain*

Taking $\{p^A, v_k^A\} = \{p^B, v_k^B\} = \{p, v_k\}(\boldsymbol{x}, t)$, $\rho^A = \rho^B = \rho(\boldsymbol{x})$, $\kappa^A = \kappa^B = \kappa(\boldsymbol{x})$ and $t = 0$ in Eq. (5.23), we arrive at the local form of the power conservation theorem in the time domain as

$$\int_{t'\in\mathbb{R}} \partial_k(pv_k)\mathrm{d}t' = \int_{t'\in\mathbb{R}} (f_k v_k + qp)\mathrm{d}t' \,. \tag{5.25}$$

Integration of Eq. (5.25) over the domain $\mathbb{D}$ with boundary $\partial\mathbb{D}$ and the use of Gauss' integral theorem in the resulting left-hand side lead to

$$\int_{\boldsymbol{x}\in\partial\mathbb{D}} \int_{t'\in\mathbb{R}} pv_k\nu_k \mathrm{dA}\mathrm{d}t' = \int_{\boldsymbol{x}\in\mathbb{D}} \int_{t'\in\mathbb{R}} (f_k v_k + qp)\mathrm{dV}\mathrm{d}t' \,. \tag{5.26}$$

Equation (5.26) is the power conservation theorem in the time domain in its global form for the domain $\mathbb{D}$.

A clear distinction between convolution-type and correlation-type reciprocity relations is made by BOJARSKI (1983), who has presented the theorem for homogeneous, isotropic and lossless media. The pertinent theorems for inhomogeneous, anisotropic fluids with relaxation is discussed by DE HOOP (1988). The seismic applications of the various reciprocity theorems are discussed in subsequent chapters.

# Chapter 6

# Field Reciprocity between Transmitter and Receiver

A first set of corollaries of the global reciprocity theorem of Section 5.1 are the transducer reciprocity properties of transmitting and receiving transducers.

In particular, we consider these experiments where the transducers interchange their position. We model the transducers as distributions of point transducers, surface transducers and volume transducers.

## 6.1. Point-transducer description

*The s-domain relations*

First the class is considered where the action of the transducers may be represented by equivalent point-source densities. The action of the sources is represented by its known volume densities of external force and/or volume injection. Let transducer $A$ be located at $\boldsymbol{x} = \boldsymbol{x}^A$ and transducer $B$ be located at $\boldsymbol{x} = \boldsymbol{x}^B$ (Fig. 6.1). We apply the reciprocity theorem of Section 5.1. Let State $A$ be identified with the state where transducer $A$ is transmitting and transducer $B$ is receiving. The acoustic wavefield equations pertaining

transducer $A$ transducer $B$

$\boldsymbol{x}^A$ $\boldsymbol{x}^B$

Figure 6.1. The point transducers $A$ and $B$.

Table 6.1. States in the reciprocity theorem

| | State $A$<br>transducer $A$ is transmitting | State $B$<br>transducer $B$ is transmitting |
|---|---|---|
| Field state | $\{\hat{p}^{T_A}, \hat{v}_k^{T_A}\}(\boldsymbol{x}, s)$ | $\{\hat{p}^{T_B}, \hat{v}_k^{T_B}\}(\boldsymbol{x}, s)$ |
| Material state | $\{\rho, \kappa\}(\boldsymbol{x})$ | $\{\rho, \kappa\}(\boldsymbol{x})$ |
| Source state | $\{\hat{q}^{T_A}, \hat{f}_k^{T_A}\}\delta(\boldsymbol{x} - \boldsymbol{x}^A)$ | $\{\hat{q}^{T_B}, \hat{f}_k^{T_B}\}\delta(\boldsymbol{x} - \boldsymbol{x}^B)$ |
| Domain $\mathbb{R}^3$ (see Fig. 6.1) | | |

to State $A$ (cf. Table 6.1) are then

$$\partial_k \hat{p}^{T_A} + s\rho \hat{v}_k^{T_A} = \hat{f}_k^{T_A} \delta(\boldsymbol{x} - \boldsymbol{x}^A) \, , \tag{6.1}$$

$$\partial_k \hat{v}_k^{T_A} + s\kappa \hat{p}^{T_A} = \hat{q}^{T_A} \delta(\boldsymbol{x} - \boldsymbol{x}^A) \, . \tag{6.2}$$

The wavefield in State $B$ is identified with the state where transducer $B$ is transmitting and transducer $A$ is receiving. The wavefield quantities in this state satisfies (cf. Table 6.1)

$$\partial_k \hat{p}^{T_B} + s\rho \hat{v}_k^{T_B} = \hat{f}_k^{T_B} \delta(\boldsymbol{x} - \boldsymbol{x}^B) \, , \tag{6.3}$$

$$\partial_k \hat{v}_k^{T_B} + s\kappa \hat{p}^{T_B} = \hat{q}^{T_B} \delta(\boldsymbol{x} - \boldsymbol{x}^B) \, . \tag{6.4}$$

Now Eq. (5.7) is applied to the domain interior to the sphere $\mathrm{S}_\Delta$ with radius $\Delta$ and center at the origin of the chosen coordinate system; $\Delta$ is so large that $\mathrm{S}_\Delta$ completely surrounds the two point transducers. In view of the causality condition we have

$$\int_{\boldsymbol{x} \in \mathrm{S}_\Delta} (\hat{p}^{T_A} \hat{v}_k^{T_B} - \hat{p}^{T_B} \hat{v}_k^{T_A}) \nu_k \mathrm{dA} = \mathrm{Order}\ (\Delta^{-1}) \ \text{ as } \ \Delta \to \infty \, . \tag{6.5}$$

Hence we have

$$\int_{\boldsymbol{x} \in \mathbb{R}^3} [\hat{f}_k^{T_A} \delta(\boldsymbol{x} - \boldsymbol{x}^A) \hat{v}_k^{T_B} + \hat{q}^{T_B} \delta(\boldsymbol{x} - \boldsymbol{x}^B) \hat{p}^{T_A}$$

$$- \hat{f}_k^{T_B} \delta(\boldsymbol{x} - \boldsymbol{x}^B) \hat{v}_k^{T_A} - \hat{q}^{T_A} \delta(\boldsymbol{x} - \boldsymbol{x}^A) \hat{p}^{T_B}] \mathrm{dV} = 0 \, , \tag{6.6}$$

with the result that the reciprocity between two point transducers is given by

$$\hat{q}^{T_B} \hat{p}^{T_A}(\boldsymbol{x}^B, s) - \hat{f}_k^{T_B} \hat{v}_k^{T_A}(\boldsymbol{x}^B, s) = \hat{q}^{T_A} \hat{p}^{T_B}(\boldsymbol{x}^A, s) - \hat{f}_k^{T_A} \hat{v}_k^{T_B}(\boldsymbol{x}^A, s) \, . \tag{6.7}$$

We now consider three special cases:

- Transducer $A$ is of the volume-injection type and transducer $B$ is of the volume-injection type, viz.,

$$\hat{f}_k^{T_A} = 0, \quad \hat{f}_k^{T_B} = 0 \, . \tag{6.8}$$

Then, the reciprocity theorem between two monopole transducers becomes (see e.g., Kinsler *et al.*, 1982, p. 168)

$$\hat{q}^{T_B} \hat{p}^{T_A}(\boldsymbol{x}^B, s) = \hat{q}^{T_A} \hat{p}^{T_B}(\boldsymbol{x}^A, s) \, . \tag{6.9}$$

- Transducer $A$ is of the volume-injection type and transducer $B$ is of the volume-force type, viz.,

$$\hat{f}_k^{T_A} = 0, \quad \hat{q}^{T_B} = 0\,. \tag{6.10}$$

Then, the reciprocity theorem between a monopole and a dipole transducer becomes

$$-\,\hat{f}_k^{T_B}\hat{v}_k^{T_A}(\boldsymbol{x}^B, s) = \hat{q}^{T_A}\hat{p}^{T_B}(\boldsymbol{x}^A, s)\,. \tag{6.11}$$

- Transducer $A$ is of the volume-force type and transducer $B$ is of the volume-force type, viz.,

$$\hat{q}^{T_A} = 0, \quad \hat{q}^{T_B} = 0\,. \tag{6.12}$$

Then, the reciprocity theorem between two dipole transducer becomes

$$\hat{f}_k^{T_B}\hat{v}_k^{T_A}(\boldsymbol{x}^B, s) = \hat{f}_k^{T_A}\hat{v}_k^{T_B}(\boldsymbol{x}^A, s)\,. \tag{6.13}$$

*The time-domain relations*

The time-domain reciprocity relations can be derived by either inversion of the $s$-domain results or directly by application of the time-domain reciprocity of the convolution type. We then arrive at

$$C_t\{q^{T_B}, p^{T_A}\}(\boldsymbol{x}^B, t) - C_t\{f_k^{T_B}, v_k^{T_A}\}(\boldsymbol{x}^B, t)$$

$$= C_t\{q^{T_A}, p^{T_B}\}(\boldsymbol{x}^A, t) - C_t\{f_k^{T_A}, v_k^{T_B}\}(\boldsymbol{x}^A, t)\,. \tag{6.14}$$

We again consider three special cases:

- Transducer $A$ is of the volume-injection type and transducer $B$ is of the volume-injection type, viz.,

$$f_k^{T_A} = 0, \quad f_k^{T_B} = 0\,. \tag{6.15}$$

Then, the reciprocity theorem between two monopole transducers becomes

$$C_t\{q^{T_B}, p^{T_A}\}(\boldsymbol{x}^B, t) = C_t\{q^{T_A}, p^{T_B}\}(\boldsymbol{x}^A, t)\,. \tag{6.16}$$

- Transducer $A$ is of the volume-injection type and transducer $B$ is of the volume-force type, viz.,

$$f_k^{T_A} = 0, \quad q^{T_B} = 0 \,. \tag{6.17}$$

Then, the reciprocity theorem between a monopole and a dipole transducer becomes

$$-\,C_t\{f_k^{T_B}, v_k^{T_A}\}(\boldsymbol{x}^B, t) = C_t\{q^{T_A}, p^{T_B}\}(\boldsymbol{x}^A, t) \,. \tag{6.18}$$

- Transducer $A$ is of the volume-force type and transducer $B$ is of the volume-force type, viz.,

$$q^{T_A} = 0, \quad q^{T_B} = 0 \,. \tag{6.19}$$

Then, the reciprocity theorem between two dipole transducers becomes

$$C_t\{f_k^{T_B}, v_k^{T_A}\}(\boldsymbol{x}^B, t) = C_t\{f_k^{T_A}, v_k^{T_B}\}(\boldsymbol{x}^A, t) \,. \tag{6.20}$$

## 6.2. Volume-transducer description

Second the class is considered where the action of the transducers may be represented by equivalent volume-source densities, distributed over the domains occupied by the transducers. The action of the sources is represented by its known volume densities of external force and/or volume injection. Let transducer $A$ occupy the bounded domain $T_A$ and transducer $B$ occupy the domain $T_B$ (Fig. 6.2). We apply the reciprocity theorem of Section 5.1. Let State $A$ be identified with the state where transducer $A$ is transmitting and transducer $B$ is receiving. The acoustic wavefield equations pertaining to State $A$ (cf. Table 6.2) are then

$$\partial_k \hat{p}^{T_A} + s\rho \hat{v}_k^{T_A} = \hat{f}_k^{T_A} \,, \tag{6.21}$$

$$\partial_k \hat{v}_k^{T_A} + s\kappa \hat{p}^{T_A} = \hat{q}^{T_A} \,. \tag{6.22}$$

The wavefield in State $B$ is identified with the state where transducer $B$ is transmitting and transducer $A$ is receiving. The wavefield quantities in this state satisfy (cf. Table 6.2)

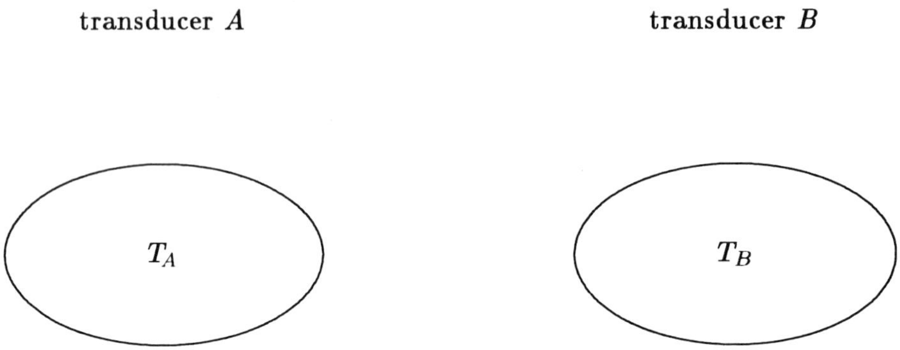

transducer $A$ transducer $B$

$T_A$ $T_B$

Figure 6.2. The volume transducers $A$ and $B$.

Table 6.2. States in the reciprocity theorem

| | State $A$<br>transducer $A$ is transmitting | State $B$<br>transducer $B$ is transmitting |
|---|---|---|
| Field state | $\{\hat{p}^{T_A}, \hat{v}_k^{T_A}\}(\boldsymbol{x}, s)$ | $\{\hat{p}^{T_B}, \hat{v}_k^{T_B}\}(\boldsymbol{x}, s)$ |
| Material state | $\{\rho, \kappa\}(\boldsymbol{x})$ | $\{\rho, \kappa\}(\boldsymbol{x})$ |
| Source state | $\{\hat{q}^{T_A}, \hat{f}_k^{T_A}\}(\boldsymbol{x}, s)$ | $\{\hat{q}^{T_B}, \hat{f}_k^{T_B}\}(\boldsymbol{x}, s)$ |
| Domain $\mathbb{R}^3$ (see Fig. 6.2) | | |

$$\partial_k \hat{p}^{T_B} + s\rho \hat{v}_k^{T_B} = \hat{f}_k^{T_B} , \tag{6.23}$$

$$\partial_k \hat{v}_k^{T_B} + s\kappa \hat{p}^{T_B} = \hat{q}^{T_B} . \tag{6.24}$$

Now Eq. (5.7) is applied to the domain interior to the sphere $\mathrm{S}_\Delta$ with radius $\Delta$ and center at the origin of the chosen coordinate system; $\Delta$ is so large that $\mathrm{S}_\Delta$ completely surrounds the two transducers domains. In view of the causality condition we have

$$\int_{\boldsymbol{x}\in \mathrm{S}_\Delta} (\hat{p}^{T_A}\hat{v}_k^{T_B} - \hat{p}^{T_B}\hat{v}_k^{T_A})\nu_k \mathrm{dA} = \text{Order}\,(\Delta^{-1}) \;\text{ as }\; \Delta \to \infty . \tag{6.25}$$

Hence, the reciprocity between the two volume transducers is given by

$$\int_{\boldsymbol{x}\in T_B} (\hat{q}^{T_B}\hat{p}^{T_A} - \hat{f}_k^{T_B}\hat{v}_k^{T_A})\mathrm{dV} = \int_{\boldsymbol{x}\in T_A} (\hat{q}^{T_A}\hat{p}^{T_B} - \hat{f}_k^{T_A}\hat{v}_k^{T_B})\mathrm{dV} , \tag{6.26}$$

where we have taken into account that the source states are only operative in $T_A$ and $T_B$, respectively.

In the special case that transducer $T_B$ can be considered as a concentrated point source the left-hand side of Eq. (6.26) has to be replaced by the left-hand side of Eq. (6.7).

The time-domain counterparts of the present reciprocity results are obtained in a similar way as before.

## 6.3. Surface-transducer description

Finally, the case is considered where the action of the transducers may be represented by equivalent surface-source densities, distributed over the boundary surfaces $\partial T_A$ and $\partial T_B$ of the respective domains occupied by the transducers (Fig. 6.3). We apply the reciprocity theorem of Section 5.1. Let State $A$ be identified with the state where transducer $A$ is transmitting and transducer $B$ is receiving. The acoustic wavefield equations pertaining to State $A$ (cf. Table 6.3) are then

$$\partial_k \hat{p}^{T_A} + s\rho \hat{v}_k^{T_A} = 0 , \tag{6.27}$$

$$\partial_k \hat{v}_k^{T_A} + s\kappa \hat{p}^{T_A} = 0 , \tag{6.28}$$

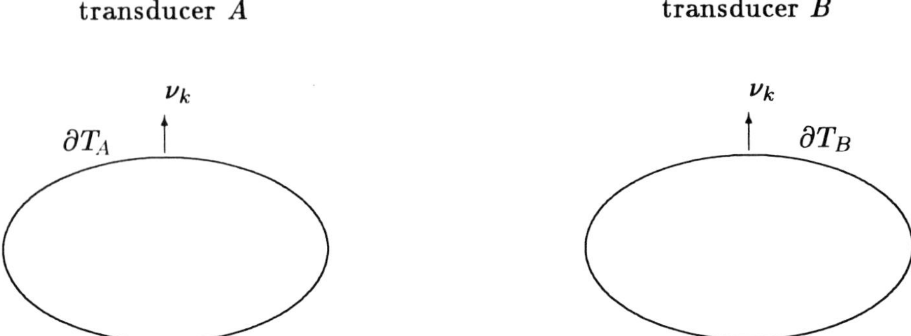


Figure 6.3. The surface transducers $A$ and $B$.

Table 6.3. States in the reciprocity theorem

| | State $A$<br>transducer $A$ is transmitting | State $B$<br>transducer $B$ is transmitting |
|---|---|---|
| Field state | $\{\hat{p}^{T_A}, \hat{v}_k^{T_A}\}(\boldsymbol{x}, s)$ | $\{\hat{p}^{T_B}, \hat{v}_k^{T_B}\}(\boldsymbol{x}, s)$ |
| Material state | $\{\rho, \kappa\}(\boldsymbol{x})$ | $\{\rho, \kappa\}(\boldsymbol{x})$ |
| Source state | $\{0, 0\}$ | $\{0, 0\}$ |
| Domain $\mathbb{R}^3 \backslash \{T_A \cup \partial T_A \cup \ T_B \cup \partial T_B\}$ (see Fig. 6.3) | | |

where $\boldsymbol{x}$ is located outside $\partial T_A$. The wavefield in State $B$ is identified with the state where transducer $B$ is transmitting and transducer $A$ is receiving. The wavefield quantities in this state satisfies (cf. Table 6.3)

$$\partial_k \hat{p}^{T_B} + s\rho \hat{v}_k^{T_B} = 0 , \tag{6.29}$$

$$\partial_k \hat{v}_k^{T_B} + s\kappa \hat{p}^{T_B} = 0 , \tag{6.30}$$

where $\boldsymbol{x}$ is located outside $\partial T_B$. Now Eq. (5.7) is applied to the domain exterior to the transducer domains and interior to the sphere $\mathrm{S}_\Delta$ with radius $\Delta$ and center at the origin of the chosen coordinate system; $\Delta$ is so large that $\mathrm{S}_\Delta$ completely surrounds the two transducers domains. In view of the causality condition we have

$$\int_{\boldsymbol{x}\in \mathrm{S}_\Delta} (\hat{p}^{T_A} \hat{v}_k^{T_B} - \hat{p}^{T_B} \hat{v}_k^{T_A}) \nu_k \mathrm{dA} = \mathrm{Order}\,(\Delta^{-1}) \ \text{ as } \ \Delta \to \infty . \tag{6.31}$$

Hence, the reciprocity between the two surface transducers is given by

$$-\int_{\boldsymbol{x}\in \partial T_B} (\hat{p}^{T_A} \hat{v}_k^{T_B} - \hat{p}^{T_B} \hat{v}_k^{T_A}) \nu_k \mathrm{dA} = -\int_{\boldsymbol{x}\in \partial T_A} (\hat{p}^{T_B} \hat{v}_k^{T_A} - \hat{p}^{T_A} \hat{v}_k^{T_B}) \nu_k \mathrm{dA} , \tag{6.32}$$

in which $\nu_k$ is the outward normal to the pertaining transducer boundary surface.

In the special case that transducer $T_B$ can be considered as a volume source, the left-hand side of Eq. (6.32) has to be replaced by the left-hand side of Eq. (6.26).

The time-domain counterparts of the present reciprocity results are obtained in a similar way as before.

The reciprocity properties derived in this chapter, are important for understanding the redundancy in the physical experiment. Moreover, these relations may serve as a numerical check in computational acoustics.

# Chapter 7

# Radiation in an Unbounded, Inhomogeneous Medium

The reciprocity theorem derived in Section 5.1 can serve to analyze the acoustic radiation generated by a known volume-source distribution in a known inhomogeneous medium; the problem of calculating the wavefield under these circumstances is commonly denoted as the direct, acoustic volume-source problem. The reciprocity theorem derived in Section 5.1 can also serve to analyze the acoustic radiation generated by a known surface-source distribution; the problem of calculating the wavefield under these circumstances is commonly denoted as the direct, acoustic surface-source problem.

## 7.1. The volume-source problem

### 7.1.1. Volume-source representations in the $s$-domain

*The $s$-domain pressure representation*

The configuration consists of a medium of infinite extent with known acoustic properties. In this medium a source is present that occupies the bounded domain $\mathbb{D}_{source}$ (Fig. 7.1). The action of the source is represented by known volume densities of external volume force and/or volume injection

$\{\hat{q}, \hat{f}_k\}$. Our aim is to arrive at a representation that expresses the acoustic pressure $\hat{p}$ in all space in terms of the source distributions. We apply the reciprocity theorem of Section 5.1. To this end, State $A$ is taken to be the actual wavefield that is generated by the sources and is causally related to them. The acoustic wavefield equations pertaining to State $A$ (cf. Table 7.1) are then

$$\partial_k \hat{p} + s\rho\hat{v}_k = \hat{f}_k , \tag{7.1}$$

$$\partial_k \hat{v}_k + s\kappa\hat{p} = \hat{q} . \tag{7.2}$$

The wavefield in State $B$ is chosen such that the application of Eq. (5.7) leads to the value of the acoustic pressure at any point in space. Inspection of the right-hand side of Eq. (5.7) reveals that this is accomplished if we take for the source distributions in State $B$ a point source of volume injection. This wavefield satisfies (cf. Table 7.1)

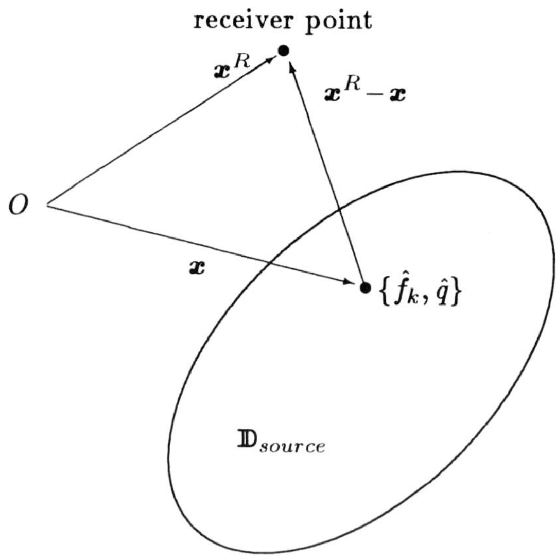


Figure 7.1. The source domain and the receiver location.

Table 7.1. States in the reciprocity theorem

| | State $A$ (actual state) | State $B$ (volume-injection Green's state) |
|---|---|---|
| Field state | $\{\hat{p}, \hat{v}_k\}(\boldsymbol{x}, s)$ | $\{\hat{p}^q, \hat{v}_k^q\}(\boldsymbol{x}\vert\boldsymbol{x}^R, s)$ |
| Material state | $\{\rho, \kappa\}(\boldsymbol{x})$ | $\{\rho, \kappa\}(\boldsymbol{x})$ |
| Source state | $\{\hat{q}, \hat{f}_k\}(\boldsymbol{x}, s)$ | $\{\hat{q}^B(s)\delta(\boldsymbol{x}-\boldsymbol{x}^R), 0\}$ |
| Domain $\mathbb{R}^3$ (see Fig. 7.1) | | |

$$\partial_k \hat{p}^q + s\rho \hat{v}_k^q = 0 , \tag{7.3}$$

$$\partial_k \hat{v}_k^q + s\kappa \hat{p}^q = \hat{q}^B \delta(\boldsymbol{x}-\boldsymbol{x}^R) , \tag{7.4}$$

where $\delta(\boldsymbol{x}-\boldsymbol{x}^R)$ denotes the three-dimensional spatial Dirac distribution (impulse function) operative at the receiver point with position vector $\boldsymbol{x}^R$, while $\hat{q}^B$ is an arbitrary constant only depending on $s$. This source is placed in the medium with the same material parameters $\rho$ and $\kappa$ as the actual ones. For the choice of Eqs. (7.3) and (7.4), State $B$ is denoted as the volume-injection Green's state. Now Eq. (5.7) is applied to the domain interior to the sphere $S_\Delta$ with radius $\Delta$ and center at the origin of the chosen coordinate system; $\Delta$ is so large that $S_\Delta$ completely surrounds $\mathbb{D}_{source}$. In view of the causality condition of Eq. (5.9) we have

$$\int_{\boldsymbol{x}\in S_\Delta} (\hat{p}\hat{v}_k^q - \hat{p}^q \hat{v}_k)\nu_k \mathrm{dA} = \text{Order}\ (\Delta^{-1}) \ \text{ as } \ \Delta \to \infty . \tag{7.5}$$

Hence we have

$$\int_{\boldsymbol{x}\in \mathbb{R}^3} [\hat{f}_k \hat{v}_k^q + \hat{q}^B \delta(\boldsymbol{x}-\boldsymbol{x}^R)\hat{p} - \hat{q}\hat{p}^q]\mathrm{dV} = 0 , \tag{7.6}$$

with the result that

$$\hat{q}^B \hat{p}(\boldsymbol{x}^R, s) = \int_{\boldsymbol{x} \in \mathbb{D}_{source}} [\hat{p}^q(\boldsymbol{x}|\boldsymbol{x}^R, s)\hat{q}(\boldsymbol{x}, s) - \hat{v}_k^q(\boldsymbol{x}|\boldsymbol{x}^R, s)\hat{f}_k(\boldsymbol{x}, s)]\mathrm{dV} \,. \quad (7.7)$$

From Eq. (7.7) a representation for $\hat{p}(\boldsymbol{x}^R, s)$ is obtained when we take into account that $\hat{p}^q$ and $\hat{v}_k^q$ are linearly related to $\hat{q}^B$. Expressing this as

$$\{\hat{p}^q, \hat{v}_k^q\}(\boldsymbol{x}|\boldsymbol{x}^R, s) = \hat{q}^B(s)\{\hat{G}^q, -\hat{\Gamma}_k^q\}(\boldsymbol{x}^R|\boldsymbol{x}, s) \,, \quad (7.8)$$

we arrive at the desired representation of the acoustic pressure

$$\hat{p}(\boldsymbol{x}^R, s) = \int_{\boldsymbol{x} \in \mathbb{D}_{source}} [\hat{G}^q(\boldsymbol{x}^R|\boldsymbol{x}, s)\hat{q}(\boldsymbol{x}, s) + \hat{\Gamma}_k^q(\boldsymbol{x}^R|\boldsymbol{x}, s)\hat{f}_k(\boldsymbol{x}, s)]\mathrm{dV} \,, \quad (7.9)$$

where $\boldsymbol{x}^R \in \mathbb{R}^3$. The first term in the integrand of Eq. (7.9) represents a monopole contribution with volume density $\hat{q}$, while the second term in the integrand of Eq. (7.9) represents a dipole contribution with volume density $\hat{f}_k$.

*The s-domain particle velocity representation*

In order to arrive at a representation that expresses the acoustic particle velocity $\hat{v}_k$ in all space in terms of the source distributions, the wavefield in State $B$ is chosen such that the application of Eq. (5.7) leads to the value of the acoustic particle velocity at any point in space. Inspection of the right-hand side of Eq. (5.7) reveals that this is accomplished if we take for the source distributions in State $B$ a point source of volume force. This wavefield satisfies (cf. Table 7.2)

$$\partial_k \hat{p}^f + s\rho \hat{v}_k^f = \hat{f}_k^B \delta(\boldsymbol{x} - \boldsymbol{x}^R) \,, \quad (7.10)$$

$$\partial_k \hat{v}_k^f + s\kappa \hat{p}^f = 0 \,, \quad (7.11)$$

where $\hat{f}_k^B$ is an arbitrary constant vector only depending on $s$. The source is placed in the medium with the same material parameters $\rho$ and $\kappa$ as the actual ones. For the choice of Eqs. (7.10) and (7.11), State $B$ is denoted as the volume-force Green's state. Now Eq. (5.7) is applied to the domain interior to the sphere $\mathrm{S}_\Delta$ with radius $\Delta$ and center at the origin of the chosen coordinate system; $\Delta$ is so large that $\mathrm{S}_\Delta$ completely surrounds $\mathbb{D}_{source}$. In view of the causality condition we have

Table 7.2. States in the reciprocity theorem

| | State $A$ (actual state) | State $B$ (volume-force Green's state) |
|---|---|---|
| Field state | $\{\hat{p}, \hat{v}_k\}(\boldsymbol{x}, s)$ | $\{\hat{p}^f, \hat{v}_k^f\}(\boldsymbol{x}\|\boldsymbol{x}^R, s)$ |
| Material state | $\{\rho, \kappa\}(\boldsymbol{x})$ | $\{\rho, \kappa\}(\boldsymbol{x})$ |
| Source state | $\{\hat{q}, \hat{f}_k\}(\boldsymbol{x}, s)$ | $\{0, \hat{f}_k^B(s)\delta(\boldsymbol{x}-\boldsymbol{x}^R)\}$ |
| Domain $\mathbb{R}^3$ (see Fig. 7.1) | | |

$$\int_{\boldsymbol{x}\in S_\Delta} (\hat{p}\hat{v}_k^f - \hat{p}^f\hat{v}_k)\nu_k \mathrm{dA} = \text{Order}\,(\Delta^{-1}) \text{ as } \Delta \to \infty \, . \tag{7.12}$$

Hence we have

$$\int_{\boldsymbol{x}\in\mathbb{R}^3} [\hat{f}_k\hat{v}_k^f - \hat{f}_k^B\delta(\boldsymbol{x}-\boldsymbol{x}^R)\hat{v}_k - \hat{q}\hat{p}^f]\mathrm{dV} = 0 \, , \tag{7.13}$$

with the result that

$$\hat{f}_l^B\hat{v}_l(\boldsymbol{x}^R, s) = \int_{\boldsymbol{x}\in\mathbb{D}_{source}} [-\hat{p}^f(\boldsymbol{x}|\boldsymbol{x}^R, s)\hat{q}(\boldsymbol{x}, s) + \hat{v}_k^f(\boldsymbol{x}|\boldsymbol{x}^R, s)\hat{f}_k(\boldsymbol{x}, s)]\mathrm{dV} \, . \tag{7.14}$$

From Eq. (7.14) a representation for $\hat{v}_l(\boldsymbol{x}^R, s)$ is obtained when we take into account that $\hat{p}^f$ and $\hat{v}_k^f$ are linearly related to $\hat{f}_l^B$. Expressing this as

$$\{\hat{p}^f, \hat{v}_k^f\}(\boldsymbol{x}|\boldsymbol{x}^R, s) = \hat{f}_l^B(s)\{-\hat{G}_l^f, \hat{\Gamma}_{l,k}^f\}(\boldsymbol{x}^R|\boldsymbol{x}, s) \, , \tag{7.15}$$

and taking into account that $\hat{f}_k^B$ is arbitrary, we arrive at the desired representation of the particle velocity

$$\hat{v}_l(\boldsymbol{x}^R, s) = \int_{\boldsymbol{x}\in\mathbb{D}_{source}} [\hat{G}_l^f(\boldsymbol{x}^R|\boldsymbol{x}, s)\hat{q}(\boldsymbol{x}, s) + \hat{\Gamma}_{l,k}^f(\boldsymbol{x}^R|\boldsymbol{x}, s)\hat{f}_k(\boldsymbol{x}, s)]\mathrm{dV} \, , \tag{7.16}$$

where $\boldsymbol{x}^R \in \mathbb{R}^3$. The first term in the integrand of Eq. (7.16) represents a monopole contribution with volume density $\hat{q}$, while the second term represents a dipole contribution with volume density $\hat{f}_k$.

Equations (7.9) and (7.16) show that the acoustic wavefield from known sources in a known medium can be calculated in all space once the wavefields radiated by appropriate point sources have been calculated. The right-hand sides of the representations express the wavefield values as a superposition of the wavefields radiated by the elementary volume sources out of which the distributed sources can be envisaged to be composed.

### 7.1.2. Green's states

In Eqs. (7.8) and (7.15), we have introduced the Green's states in an unbounded medium. In this section we shall derive some general properties of the Green's states in an (unbounded) inhomogeneous medium.

Comparing Eqs. (7.3) and (7.8), it follows that

$$\hat{\Gamma}^q_k(\boldsymbol{x}^R|\boldsymbol{x},s) = \frac{1}{s\rho(\boldsymbol{x})}\partial_k \hat{G}^q(\boldsymbol{x}^R|\boldsymbol{x},s)\,, \tag{7.17}$$

while from Eqs. (7.10) and (7.15) it follows that

$$\hat{\Gamma}^f_{l,k}(\boldsymbol{x}^R|\boldsymbol{x},s) = \frac{1}{s\rho(\boldsymbol{x})}\left[\partial_k \hat{G}^f_l(\boldsymbol{x}^R|\boldsymbol{x},s) + \delta(\boldsymbol{x}^R - \boldsymbol{x})\delta_{l,k}\right]\,, \tag{7.18}$$

where $\delta_{l,k}$ is Kronecker symbol, $\delta_{l,k} = 1$ when $k = l$ and $\delta_{l,k} = 0$ when $k \neq l$.

From Eq. (7.8) and the reciprocity relation of Eq. (6.9) we obtain the following symmetry property (see also FRIEDLANDER, 1958, p. 143)

$$\hat{G}^q(\boldsymbol{x}^R|\boldsymbol{x},s) = \hat{G}^q(\boldsymbol{x}|\boldsymbol{x}^R,s)\,. \tag{7.19}$$

From Eqs. (7.8), (7.15) and the reciprocity relation of Eq. (6.11) we obtain the following symmetry property

$$\hat{\Gamma}^q_l(\boldsymbol{x}^R|\boldsymbol{x},s) = -\hat{G}^f_l(\boldsymbol{x}|\boldsymbol{x}^R,s)\,. \tag{7.20}$$

As a consequence of Eqs. (7.17), (7.19) and (7.20), we have

$$\hat{G}_l^f(\boldsymbol{x}^R|\boldsymbol{x},s) = \frac{-1}{s\rho(\boldsymbol{x}^R)}\partial_l^R \hat{G}^q(\boldsymbol{x}^R|\boldsymbol{x},s)\,, \tag{7.21}$$

where $\partial_l^R$ denotes the spatial derivative with respect to $x_l^R$. Further, using Eqs. (7.18) and (7.21) we have

$$\hat{\Gamma}_{l,k}^f(\boldsymbol{x}^R|\boldsymbol{x},s) = \frac{1}{s\rho(\boldsymbol{x}^R)}\left[-\frac{1}{s\rho(\boldsymbol{x})}\partial_l^R\partial_k \hat{G}^q(\boldsymbol{x}^R|\boldsymbol{x},s) + \delta(\boldsymbol{x}^R-\boldsymbol{x})\delta_{l,k}\right]. \tag{7.22}$$

From Eq. (7.15) and the reciprocity relation of Eq. (6.13) we obtain the following symmetry property

$$\hat{\Gamma}_{l,k}^f(\boldsymbol{x}^R|\boldsymbol{x},s) = \hat{\Gamma}_{k,l}^f(\boldsymbol{x}|\boldsymbol{x}^R,s)\,. \tag{7.23}$$

*The s-domain Green's states in an unbounded, homogeneous medium*

The construction of the Green's states is complicated in inhomogeneous media, but is fairly straightforward in a homogeneous medium with constants $\rho$ and $\kappa$ (see Section 4.7). In the latter case they are given by

$$\hat{G}^q(\boldsymbol{x}^R|\boldsymbol{x},s) = s\rho\hat{G}(\boldsymbol{x}^R-\boldsymbol{x},s)\,, \tag{7.24}$$

$$\hat{\Gamma}_k^q(\boldsymbol{x}^R|\boldsymbol{x},s) = -\partial_k^R\hat{G}(\boldsymbol{x}^R-\boldsymbol{x},s)\,, \tag{7.25}$$

$$\hat{G}_l^f(\boldsymbol{x}^R|\boldsymbol{x},s) = -\partial_l^R\hat{G}(\boldsymbol{x}^R-\boldsymbol{x},s)\,, \tag{7.26}$$

$$\hat{\Gamma}_{l,k}^f(\boldsymbol{x}^R|\boldsymbol{x},s) = \frac{1}{s\rho}\left[\partial_l^R\partial_k^R\hat{G}(\boldsymbol{x}^R-\boldsymbol{x},s) + \delta(\boldsymbol{x}^R-\boldsymbol{x})\delta_{l,k}\right], \tag{7.27}$$

where $\partial_k^R$ denotes the spatial derivative with respect to $x_k^R$. In Eqs. (7.24) - (7.27), the scalar Green's function $\hat{G}(\boldsymbol{x},s)$ is given by (cf. Chapter 4)

$$\hat{G}(\boldsymbol{x},s) = \frac{\exp(-\hat{\gamma}|\boldsymbol{x}|)}{4\pi|\boldsymbol{x}|} \quad \text{with } \hat{\gamma} = s(\kappa\rho)^{\frac{1}{2}} = \frac{s}{c}\,. \tag{7.28}$$

The time-domain properties of the Green's states follow directly from an inversion of the $s$-domain results of Eqs. (7.17) - (7.28).

*Limiting values as* $\boldsymbol{x} \rightarrow \boldsymbol{x}^R$

To investigate the singular behavior of the Green's states in inhomogeneous media, we consider the limiting values as $\boldsymbol{x} \rightarrow \boldsymbol{x}^R$. We assume that the medium is locally homogeneous at $\boldsymbol{x}^R$. In the limiting case that $\boldsymbol{x}$ approaches $\boldsymbol{x}^R$, the solution of Eqs. (7.3) and (7.4) is given by

$$\hat{p}^q(\boldsymbol{x}|\boldsymbol{x}^R, s) = s\rho(\boldsymbol{x}^R)\hat{q}^B \frac{1}{4\pi|\boldsymbol{x}-\boldsymbol{x}^R|} \quad \text{as } \boldsymbol{x} \rightarrow \boldsymbol{x}^R , \tag{7.29}$$

and

$$\hat{v}_k^q(\boldsymbol{x}|\boldsymbol{x}^R, s) = -\hat{q}^B \partial_k \frac{1}{4\pi|\boldsymbol{x}-\boldsymbol{x}^R|} = \hat{q}^B \frac{x_k - x_k^R}{4\pi|\boldsymbol{x}-\boldsymbol{x}^R|^3} \quad \text{as } \boldsymbol{x} \rightarrow \boldsymbol{x}^R . \tag{7.30}$$

We directly observe that this solution satisfies Eq. (7.3) around $\boldsymbol{x} = \boldsymbol{x}^R$. In order to show that the limiting values of Eqs. (7.29) and (7.30) satisfy Eq. (7.4) around $\boldsymbol{x} = \boldsymbol{x}^R$, we rewrite the latter equation in its global form

$$\int_{\boldsymbol{x}\in\mathbb{D}_\delta} (\partial_k \hat{v}_k^q + s\kappa\hat{p}^q)\mathrm{dV} = \hat{q}^B , \tag{7.31}$$

where $\mathbb{D}_\delta$ is a spherical domain around $\boldsymbol{x}^R$ with vanishing radius $\delta = |\boldsymbol{x}-\boldsymbol{x}^R|$. Using Eq. (7.29), we observe that the contribution of the second term in the integrand of Eq. (7.31) vanishes as $\delta \downarrow 0$. Using Gauss' integral theorem we rewrite Eq. (7.31) as

$$\int_{\boldsymbol{x}\in\partial\mathbb{D}_\delta} \nu_k \hat{v}_k^q \mathrm{dA} = \hat{q}^B \quad \text{as } \delta \downarrow 0 , \tag{7.32}$$

where the normal vector $\nu_k$, directed away from $\mathbb{D}_\delta$, is given by

$$\nu_k = \frac{x_k - x_k^R}{|\boldsymbol{x}-\boldsymbol{x}^R|} , \tag{7.33}$$

and where $\partial\mathbb{D}_\delta$ denotes the spherical surface around the point $\boldsymbol{x} = \boldsymbol{x}^R$. Using Eq. (7.30) we observe that

$$\nu_k \hat{v}_k^q = \frac{\hat{q}^B}{4\pi|\boldsymbol{x}-\boldsymbol{x}^R|^2} = \frac{\hat{q}^B}{4\pi\delta^2} \tag{7.34}$$

and that the left-hand side of Eq. (7.32) is equal to $\hat{q}^B$. Thus, Eq. (7.31) is satisfied. Hence, Eqs. (7.29) and (7.30) represent the limiting values of the acoustic quantities $\{\hat{p}^q, \hat{v}_k^q\}$ as $\boldsymbol{x}$ approaches $\boldsymbol{x}^R$.

From Eqs. (7.8), (7.29) and (7.30) the limiting values of the Green's states are obtained as

$$\hat{G}^q(\boldsymbol{x}^R|\boldsymbol{x},s) = s\rho(\boldsymbol{x}^R)\frac{1}{4\pi|\boldsymbol{x}^R-\boldsymbol{x}|} \quad \text{as } \boldsymbol{x}\to\boldsymbol{x}^R\,, \tag{7.35}$$

$$\hat{\Gamma}^q_k(\boldsymbol{x}^R|\boldsymbol{x},s) = -\partial^R_k\frac{1}{4\pi|\boldsymbol{x}^R-\boldsymbol{x}|} \quad \text{as } \boldsymbol{x}\to\boldsymbol{x}^R\,, \tag{7.36}$$

while using Eqs. (7.21) and (7.18),

$$\hat{G}^f_l(\boldsymbol{x}^R|\boldsymbol{x},s) = -\partial^R_l\frac{1}{4\pi|\boldsymbol{x}^R-\boldsymbol{x}|} \quad \text{as } \boldsymbol{x}\to\boldsymbol{x}^R\,, \tag{7.37}$$

$$\hat{\Gamma}^f_{l,k}(\boldsymbol{x}^R|\boldsymbol{x},s) = \frac{1}{s\rho(\boldsymbol{x}^R)}\left[\partial^R_l\partial^R_k\frac{1}{4\pi|\boldsymbol{x}^R-\boldsymbol{x}|} + \delta(\boldsymbol{x}^R-\boldsymbol{x})\delta_{l,k}\right] \quad \text{as } \boldsymbol{x}\to\boldsymbol{x}^R\,. \tag{7.38}$$

It is evident that the limiting values of the Green's states also hold for the case of a homogeneous medium, but in this case they directly follow from Eqs. (7.24) - (7.28).

### 7.1.3. Cauchy's domain integrals

When $\boldsymbol{x}^R \in \mathbb{D}_{source}$, the evaluation of the domain integrals of Eqs. (7.9) and (7.16) have to be interpreted as their Cauchy principal value, where the contribution around the singular point $\boldsymbol{x} = \boldsymbol{x}^R$ is excluded symmetrically and calculated analytically (Lee *at al.*, 1980). To this end we consider the contribution of the integrations over a vanishing spherical domain $\mathbb{D}_\delta$ with radius $\delta$ and center some point $\boldsymbol{x}' \in \mathbb{D}_{source}$. We assume that $\boldsymbol{x}^R \in \mathbb{D}_\delta$. Note that $\boldsymbol{x}^R \to \boldsymbol{x}'$ when $\delta \downarrow 0$. Using the expressions for the limiting values of the Green's states of Eqs. (7.35) - (7.38), we may write

$$\int_{\boldsymbol{x}\in\mathbb{D}_\delta} \hat{G}^q(\boldsymbol{x}^R|\boldsymbol{x},s)\hat{q}(\boldsymbol{x},s)\mathrm{dV} = s\rho(\boldsymbol{x}^R)\hat{q}(\boldsymbol{x}^R,s)\int_{\boldsymbol{x}\in\mathbb{D}_\delta}\frac{1}{4\pi|\boldsymbol{x}^R-\boldsymbol{x}|}\mathrm{dV}\,, \tag{7.39}$$

$$\int_{\boldsymbol{x}\in\mathbb{D}_\delta} \hat{\Gamma}^q_k(\boldsymbol{x}^R|\boldsymbol{x},s)\hat{f}_k(\boldsymbol{x},s)\mathrm{dV} = -\hat{f}_k(\boldsymbol{x}^R,s)\partial^R_k\int_{\boldsymbol{x}\in\mathbb{D}_\delta}\frac{1}{4\pi|\boldsymbol{x}^R-\boldsymbol{x}|}\mathrm{dV}\,, \tag{7.40}$$

$$\int_{\boldsymbol{x}\in\mathbb{D}_\delta} \hat{G}_l^f(\boldsymbol{x}^R|\boldsymbol{x},s)\hat{q}(\boldsymbol{x},s)\mathrm{dV} = -\hat{q}(\boldsymbol{x}^R,s)\partial_l^R\int_{\boldsymbol{x}\in\mathbb{D}_\delta} \frac{1}{4\pi|\boldsymbol{x}^R-\boldsymbol{x}|}\mathrm{dV}\,, \quad (7.41)$$

$$\int_{\boldsymbol{x}\in\mathbb{D}_\delta} \hat{\Gamma}_{l,k}^f(\boldsymbol{x}^R|\boldsymbol{x},s)\hat{f}_k(\boldsymbol{x},s)\mathrm{dV}$$
$$= \frac{1}{s\rho(\boldsymbol{x}^R)}\left[\hat{f}_k(\boldsymbol{x}^R,s)\partial_l^R\partial_k^R\int_{\boldsymbol{x}\in\mathbb{D}_\delta}\frac{1}{4\pi|\boldsymbol{x}^R-\boldsymbol{x}|}\mathrm{dV} + \hat{f}_l(\boldsymbol{x}^R,s)\right]\,, \quad (7.42)$$

when $\delta \downarrow 0$. It is assumed that the domain $\mathbb{D}_\delta$ does not contain a discontinuity of $\hat{q}$ and/or $\hat{f}_k$.

First we turn our attention to the integral at the right-hand side of Eq. (7.39). The simplest way to evaluate this integral is to introduce spherical coordinates in the $\boldsymbol{x}$-space with center at $\boldsymbol{x} = \boldsymbol{x}'$ and the direction $\boldsymbol{x}^R - \boldsymbol{x}'$ as polar axis. Let $r = |\boldsymbol{x}-\boldsymbol{x}'|$ and $\theta$ the polar angle between $\boldsymbol{x}^R - \boldsymbol{x}'$ and $\boldsymbol{x}-\boldsymbol{x}'$, then the range of integration is $0 \le r \le \delta$, $0 \le \theta \le \pi$, $0 \le \phi < 2\pi$, where $\phi$ is the azimuth angle in the plane perpendicular to $\boldsymbol{x}^R - \boldsymbol{x}'$. Let further $r^R = |\boldsymbol{x}^R - \boldsymbol{x}'|$. Then in the integral we have

$$|\boldsymbol{x}^R - \boldsymbol{x}| = \left[(r^R)^2 + r^2 - 2r^R r\cos(\theta)\right]^{\frac{1}{2}}\,, \quad (7.43)$$

and $\mathrm{dV} = r^2\sin(\theta)\mathrm{d}r\mathrm{d}\theta\mathrm{d}\phi$. In the resulting integral we first carry out the integration with respect to $\phi$. Since the integrand is independent of $\phi$, this merely amounts to a multiplication by a function of $2\pi$. Next we carry out the integration with respect to $\theta$, which is elementary. After this we have

$$\begin{aligned}\int_{\boldsymbol{x}\in\mathbb{D}_\delta}\frac{1}{4\pi|\boldsymbol{x}^R-\boldsymbol{x}|}\mathrm{dV} &= \frac{1}{2r^R}\int_0^\delta [(r^R + r) - |r^R - r|]r\mathrm{d}r \\ &= \frac{1}{2}\delta^2 - \frac{1}{6}(r^R)^2 \\ &= \frac{1}{2}\delta^2 - \frac{1}{6}|\boldsymbol{x}^R - \boldsymbol{x}'|^2\,. \end{aligned} \quad (7.44)$$

Choosing $\boldsymbol{x}^R = \boldsymbol{x}'$ and letting $\delta \downarrow 0$, we arrive at

$$\int_{\boldsymbol{x}\in\mathbb{D}_\delta}\frac{1}{4\pi|\boldsymbol{x}^R-\boldsymbol{x}|}\mathrm{dV} = 0 \quad \text{as } \delta \downarrow 0\,. \quad (7.45)$$

The integrals on the right-hand sides of Eqs. (7.40) and (7.41) are calculated by taking the derivative of Eq. (7.44) with respect to $x_k^R$, viz.,

$$\partial_k^R \int_{\boldsymbol{x}\in\mathbb{D}_\delta}\frac{1}{4\pi|\boldsymbol{x}^R-\boldsymbol{x}'|}\mathrm{dV} = -\frac{1}{3}(x_k^R - x_k')\,. \quad (7.46)$$

Choosing $\boldsymbol{x}^R = \boldsymbol{x}'$, we arrive at

$$\partial_k^R \int_{\boldsymbol{x}\in\mathbb{D}_\delta} \frac{1}{4\pi|\boldsymbol{x}^R - \boldsymbol{x}|} \mathrm{dV} = 0 \,. \tag{7.47}$$

Sofar, the integrals of Eqs. (7.39) - (7.41) vanish when $\delta \downarrow 0$.

Finally, the integral on the right-hand side of Eq. (7.42) is calculated by taking the derivative of Eq. (7.46) with respect to $x_l^R$, viz.

$$\partial_l^R \partial_k^R \int_{\boldsymbol{x}\in\mathbb{D}_\delta} \frac{1}{4\pi|\boldsymbol{x}^R - \boldsymbol{x}|} \mathrm{dV} = -\frac{1}{3}\delta_{l,k} \,. \tag{7.48}$$

Observe that the latter integral does not vanish for vanishing $\delta$. Hence, the integral of Eq. (7.42) becomes

$$\int_{\boldsymbol{x}\in\mathbb{D}_\delta} \hat{\Gamma}_{l,k}^f(\boldsymbol{x}^R|\boldsymbol{x}, s)\hat{f}_k(\boldsymbol{x}, s)\mathrm{dV} = \frac{2}{3}\frac{1}{s\rho(\boldsymbol{x}^R)}\hat{f}_l(\boldsymbol{x}^R, s) \,, \tag{7.49}$$

when $\delta \downarrow 0$.

In conclusion we state that Eqs. (7.9) and (7.16) have to be interpreted as the Cauchy integral representations

$$\hat{p}(\boldsymbol{x}^R, s) = \fint_{\boldsymbol{x}\in\mathbb{D}_{source}} [\hat{G}^q(\boldsymbol{x}^R|\boldsymbol{x}, s)\hat{q}(\boldsymbol{x}, s) + \hat{\Gamma}_k^q(\boldsymbol{x}^R|\boldsymbol{x}, s)\hat{f}_k(\boldsymbol{x}, s)]\mathrm{dV} \,, \tag{7.50}$$

and

$$\hat{v}_l(\boldsymbol{x}^R, s) = \fint_{\boldsymbol{x}\in\mathbb{D}_{source}} [\hat{G}_l^f(\boldsymbol{x}^R|\boldsymbol{x}, s)\hat{q}(\boldsymbol{x}, s) + \hat{\Gamma}_{l,k}^f(\boldsymbol{x}^R|\boldsymbol{x}, s)\hat{f}_k(\boldsymbol{x}, s)]\mathrm{dV} + \frac{2}{3}\frac{1}{s\rho(\boldsymbol{x}^R)}\hat{f}_l(\boldsymbol{x}^R, s) \,, \tag{7.51}$$

when $\boldsymbol{x}^R \in \mathbb{R}^3$. The integral sign $\fint$ means the integration over the pertaining domain with a symmetric exclusion of the singular point, if necessary. It is noted that at any surface of discontinuity in $\hat{q}$ and/or $\hat{f}_k$ the right-hand sides of Eqs. (7.50) and (7.51) yield half the sum of the limiting values of the left-hand sides at either side of the relevant surface of discontinuity (consistent with the definition of a discontinuity in a domain, see Eq. (1.30)), provided that the integrals are interpreted as their Cauchy principal value.

### 7.1.4. Volume-source representations in the time domain

By applying the standard rules for inversion from the $s$-domain to the time domain, the time-domain representations are directly obtained from Eqs. (7.50) and (7.51) as

$$\int_{\boldsymbol{x}\in\mathbb{D}_{source}} [C_t\{G^q, q\} + C_t\{\Gamma^q_k, f_k\}]\mathrm{dV} = \chi_{\mathrm{T}}(t)p(\boldsymbol{x}^R, t)\,, \tag{7.52}$$

$$\int_{\boldsymbol{x}\in\mathbb{D}_{source}} [C_t\{G^f_l, q\} + C_t\{\Gamma^f_{l,k}, f_k\}]\mathrm{dV} + \frac{2}{3\rho(\boldsymbol{x}^R)}\mathcal{I}_t f_l(\boldsymbol{x}^R, t) = \chi_{\mathrm{T}}(t)v_l(\boldsymbol{x}^R, t)\,, \tag{7.53}$$

where $\boldsymbol{x}^R \in \mathbb{R}^3$. The Green's states $G^q = G^q(\boldsymbol{x}^R|\boldsymbol{x}, t)$, $\Gamma^q_k = \Gamma^q_k(\boldsymbol{x}^R|\boldsymbol{x}, t)$, $G^f_l = G^f_l(\boldsymbol{x}^R|\boldsymbol{x}, t)$ and $\Gamma^f_{l,k} = \Gamma^f_{l,k}(\boldsymbol{x}^R|\boldsymbol{x}, t)$ are the time-domain counterparts of the $s$-domain Green's states $\hat{G}^q$, $\hat{\Gamma}^q_k$, $\hat{G}^f_l$ and $\hat{\Gamma}^f_{l,k}$. Further, $C_t\{\cdot,\cdot\} = C_t\{\cdot,\cdot\}(\boldsymbol{x}, t)$ denotes the temporal convolution (see Eq. (1.19)).

The time-domain representations of Eqs. (7.52) - (7.53) can also be derived by remaining in the time domain. Performing a similar analysis using the time-domain reciprocity theorem of convolution type of Section 5.2 we obtain the same results.

For the special case of a homogeneous background medium, the pertaining relations of Eqs. (7.52) and (7.53) reduce to

$$\int_{\boldsymbol{x}\in\mathbb{D}_{source}} \left[\rho\frac{\partial_t q(\boldsymbol{x}, t - \frac{|\boldsymbol{x}^R - \boldsymbol{x}|}{c})}{4\pi|\boldsymbol{x}^R - \boldsymbol{x}|} - \partial^R_k \frac{f_k(\boldsymbol{x}, t - \frac{|\boldsymbol{x}^R - \boldsymbol{x}|}{c})}{4\pi|\boldsymbol{x}^R - \boldsymbol{x}|}\right]\mathrm{dV} = \chi_{\mathrm{T}}(t)p(\boldsymbol{x}^R, t) \tag{7.54}$$

(cf. Eqs. (4.85) and (4.93)), and

$$\int_{\boldsymbol{x}\in\mathbb{D}_{source}} \left[-\partial^R_l \frac{q(\boldsymbol{x}, t - \frac{|\boldsymbol{x}^R - \boldsymbol{x}|}{c})}{4\pi|\boldsymbol{x}^R - \boldsymbol{x}|} + \frac{1}{\rho}\partial^R_l\partial^R_k \frac{\mathcal{I}_t f_k(\boldsymbol{x}, t - \frac{|\boldsymbol{x}^R - \boldsymbol{x}|}{c})}{4\pi|\boldsymbol{x}^R - \boldsymbol{x}|}\right]\mathrm{dV} + \frac{2}{3\rho}\mathcal{I}_t f_l(\boldsymbol{x}^R, t) = \chi_{\mathrm{T}}(t)v_l(\boldsymbol{x}^R, t) \tag{7.55}$$

(cf. Eqs. (4.86) and (4.94)), where $\boldsymbol{x}^R \in \mathbb{R}^3$. The first term in the integrand of Eq. (7.54) and (7.55) represents the monopole contribution distributed over the source domain $\mathbb{D}_{source}$, while the second term represents the dipole contribution distributed over the source domain $\mathbb{D}_{source}$.

## 7.2. The surface-source problem

### 7.2.1. Surface-source representations in the $s$-domain

*The $s$-domain pressure representation*

The configuration consists of a medium of infinite extent with known acoustic properties. In this medium a source is present that occupies a bounded domain $\mathbb{D}_{source}$ with enclosing boundary $\partial\mathbb{D}_{source}$ (Fig. 7.2). Our aim is to arrive at a representation that expresses the acoustic pressure $\hat{p}$ in the domain $\mathbb{D}'_{source}$ outside $\partial\mathbb{D}_{source}$ in terms of the source distributions on the surface $\partial\mathbb{D}_{source}$. We apply the reciprocity theorem of Section 5.1. To this end, State $A$ is taken to be the actual wavefield that is generated by the sources and is causally related to them. The acoustic wavefield equations pertaining to State $A$ (cf. Table 7.3) are then

$$\partial_k\hat{p} + s\rho\hat{v}_k = 0, \quad \boldsymbol{x} \in \mathbb{D}'_{source}, \tag{7.56}$$

$$\partial_k\hat{v}_k + s\kappa\hat{p} = 0, \quad \boldsymbol{x} \in \mathbb{D}'_{source}. \tag{7.57}$$

The wavefield in State $B$ is chosen such that the application of Eq. (5.7) leads to the value of the acoustic pressure at any point in the domain $\mathbb{D}'_{source}$. Inspection of the right-hand side of Eq. (5.7) reveals that this is accomplished if we take for the source distributions in State $B$ a point source of volume injection. This wavefield satisfies (cf. Table 7.3)

$$\partial_k\hat{p}^q + s\rho\hat{v}^q_k = 0, \tag{7.58}$$

$$\partial_k\hat{v}^q_k + s\kappa\hat{p}^q = \hat{q}^B\delta(\boldsymbol{x}-\boldsymbol{x}^R), \tag{7.59}$$

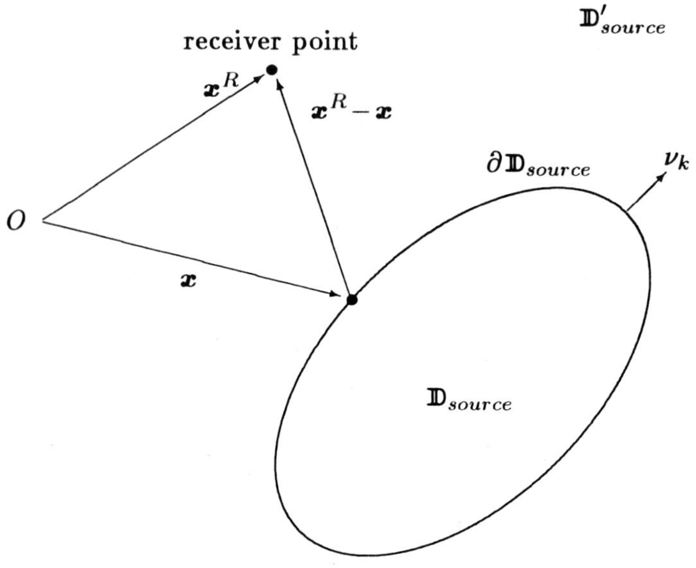


Figure 7.2. The source domain and the receiver location.

Table 7.3. States in the reciprocity theorem

| | State $A$ (actual state) | State $B$ (volume-injection Green's state) |
|---|---|---|
| Field state | $\{\hat{p}, \hat{v}_k\}(\boldsymbol{x}, s)$ | $\{\hat{p}^q, \hat{v}_k^q\}(\boldsymbol{x}\vert\boldsymbol{x}^R, s)$ |
| Material state | $\{\rho, \kappa\}(\boldsymbol{x})$ | $\{\rho, \kappa\}(\boldsymbol{x})$ |
| Source state | $\{0, 0\}$ | $\{\hat{q}^B(s)\delta(\boldsymbol{x}-\boldsymbol{x}^R), 0\}$ |
| Domain $\mathbb{D}'_{source}$ (see Fig. 7.2) | | |

where $\delta(\boldsymbol{x} - \boldsymbol{x}^R)$ denotes the three-dimensional spatial Dirac distribution (impulse function) operative at the receiver point with position vector $\boldsymbol{x}^R$, while $\hat{q}^B$ is an arbitrary constant only depending on $s$. This source is placed in the medium with the same material parameters $\rho$ and $\kappa$ as the actual ones. For the choice of Eqs. (7.58) and (7.59), State $B$ is denoted as the volume-injection Green's state. Now Eq. (5.7) is applied to the domain exterior to $\partial\mathbb{D}_{source}$ and interior to the sphere $\mathrm{S}_\Delta$ with radius $\Delta$ and center at the origin of the chosen coordinate system; $\Delta$ is so large that $\mathrm{S}_\Delta$ completely surrounds $\mathbb{D}_{source}$. In view of the causality condition we have

$$\int_{\boldsymbol{x}\in \mathrm{S}_\Delta} (\hat{p}\hat{v}_k^q - \hat{p}^q\hat{v}_k)\nu_k \mathrm{dA} = \mathrm{Order}\ (\Delta^{-1}) \ \text{ as } \ \Delta \to \infty \, . \tag{7.60}$$

Hence, when $\boldsymbol{x}^R \in \mathbb{D}'_{source}$, we have

$$\hat{q}^B\hat{p}(\boldsymbol{x}^R, s) = \int_{\boldsymbol{x}\in\partial\mathbb{D}_{source}} [\hat{p}^q(\boldsymbol{x}|\boldsymbol{x}^R, s)\hat{v}_k(\boldsymbol{x}, s) - \hat{v}_k^q(\boldsymbol{x}|\boldsymbol{x}^R, s)\hat{p}(\boldsymbol{x}, s)]\nu_k \mathrm{dA} \, , \tag{7.61}$$

in which the normal vector $\nu_k$ is directed away from $\mathbb{D}_{source}$. From Eq. (7.61) a representation for $\hat{p}(\boldsymbol{x}^R, s)$ is obtained when we take into account that $\hat{p}^q$ and $\hat{v}_k^q$ are linearly related to $\hat{q}^B$. Expressing this as

$$\{\hat{p}^q, \hat{v}_k^q\}(\boldsymbol{x}|\boldsymbol{x}^R, s) = \hat{q}^B(s)\{\hat{G}^q, -\hat{\Gamma}_k^q\}(\boldsymbol{x}^R|\boldsymbol{x}, s) \, , \tag{7.62}$$

we arrive at the desired representation of the acoustic pressure

$$\hat{p}(\boldsymbol{x}^R, s) = \int_{\boldsymbol{x}\in\partial\mathbb{D}_{source}} [\hat{G}^q(\boldsymbol{x}^R|\boldsymbol{x}, s)\hat{v}_k(\boldsymbol{x}, s) + \hat{\Gamma}_k^q(\boldsymbol{x}^R|\boldsymbol{x}, s)\hat{p}(\boldsymbol{x}, s)]\nu_k \mathrm{dA} \quad \text{when } \ \boldsymbol{x}^R \in \mathbb{D}'_{source} \, . \tag{7.63}$$

The first term in the integrand of Eq. (7.63) represents a monopole contribution with surface density $\nu_k\hat{v}_k$, while the second term represents a dipole contribution with surface density $\hat{p}\nu_k$.

### *The $s$-domain particle velocity representation*

In order to arrive at a representation that expresses the acoustic particle velocity $\hat{v}_k$ in the domain $\mathbb{D}'_{source}$ outside $\partial\mathbb{D}_{source}$ in terms of the source distributions on the surface $\partial\mathbb{D}_{source}$, the wavefield in State $B$ is chosen such that the application of Eq. (5.7) leads to the value of the acoustic particle velocity at any point in the domain $\mathbb{D}'_{source}$. Inspection of the right-hand side of Eq. (5.7) reveals that this is accomplished if we take for the source distributions in State $B$ a point source of volume force. This wavefield satisfies (cf. Table 7.4)

$$\partial_k \hat{p}^f + s\rho \hat{v}_k^f = \hat{f}_k^B \delta(\boldsymbol{x} - \boldsymbol{x}^R) , \tag{7.64}$$

$$\partial_k \hat{v}_k^f + s\kappa \hat{p}^f = 0 , \tag{7.65}$$

where $\hat{f}_k^B$ is an arbitrary constant vector only depending on $s$. The source is placed in the medium with the same material parameters $\rho$ and $\kappa$ as the actual ones. For the choice of Eqs. (7.64) and (7.65), State $B$ is denoted as the volume-force Green's state. Now Eq. (5.7) is applied to the domain exterior to $\partial \mathbb{D}_{source}$ and interior to the sphere $\mathrm{S}_\Delta$ with radius $\Delta$ and center at the origin of the chosen coordinate system; $\Delta$ is so large that $\mathrm{S}_\Delta$ completely surrounds $\mathbb{D}_{source}$. In view of the causality condition we have

$$\int_{\boldsymbol{x} \in \mathrm{S}_\Delta} (\hat{p} \hat{v}_k^f - \hat{p}^f \hat{v}_k) \nu_k \mathrm{dA} = \text{Order}\ (\Delta^{-1}) \ \text{ as } \ \Delta \to \infty . \tag{7.66}$$

Table 7.4. States in the reciprocity theorem

| | State $A$ (actual state) | State $B$ (volume-force Green's state) |
|---|---|---|
| Field state | $\{\hat{p}, \hat{v}_k\}(\boldsymbol{x}, s)$ | $\{\hat{p}^f, \hat{v}_k^f\}(\boldsymbol{x} \vert \boldsymbol{x}^R, s)$ |
| Material state | $\{\rho, \kappa\}(\boldsymbol{x})$ | $\{\rho, \kappa\}(\boldsymbol{x})$ |
| Source state | $\{0, 0\}$ | $\{0, \hat{f}_k^B(s) \delta(\boldsymbol{x} - \boldsymbol{x}^R)\}$ |
| Domain $\mathbb{D}'_{source}$ (see Fig. 7.2) | | |

Hence, when $\boldsymbol{x}^R \in \mathbb{D}'_{source}$, we have

$$\hat{f}_l^B \hat{v}_l(\boldsymbol{x}^R, s) = \int_{\boldsymbol{x}\in\partial\mathbb{D}_{source}} [-\hat{p}^f(\boldsymbol{x}|\boldsymbol{x}^R, s)\hat{v}_k(\boldsymbol{x}, s) + \hat{v}_k^f(\boldsymbol{x}|\boldsymbol{x}^R, s)\hat{p}(\boldsymbol{x}, s)]\nu_k \mathrm{dA}, \tag{7.67}$$

in which the normal vector $\nu_k$ is directed away from $\mathbb{D}_{source}$. From Eq. (7.67) a representation for $\hat{v}_l(\boldsymbol{x}^R, s)$ is obtained when we take into account that $\hat{p}^f$ and $\hat{v}_k^f$ are linearly related to $\hat{f}_l^B$. Expressing this as

$$\{\hat{p}^f, \hat{v}_k^f\}(\boldsymbol{x}|\boldsymbol{x}^R, s) = \hat{f}_l^B(s)\{-\hat{G}_l^f, \hat{\Gamma}_{l,k}^f\}(\boldsymbol{x}^R|\boldsymbol{x}, s)\,, \tag{7.68}$$

we arrive at the desired representation of the particle velocity

$$\hat{v}_l(\boldsymbol{x}^R, s) = \int_{\boldsymbol{x}\in\partial\mathbb{D}_{source}} [\hat{G}_l^f(\boldsymbol{x}^R|\boldsymbol{x}, s)\hat{v}_k(\boldsymbol{x}, s) + \hat{\Gamma}_{l,k}^f(\boldsymbol{x}^R|\boldsymbol{x}, s)\hat{p}(\boldsymbol{x}, s)]\nu_k \mathrm{dA} \quad \text{when } \boldsymbol{x}^R \in \mathbb{D}'_{source}. \tag{7.69}$$

The first term in the integrand of Eq. (7.69) represents a monopole contribution with surface density $\nu_k \hat{v}_k$, while the second term represents a dipole contribution with surface density $\hat{p}\nu_k$.

Equations (7.63) and (7.69) show that the acoustic wavefield outside a closed surface can be calculated once both the acoustic pressure and the normal component of the particle velocity on this surface are known, and once the wavefields radiated by appropriate point sources have been calculated. The right-hand sides of the representations express the wavefield values as a superposition of the wavefields radiated by elementary surface sources out of which the distributed sources can be envisaged to be composed. The surface sources are represented by the Green's states. When the medium is homogeneous outside the pertaining surface, they are given in Eqs. (7.24) - (7.27). Note that the representation for $\hat{v}_k(\boldsymbol{x}^R, s)$, when $\boldsymbol{x}^R \in \mathbb{D}'_{source}$, can directly be obtained from the wavefield equation (7.56) as

$$\hat{v}_k(\boldsymbol{x}^R, s) = \frac{-1}{s\rho(\boldsymbol{x}^R)} \partial_k^R \hat{p}(\boldsymbol{x}^R, s)\,. \tag{7.70}$$

This can also be seen by comparing Eqs. (7.63) and (7.69), and using the properties of the Green's states as given by Eq. (7.21) and (7.22) for $\boldsymbol{x}^R \neq \boldsymbol{x}$. Therefore we only consider Eq. (7.63) as the fundamental representation and Eq. (7.69) as an auxiliary relation which can always be obtained by using Eqs. (7.63) and (7.70).

### 7.2.2. Cauchy's boundary integrals

In this section we investigate the integral representation for the acoustic pressure when $\boldsymbol{x}^R$ approaches $\partial\mathbb{D}_{source}$ from $\mathbb{D}'_{source}$ via $\boldsymbol{x}^R = \lim_{\varepsilon\downarrow 0}(\boldsymbol{x}' + \varepsilon\boldsymbol{\nu}')$, where $\boldsymbol{x}'$ is a point on $\partial\mathbb{D}_{source}$ and $\boldsymbol{\nu}'$ is the normal in $\boldsymbol{x}'$ directed away from $\mathbb{D}_{source}$. We rewrite the integral in Eq. (7.63) as a contribution from $\partial\mathbb{D}_{source}\backslash\partial\mathbb{D}_\delta$ and $\partial\mathbb{D}_\delta$, where the latter is a surface area of $\partial\mathbb{D}_{source}$ symmetrically located around $\boldsymbol{x}'$. We may write

$$\int_{\boldsymbol{x}\in\partial\mathbb{D}_{source}} [\hat{G}^q(\boldsymbol{x}^R|\boldsymbol{x},s)\hat{v}_k(\boldsymbol{x},s) + \hat{\Gamma}^q_k(\boldsymbol{x}^R|\boldsymbol{x},s)\hat{p}(\boldsymbol{x},s)]\nu_k \mathrm{dA}$$

$$= \int_{\boldsymbol{x}\in\partial\mathbb{D}_{source}\backslash\partial\mathbb{D}_\delta} [\hat{G}^q(\boldsymbol{x}^R|\boldsymbol{x},s)\hat{v}_k(\boldsymbol{x},s) + \hat{\Gamma}^q_k(\boldsymbol{x}^R|\boldsymbol{x},s)\hat{p}(\boldsymbol{x},s)]\nu_k \mathrm{dA}$$

$$+ s\rho(\boldsymbol{x}')\hat{v}_k(\boldsymbol{x}',s)\nu'_k \int_{\boldsymbol{x}\in\partial\mathbb{D}_\delta} \frac{1}{4\pi|\boldsymbol{x}'+\varepsilon\boldsymbol{\nu}'-\boldsymbol{x}|}\mathrm{dA}$$

$$- \hat{p}(\boldsymbol{x}',s)\partial_\varepsilon \int_{\boldsymbol{x}\in\partial\mathbb{D}_\delta} \frac{1}{4\pi|\boldsymbol{x}'+\varepsilon\boldsymbol{\nu}'-\boldsymbol{x}|}\mathrm{dA}\,, \tag{7.71}$$

where we have assumed that the surface area $\partial\mathbb{D}_\delta$ is small enough to approximate the acoustic quantities by their values at $\boldsymbol{x}'$ and in addition that $\varepsilon$ is small enough to approximate the Green's states by Eqs. (7.35) and (7.36). Note that in the last term of the right-hand side of Eq. (7.71) the normal derivative $\nu_k\partial^R_k$ has been replaced by the derivative with respect to $\varepsilon$. We further assume that the surface $\partial\mathbb{D}_{source}$ is a smooth surface with a continuous normal. Then, the surface area $\partial\mathbb{D}_\delta$ can be considered as a locally plane one, viz., a flat disk with vanishing radius $\delta$ and with center at $\boldsymbol{x} = \boldsymbol{x}'$, and the integrals over this surface area can be calculated analytically (see also HÖNL *et al.*, 1961, pp. 234-235).

Let us consider the second integral on the right-hand side of Eq. (7.71). This integral can be calculated by introducing polar coordinates $r = |\boldsymbol{x}'-\boldsymbol{x}|$, $0 \le r \le \delta$, and the polar angle $\phi$, $0 \le \phi < 2\pi$. Noting that, $\boldsymbol{\nu}'\cdot(\boldsymbol{x}'-\boldsymbol{x}) = 0$, we obtain

$$\int_{\boldsymbol{x}\in\partial\mathbb{D}_\delta} \frac{1}{4\pi|\boldsymbol{x}'+\varepsilon\boldsymbol{\nu}'-\boldsymbol{x}|}\mathrm{dA} = \frac{1}{2}\int_0^\delta \frac{1}{(r^2+\varepsilon^2)^{\frac{1}{2}}} r\mathrm{d}r = \frac{1}{2}(\delta^2+\varepsilon^2)^{\frac{1}{2}} - \frac{1}{2}\varepsilon\,. \tag{7.72}$$

Letting $\varepsilon \downarrow 0$ and $\delta \downarrow 0$ we arrive at

$$\int_{\boldsymbol{x}\in\partial\mathbb{D}_\delta} \frac{1}{4\pi|\boldsymbol{x}'+\varepsilon\boldsymbol{\nu}'-\boldsymbol{x}|}\mathrm{dA} = 0\,, \tag{7.73}$$

and the second term on the right-hand side of Eq. (7.71) vanishes.

Next, the normal derivative of Eq. (7.72) is obtained as

$$\partial_\varepsilon \int_{\boldsymbol{x}\in\partial\mathbb{D}_\delta} \frac{1}{4\pi|\boldsymbol{x}'+\varepsilon\boldsymbol{\nu}'-\boldsymbol{x}|}\mathrm{dA} = \frac{1}{2}\frac{\varepsilon}{(\delta^2+\varepsilon^2)^{\frac{1}{2}}} - \frac{1}{2}\,. \tag{7.74}$$

We take the limit that $\varepsilon \downarrow 0$ to arrive at

$$\partial_\varepsilon \int_{\boldsymbol{x}\in\partial\mathbb{D}_\delta} \frac{1}{4\pi|\boldsymbol{x}'+\varepsilon\boldsymbol{\nu}'-\boldsymbol{x}|}\mathrm{dA} = -\frac{1}{2}\,, \tag{7.75}$$

where $\delta$ is vanishing small, but not equal to zero. This means that the third term of Eq. (7.71) tends to the value $\frac{1}{2}\hat{p}(\boldsymbol{x}^R, s)$.

In conclusion we observe that the integral representation of Eq. (7.63) has to be replaced by

$$\tfrac{1}{2}\hat{p}(\boldsymbol{x}^R, s) = ⨍_{\boldsymbol{x}\in\partial\mathbb{D}_{source}} [\hat{G}^q(\boldsymbol{x}^R|\boldsymbol{x}, s)\hat{v}_k(\boldsymbol{x}, s) + \hat{\Gamma}^q_k(\boldsymbol{x}^R|\boldsymbol{x}, s)\hat{p}(\boldsymbol{x}, s)]\nu_k\mathrm{dA}$$
$$\text{when } \boldsymbol{x}^R \in \partial\mathbb{D}_{source}\,. \tag{7.76}$$

The integral sign $⨍$ means the integration over the pertaining boundary with a symmetric exclusion of the singular point, if necessary.

### 7.2.3. Surface-source representations in the time domain

By applying the standard rules for inversion from the $s$-domain to the time domain, the time-domain representations are directly obtained from Eqs. (7.63) and (7.76) as

$$\int_{\boldsymbol{x}\in\partial\mathbb{D}_{source}} [C_t\{G^q, \nu_k v_k\} + C_t\{\Gamma^q_k, \nu_k p\}]\mathrm{dA} = \chi_T(t)p(\boldsymbol{x}^R, t)$$
$$\text{when } \boldsymbol{x}^R \in \mathbb{D}'_{source}\,, \tag{7.77}$$

$$\oint_{\boldsymbol{x}\in\partial\mathbb{D}_{source}} [C_t\{G^q, \nu_k v_k\} + C_t\{\Gamma_k^q, \nu_k p\}]\mathrm{dA} = \tfrac{1}{2}\chi_T(t)p(\boldsymbol{x}^R, t)$$

$$\text{when } \boldsymbol{x}^R \in \partial\mathbb{D}_{source} \,. \qquad (7.78)$$

Again, the Green's states $G^q = G^q(\boldsymbol{x}^R|\boldsymbol{x}, t)$ and $\Gamma_k^q = \Gamma_k^q(\boldsymbol{x}^R|\boldsymbol{x}, t)$ are the time-domain counterparts of the $s$-domain Green's states $\hat{G}^q$ and $\hat{\Gamma}_k^q$. Further, $C_t\{\cdot,\cdot\} = C_t\{\cdot,\cdot\}(\boldsymbol{x}, t)$ denotes the temporal convolution defined by Eq. (1.19).

The time-domain representations of Eqs. (7.77) - (7.78) can also be derived in the time domain. Performing a similar analysis using the time-domain reciprocity theorem of convolution type of Section 5.2, we obtain the same results.

For the special case of a homogeneous background medium the pertaining relations of Eqs. (7.77) and (7.78) reduce to

$$\int_{\boldsymbol{x}\in\partial\mathbb{D}_{source}} \left[\rho\frac{\partial_t v_k(\boldsymbol{x}, t - \frac{|\boldsymbol{x}^R - \boldsymbol{x}|}{c})}{4\pi|\boldsymbol{x}^R - \boldsymbol{x}|} - \partial_k^R \frac{p(\boldsymbol{x}, t - \frac{|\boldsymbol{x}^R - \boldsymbol{x}|}{c})}{4\pi|\boldsymbol{x}^R - \boldsymbol{x}|}\right] \nu_k \mathrm{dA} \qquad (7.79)$$

$$= \chi_{\mathrm{T}}(t)p(\boldsymbol{x}^R, t) \quad \text{when } \boldsymbol{x}^R \in \mathbb{D}'_{source} \,,$$

and

$$\oint_{\boldsymbol{x}\in\partial\mathbb{D}_{source}} \left[\rho\frac{\partial_t v_k(\boldsymbol{x}, t - \frac{|\boldsymbol{x}^R - \boldsymbol{x}|}{c})}{4\pi|\boldsymbol{x}^R - \boldsymbol{x}|} - \partial_k^R \frac{p(\boldsymbol{x}, t - \frac{|\boldsymbol{x}^R - \boldsymbol{x}|}{c})}{4\pi|\boldsymbol{x}^R - \boldsymbol{x}|}\right] \nu_k \mathrm{dA} \qquad (7.80)$$

$$= \tfrac{1}{2}\chi_{\mathrm{T}}(t)p(\boldsymbol{x}^R, t) \quad \text{when } \boldsymbol{x}^R \in \partial\mathbb{D}_{source} \,.$$

Finally, we will discuss some useful reciprocity theorems that apply to bounded domains.

### 7.2.4. Oseen's extinction theorem

As has been shown, acoustic wavefield representations can be obtained that express the acoustic pressure and the particle velocity, at all points $\boldsymbol{x}^R \in \mathbb{D}'_{source}$, in terms of the equivalent surface-source distributions that generate the wavefield. In a number of cases, however, we are interested in the consequences of taking the point of observation inside the source domain ($\boldsymbol{x}^R \in \mathbb{D}_{source}$). Carrying out the previous application of the reciprocity of Section 7.2.1, Eq. (7.63) has to be replaced by

$$\int_{\boldsymbol{x}\in\partial\mathbb{D}_{source}} [\hat{G}^q(\boldsymbol{x}^R|\boldsymbol{x},s)\hat{v}_k(\boldsymbol{x},s) + \hat{\Gamma}_k^q(\boldsymbol{x}^R|\boldsymbol{x},s)\hat{p}(\boldsymbol{x},s)]\nu_k \mathrm{dA} = 0 \quad \text{when } \boldsymbol{x}^R \in \mathbb{D}_{source}\,. \tag{7.81}$$

This result for $\boldsymbol{x}^R \in \mathbb{D}_{source}$ shows that the left-hand side of Eq. (7.81) vanishes inside the source domain. This property is known as Oseen's extinction theorem (OSEEN, 1915). This theorem is used in computational acoustics, where the relevant numerical methods are known as null-field methods (BATES and WALL, 1977) and T-matrix methods (WATERMAN, 1969).

The time-domain counterpart of Oseen's extinction theorem is obtained in a similar way as before by either inversion of the $s$-domain results or by direct application of the time-domain reciprocity theorem of convolution type.

### 7.2.5. Representation theorem for a bounded subdomain

In some applications we need the results of the application of the reciprocity theorem with respect to the wavefields in a bounded subdomain $\mathbb{D}$ exterior to the source domain $\mathbb{D}_{source}$ (see Fig. 7.3). We apply the reciprocity theorem of Section 5.1 to the domain $\mathbb{D}$ inside the closed contour $\partial\mathbb{D}$ with outward normal $\nu_k$. To this end, State $A$ is taken to be the actual wavefield that is generated by the sources in $\mathbb{D}_{source}$. The wavefield of State $B$ is generated by a point source of volume injection (cf. Table 7.5). Using the Green's states of Eq. (7.62) we arrive at

$$\int_{\boldsymbol{x}\in\partial\mathbb{D}} [\hat{G}^q(\boldsymbol{x}^R|\boldsymbol{x},s)\hat{v}_k(\boldsymbol{x},s) + \hat{\Gamma}_k^q(\boldsymbol{x}^R|\boldsymbol{x},s)\hat{p}(\boldsymbol{x},s)]\nu_k \mathrm{dA} = -\hat{p}(\boldsymbol{x}^R,s) \quad \text{when } \boldsymbol{x}^R \in \mathbb{D}\,, \tag{7.82}$$

$$\int_{\boldsymbol{x}\in\partial\mathbb{D}} [\hat{G}^q(\boldsymbol{x}^R|\boldsymbol{x},s)\hat{v}_k(\boldsymbol{x},s) + \hat{\Gamma}_k^q(\boldsymbol{x}^R|\boldsymbol{x},s)\hat{p}(\boldsymbol{x},s)]\nu_k \mathrm{dA} = 0 \quad \text{when } \boldsymbol{x}^R \in \mathbb{D}'\,, \tag{7.83}$$

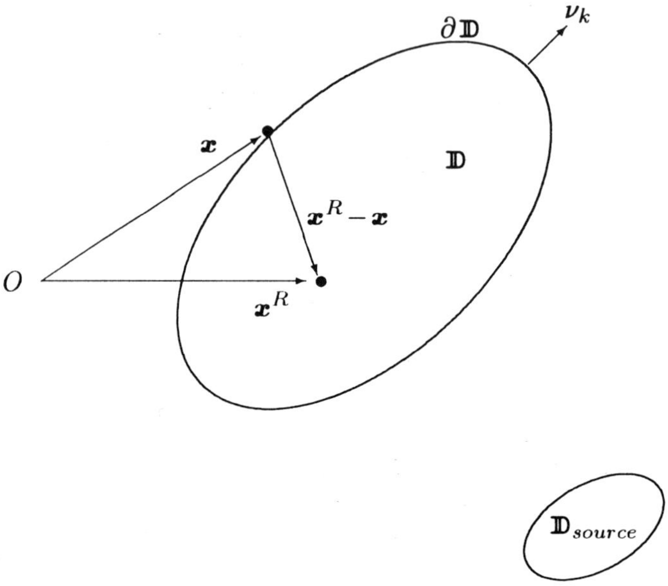


Figure 7.3. The bounded subdomain $\mathbb{D}$ exterior to the source domain $\mathbb{D}_{source}$.

Table 7.5. States in the reciprocity theorem

| | State $A$ (actual state) | State $B$ (volume-injection Green's state) |
|---|---|---|
| Field state | $\{\hat{p}, \hat{v}_k\}(\boldsymbol{x}, s)$ | $\{\hat{p}^q, \hat{v}_k^q\}(\boldsymbol{x}\vert\boldsymbol{x}^R, s)$ |
| Material state | $\{\rho, \kappa\}(\boldsymbol{x})$ | $\{\rho, \kappa\}(\boldsymbol{x})$ |
| Source state | $\{0, 0\}$ | $\{\hat{q}^B(s)\delta(\boldsymbol{x}-\boldsymbol{x}^R), 0\}$ |
| Domain $\mathbb{D}$ (see Fig. 7.3) | | |

and

$$\int_{\boldsymbol{x}\in\partial\mathbb{D}} [\hat{G}^q(\boldsymbol{x}^R|\boldsymbol{x},s)\hat{v}_k(\boldsymbol{x},s) + \hat{\Gamma}^q_k(\boldsymbol{x}^R|\boldsymbol{x},s)\hat{p}(\boldsymbol{x},s)]\nu_k \mathrm{dA} = -\tfrac{1}{2}\hat{p}(\boldsymbol{x}^R,s)$$

$$\text{when } \boldsymbol{x}^R \in \partial\mathbb{D}\,, \qquad (7.84)$$

where $\mathbb{D}'$ is the complement of $\mathbb{D}\cup\partial\mathbb{D}$ in $\mathbb{R}^3$.

The time-domain counterparts of the latter results are obtained in a similar way as before by either inversion of the $s$-domain results or by direct application of the time-domain reciprocity theorem of convolution type.

This concludes the discussion of the source-type integral representations. In the next chapter, these representations are the point of departure to formulate the scattering problem in terms of integral equations.

# Chapter 8

# Scattering by a Bounded Contrasting Domain

In this chapter, the scattering of acoustic waves by a contrasting fluid domain of bounded extent, present in an inhomogeneous, fluid embedding of infinite extent, is investigated in more detail. The integral-equation formulations of the scattering problem are presented. When we consider the scattering object as a volume scatterer a domain-integral equation formulation is the most versatile technique, while the boundary-integral equation formulation is most convenient when we treat the scatterer as a surface scatterer.

## 8.1. The domain-integral equation formulation

We investigate the direct or forward scattering of acoustic waves by a contrasting fluid domain of bounded extent, present in a fluid embedding of infinite extent. Let $\mathbb{D}_{sct}$ be the bounded domain occupied by the fluid scatterer and let $\rho^s = \rho^s(\boldsymbol{x})$ be its volume density of mass and $\kappa^s = \kappa^s(\boldsymbol{x})$ its compressibility. The domain exterior to $\mathbb{D}_{sct}$ is denoted by $\mathbb{D}'_{sct}$. In $\mathbb{D}'_{sct}$, the embedding, an inhomogeneous fluid is present; its volume density of mass is denoted by $\rho = \rho(\boldsymbol{x})$ and its compressibility by $\kappa = \kappa(\boldsymbol{x})$ (Fig. 8.1). We assume that the Green's states (cf. Chapter 7) of the embedding (or background medium) can be determined. The total acoustic wavefield in the configuration $\{\hat{p}, \hat{v}_k\}$ is decomposed into the incident wavefield $\{\hat{p}^{inc}, \hat{v}_k^{inc}\}$ and the scattered wavefield $\{\hat{p}^{sct}, \hat{v}_k^{sct}\}$. The incident wavefield is the wavefield that would be present in the entire configuration if the domain $\mathbb{D}_{sct}$ showed no contrast with the embedding. The total wavefield is generated by sources that are located outside the scattering domain. Since these sources remain present even if the scattering domain is thought to be absent, they also serve as sources for the incident wavefield. The incident wavefield quantities can be calculated with the representations obtained in Chapter 7. We start our analysis in the $s$-domain. The results in the time domain are obtained by applying the standard rules for inversion from the $s$-domain to the time domain.

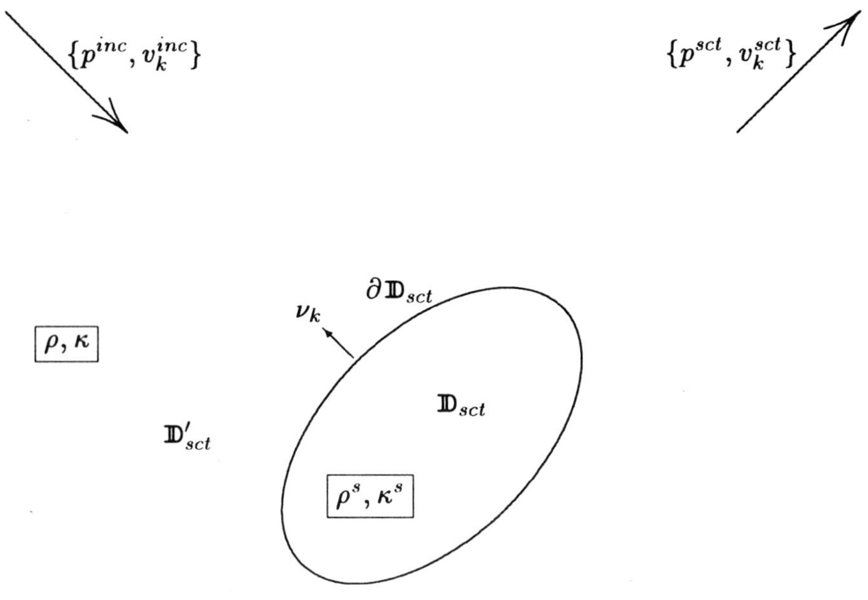


Figure 8.1. The scattering domain.

### 8.1.1. Domain-integral representations in the $s$-domain

In the $s$-domain, the scattered wavefield is also the difference between the total wavefield and the incident wavefield. Hence,

$$\{\hat{p}^{sct}, \hat{v}_k^{sct}\} = \{\hat{p} - \hat{p}^{inc}, \hat{v}_k - \hat{v}_k^{inc}\} . \tag{8.1}$$

Through a particular reasoning we now want to express that the scattered wavefield originates from the contrast in acoustic properties that the scattering object shows with respect to its embedding. First, we observe that, since the total wavefield is sourcefree in $\mathbb{D}_{sct}$,

$$\partial_k \hat{p} + s\rho^s \hat{v}_k = 0, \quad \boldsymbol{x} \in \mathbb{D}_{sct} , \tag{8.2}$$

$$\partial_k \hat{v}_k + s\kappa^s \hat{p} = 0, \quad \boldsymbol{x} \in \mathbb{D}_{sct} . \tag{8.3}$$

Secondly, the incident wavefield has no sources in $\mathbb{D}_{sct}$, while for it the constitutive parameters in $\mathbb{D}_{sct}$ have the same value as for the embedding. Hence,

$$\partial_k \hat{p}^{inc} + s\rho \hat{v}_k^{inc} = 0, \quad \boldsymbol{x} \in \mathbb{D}_{sct} , \tag{8.4}$$

$$\partial_k \hat{v}_k^{inc} + s\kappa \hat{p}^{inc} = 0, \quad \boldsymbol{x} \in \mathbb{D}_{sct} . \tag{8.5}$$

Upon rewriting Eqs. (8.2) and (8.3) as

$$\partial_k \hat{p} + s\rho \hat{v}_k = s(\rho - \rho^s)\hat{v}_k, \quad \boldsymbol{x} \in \mathbb{D}_{sct} , \tag{8.6}$$

$$\partial_k \hat{v}_k + s\kappa \hat{p} = s(\kappa - \kappa^s)\hat{p}, \quad \boldsymbol{x} \in \mathbb{D}_{sct} , \tag{8.7}$$

subtracting Eq. (8.4) from Eq. (8.6) and Eq. (8.5) from Eq (8.7) and using Eq. (8.1), we arrive at

$$\partial_k \hat{p}^{sct} + s\rho \hat{v}_k^{sct} = s(\rho - \rho^s)\hat{v}_k, \quad \boldsymbol{x} \in \mathbb{D}_{sct} , \tag{8.8}$$

$$\partial_k \hat{v}_k^{sct} + s\kappa \hat{p}^{sct} = s(\kappa - \kappa^s)\hat{p}, \quad \boldsymbol{x} \in \mathbb{D}_{sct} . \tag{8.9}$$

Thirdly, we observe that the scattered wavefield is sourcefree in the embedding (where the total wavefield and the incident wavefield have the same sources), and hence

$$\partial_k \hat{p}^{sct} + s\rho \hat{v}_k^{sct} = 0, \quad \boldsymbol{x} \in \mathbb{D}'_{sct} , \tag{8.10}$$

$$\partial_k \hat{v}_k^{sct} + s\kappa \hat{p}^{sct} = 0, \quad \boldsymbol{x} \in \mathbb{D}'_{sct} . \tag{8.11}$$

Equations (8.8) - (8.11) are now combined to

$$\partial_k \hat{p}^{sct} + s\rho \hat{v}_k^{sct} = \hat{f}_k^{sct}, \quad \boldsymbol{x} \in \mathbb{R}^3, \tag{8.12}$$

$$\partial_k \hat{v}_k^{sct} + s\kappa \hat{p}^{sct} = \hat{q}^{sct}, \quad \boldsymbol{x} \in \mathbb{R}^3, \tag{8.13}$$

where

$$\hat{f}_k^{sct} = \{s(\rho-\rho^s)\hat{v}_k, 0\}, \quad \boldsymbol{x} \in \{\mathbb{D}_{sct}, \mathbb{D}'_{sct}\}, \tag{8.14}$$

$$\hat{q}^{sct} = \{s(\kappa-\kappa^s)\hat{p}, 0\}, \quad \boldsymbol{x} \in \{\mathbb{D}_{sct}, \mathbb{D}'_{sct}\}. \tag{8.15}$$

If $\hat{f}^{sct}$ and $\hat{q}^{sct}$ were known, Eqs. (8.12) and (8.13) would constitute an acoustic radiation problem with known sources in an unbounded, inhomogeneous fluid with the same constitutive parameters as the embedding. This problem has been discussed in Chapter 7 (cf. Eqs. (7.50) and (7.51)). Hence we may write

$$\hat{p}^{sct}(\boldsymbol{x}^R, s) = \int_{\boldsymbol{x}\in\mathbb{D}_{sct}} [\hat{G}^q\, s(\kappa-\kappa^s)\hat{p}(\boldsymbol{x}, s) + \hat{\Gamma}_k^q\, s(\rho-\rho^s)\hat{v}_k(\boldsymbol{x}, s)]\mathrm{dV}, \tag{8.16}$$

and

$$\hat{v}_l^{sct}(\boldsymbol{x}^R, s) = \int_{\boldsymbol{x}\in\mathbb{D}_{sct}} [\hat{G}_l^f\, s(\kappa-\kappa^s)\hat{p}(\boldsymbol{x}, s) + \hat{\Gamma}_{l,k}^f\, s(\rho-\rho^s)\hat{v}_k(\boldsymbol{x}, s)]\mathrm{dV}$$

$$+ \frac{2}{3}\frac{\rho(\boldsymbol{x}^R) - \rho^s(\boldsymbol{x}^R)}{\rho(\boldsymbol{x}^R)}\hat{v}_l(\boldsymbol{x}^R, s), \tag{8.17}$$

where $\boldsymbol{x}^R \in \mathbb{R}^3$; the Green's states $\hat{G}^q = \hat{G}^q(\boldsymbol{x}^R|\boldsymbol{x}, s)$, $\hat{\Gamma}_k^q = \hat{\Gamma}_k^q(\boldsymbol{x}^R|\boldsymbol{x}, s)$, $\hat{G}_l^f = \hat{G}_l^f(\boldsymbol{x}^R|\boldsymbol{x}, s)$ and $\hat{\Gamma}_{l,k}^f = \hat{\Gamma}_{l,k}^f(\boldsymbol{x}^R|\boldsymbol{x}, s)$ are discussed in Chapter 7. In Eq. (8.17) and further analysis it is understood that $\rho(\boldsymbol{x}^R) - \rho^s(\boldsymbol{x}^R) = 0$ when $\boldsymbol{x}^R \in \mathbb{D}'_{sct}$.

It is noted that at any surface of discontinuity in $\kappa^s$ and/or $\rho^s$ the right-hand sides of Eqs. (8.16) and (8.17) yield half the sum of the limiting values of the left-hand sides at either side of the relevant surface of discontinuity, provided that the integrals are interpreted as their Cauchy principal value (see Section 7.1.3)

Equations (8.16) - (8.17) are the desired integral representations. Once the total wavefield quantities $\hat{p}$ and $\hat{v}_k$ in $\mathbb{D}_{sct}$ are known, these integral representations enable us to calculate the acoustic wavefield quantities in all space.

### 8.1.2. Domain-integral equations in the $s$-domain

The domain-integral equations are obtained by confining in Eqs. (8.16) and (8.17) the position of observation to the interior of the scatterer, i.e., $\boldsymbol{x}^R \in \mathbb{D}_{sct}$. Using Eq. (8.1) to express $\hat{p}^{sct}$ and $\hat{v}_k^s$ in terms of the known values of $\hat{p}^{inc}$ and $\hat{v}_k^{inc}$ and the as yet unknown values of $\hat{p}$ and $\hat{v}_k$ in $\mathbb{D}_{sct}$, we arrive at a system of linear integral equations

$$\hat{p}(\boldsymbol{x},s) - \int_{\boldsymbol{x}'\in\mathbb{D}_{sct}} [\hat{G}^q(\boldsymbol{x}|\boldsymbol{x}',s)s(\kappa-\kappa^s)\hat{p}(\boldsymbol{x}',s) + \hat{\Gamma}_k^q(\boldsymbol{x}|\boldsymbol{x}',s)s(\rho-\rho^s)\hat{v}_k(\boldsymbol{x}',s)]\mathrm{dV} = \hat{p}^{inc}(\boldsymbol{x},s), \quad \boldsymbol{x}\in\mathbb{D}_{sct}\,, \tag{8.18}$$

$$\left[\frac{1}{3}+\frac{2}{3}\frac{\rho^s(\boldsymbol{x})}{\rho(\boldsymbol{x})}\right]\hat{v}_l(\boldsymbol{x},s) - \int_{\boldsymbol{x}'\in\mathbb{D}_{sct}} [\hat{G}_l^f(\boldsymbol{x}|\boldsymbol{x}',s)s(\kappa-\kappa^s)\hat{p}(\boldsymbol{x}',s) + \hat{\Gamma}_{l,k}^f(\boldsymbol{x}|\boldsymbol{x}',s)s(\rho-\rho^s)\hat{v}_k(\boldsymbol{x}',s)]\mathrm{dV} = \hat{v}_l^{inc}(\boldsymbol{x},s)\,, \quad \boldsymbol{x}\in\mathbb{D}_{sct}\,. \tag{8.19}$$

From this system of integral equations, the acoustic pressure $\hat{p}$ and the particle velocity $\hat{v}_k$ in $\mathbb{D}_{sct}$ can, in principle, be determined. In practice, the integral equations associated with the direct scattering problem have to be solved with the aid of numerical methods. In special cases, analytical approximations to the values of the wavefield in the interior of the scatterer can be made. Specifically, the integrals contain a singular point $\boldsymbol{x} = \boldsymbol{x}'$. These integrals have to be interpreted as their Cauchy principal value and around the singular point analytical computations have to be performed. The iterative solution of integral equations has been discussed in Chapter 2. Once the solution of the integral equations has been obtained, the scattered wavefield in all space follows by evaluating the right-hand sides of Eqs. (8.16) and (8.17).

For the special case of a homogeneous background medium, the system of integral equations of Eqs. (8.18) and (8.19) reduces to

$$\hat{p}(\boldsymbol{x},s) - \int_{\boldsymbol{x}'\in\mathbb{D}_{sct}} \Big[s^2\rho(\kappa-\kappa^s)\hat{G}(\boldsymbol{x}-\boldsymbol{x}',s)\hat{p}(\boldsymbol{x}',s) - s(\rho-\rho^s)\partial_k\hat{G}(\boldsymbol{x}-\boldsymbol{x}',s)\hat{v}_k(\boldsymbol{x}',s)\Big]\,\mathrm{dV} = \hat{p}^{inc}(\boldsymbol{x},s), \quad \boldsymbol{x}\in\mathbb{D}_{sct}\,, \tag{8.20}$$

$$\left[\frac{1}{3}+\frac{2}{3}\frac{\rho^s(\boldsymbol{x})}{\rho}\right]\hat{v}_l(\boldsymbol{x},s) - \int_{\boldsymbol{x}'\in\mathbb{D}_{sct}} \Big[-s(\kappa-\kappa^s)\partial_l\hat{G}(\boldsymbol{x}-\boldsymbol{x}',s)\hat{p}(\boldsymbol{x}',s) + \frac{(\rho-\rho^s)}{\rho}\partial_l\partial_k\hat{G}(\boldsymbol{x}-\boldsymbol{x}',s)\hat{v}_k(\boldsymbol{x}',s)\Big]\,\mathrm{dV} = \hat{v}_l^{inc}(\boldsymbol{x},s)\,, \quad \boldsymbol{x}\in\mathbb{D}_{sct}\,, \tag{8.21}$$

where

$$\hat{G}(\boldsymbol{x},s) = \frac{\exp(-\hat{\gamma}|\boldsymbol{x}|)}{4\pi|\boldsymbol{x}|} \quad \text{with } \hat{\gamma} = s(\kappa\rho)^{\frac{1}{2}} = \frac{s}{c}\,. \tag{8.22}$$

When there is no contrast in the volume density of mass ($\rho^s = \rho$) we observe that we are left with one integral equation for the acoustic pressure only, viz.,

$$\hat{p}(\boldsymbol{x},s) + \int_{\boldsymbol{x}'\in\mathbb{D}_{sct}} s^2[(c^s)^{-2}-c^{-2}]\hat{G}(\boldsymbol{x}-\boldsymbol{x}',s)\hat{p}(\boldsymbol{x}',s)\mathrm{dV} = \hat{p}^{inc}(\boldsymbol{x},s)\,, \quad \boldsymbol{x}\in\mathbb{D}_{sct}\,, \tag{8.23}$$

where $c = (\rho\kappa)^{-\frac{1}{2}}$ is the wave speed of the embedding and $c^s = c^s(\boldsymbol{x}') = (\rho\kappa^s)^{-\frac{1}{2}}$ is the wave speed in a point $\boldsymbol{x}'$ of the scatterer. In mathematical physics, this integral equation has been discussed extensively, e.g., MORSE and FESHBACH (1953, p. 1069).

*The Born approximation in the s-domain*

For small values of $\kappa - \kappa^s$ and $\rho - \rho^s$, the system of integral equations can be solved iteratively by using the Neumann expansion of Section 2.5. The first term of this expansion yields

$$\hat{p}(\boldsymbol{x},s) \cong \hat{p}^{inc}(\boldsymbol{x},s)\,, \tag{8.24}$$

$$\hat{v}_l(\boldsymbol{x}, s) \cong \hat{v}_l^{inc}(\boldsymbol{x}, s) \,. \tag{8.25}$$

This is the so-called first Born approximation (see BORN and WOLF, 1965, p. 452).

For the special case of a homogeneous background medium and no contrast in the volume density of mass, DE HOOP (1991) has shown that the Neumann iterative solution of integral equation (8.23) converges for all real and positive values of $s$, provided that $|(c^s)^{-2} - c^{-2}| < c^{-2}$, independently of the size of the scatterer. An alternative analysis valid for all imaginary values of $s$ ($s \to j\omega$) shows that the Neumann expansion is convergent, if the criterion $|\omega^2|\,|(c^s)^{-2} - c^{-2}|\Delta_{sct}^2 < 2$ holds, where $\Delta_{sct}$ is the radius of the smallest ball in which the scatterer is contained. Note that the size of the object is now included in the convergence criterion.

### 8.1.3. Domain-integral representations in the time domain

By applying the standard rules for inversion from the $s$-domain to the time domain, the time-domain representations are directly obtained from Eqs. (8.16) and (8.17) as

$$\int_{\boldsymbol{x}\in\mathbb{D}_{sct}} [(\kappa-\kappa^s)C_t\{G^q, \partial_t p\} + (\rho-\rho^s)C_t\{\Gamma_k^q, \partial_t v_k\}]\mathrm{dV} = \chi_{\mathrm{T}}(t)p^{sct}(\boldsymbol{x}^R, t) \tag{8.26}$$

and

$$\int_{\boldsymbol{x}\in\mathbb{D}_{sct}} [(\kappa-\kappa^s)C_t\{G_l^f, \partial_t p\} + (\rho-\rho^s)C_t\{\Gamma_{l,k}^f, \partial_t v_k\}]\mathrm{dV} + \frac{2}{3}\frac{\rho(\boldsymbol{x}^R)-\rho^s(\boldsymbol{x}^R)}{\rho(\boldsymbol{x}^R)}\chi_{\mathrm{T}}(t)\, v_l(\boldsymbol{x}^R, t) = \chi_{\mathrm{T}}(t)v_l^{sct}(\boldsymbol{x}^R, t)\,, \tag{8.27}$$

where $\boldsymbol{x}^R \in \mathbb{R}^3$. The Green's states $G^q = G^q(\boldsymbol{x}^R|\boldsymbol{x}, t)$, $\Gamma_k^q = \Gamma_k^q(\boldsymbol{x}^R|\boldsymbol{x}, t)$, $G_l^f = G_l^f(\boldsymbol{x}^R|\boldsymbol{x}, t)$ and $\Gamma_{l,k}^f = \Gamma_{l,k}^f(\boldsymbol{x}^R|\boldsymbol{x}, t)$ are the time-domain counterparts of the $s$-domain Green's states $\hat{G}^q$, $\hat{\Gamma}_k^q$, $\hat{G}_l^f$ and $\hat{\Gamma}_{l,k}^f$. Further, $C_t\{\cdot,\cdot\} = C_t\{\cdot,\cdot\}(\boldsymbol{x}, t)$ denotes the temporal convolution (see Eq. (1.19)).

The time-domain representations of Eqs. (8.26) and (8.27) can also be derived by remaining in the time domain. Performing a similar analysis using the time-domain reciprocity theorem of convolution type of Section 5.2 we obtain the same results.

For the special case of a homogeneous background, the pertaining relations of Eqs. (8.26) and (8.27) reduce to

$$\int_{\boldsymbol{x}\in\mathbb{D}_{sct}} \left[ \rho(\kappa-\kappa^s)\frac{\partial_t^2 p(\boldsymbol{x},t-\frac{|\boldsymbol{x}^R-\boldsymbol{x}|}{c})}{4\pi|\boldsymbol{x}^R-\boldsymbol{x}|} - (\rho-\rho^s)\partial_k^R\frac{\partial_t v_k(\boldsymbol{x},t-\frac{|\boldsymbol{x}^R-\boldsymbol{x}|}{c})}{4\pi|\boldsymbol{x}^R-\boldsymbol{x}|}\right]\mathrm{dV} = \chi_{\mathrm{T}}(t)p^{sct}(\boldsymbol{x}^R,t) \tag{8.28}$$

and

$$\int_{\boldsymbol{x}\in\mathbb{D}_{sct}} \left[ -(\kappa-\kappa^s)\partial_l^R\frac{\partial_t p(\boldsymbol{x},t-\frac{|\boldsymbol{x}^R-\boldsymbol{x}|}{c})}{4\pi|\boldsymbol{x}^R-\boldsymbol{x}|} + \frac{(\rho-\rho^s)}{\rho}\partial_l^R\partial_k^R\frac{v_k(\boldsymbol{x},t-\frac{|\boldsymbol{x}^R-\boldsymbol{x}|}{c})}{4\pi|\boldsymbol{x}^R-\boldsymbol{x}|}\right]\mathrm{dV} + \frac{2}{3}\frac{\rho-\rho^s(\boldsymbol{x}^R)}{\rho}\chi_{\mathrm{T}}(t)\,v_l(\boldsymbol{x}^R,t) = \chi_{\mathrm{T}}(t)v_l^{sct}(\boldsymbol{x}^R,t)\,. \tag{8.29}$$

### 8.1.4. Domain-integral equations in the time domain

In a similar way as in the previous section, the time-domain counterparts of the integral equations (8.18) and (8.19) are obtained by applying the standard rules for inversion from the $s$-domain to the time domain. We arrive at

$$p(\boldsymbol{x},t) - \int_{\boldsymbol{x}'\in\mathbb{D}_{sct}} [(\kappa-\kappa^s)C_t\{G^q,\partial_t p\} + (\rho-\rho^s)C_t\{\Gamma_k^q,\partial_t v_k\}]\mathrm{dV} = p^{inc}(\boldsymbol{x},t), \quad \boldsymbol{x}\in\mathbb{D}_{sct}, \quad t\in\mathrm{T}\,, \tag{8.30}$$

$$\left[\frac{1}{3}+\frac{2}{3}\frac{\rho^s(\boldsymbol{x})}{\rho(\boldsymbol{x})}\right] v_l(\boldsymbol{x},t) - \int_{\boldsymbol{x}'\in\mathbb{D}_{sct}} [(\kappa-\kappa^s)C_t\{G_l^f, \partial_t p\}$$

$$+ (\rho-\rho^s)C_t\{\Gamma_{l,k}^f, \partial_t v_k\}]\mathrm{dV}$$

$$= v_l^{inc}(\boldsymbol{x},t), \quad \boldsymbol{x}\in\mathbb{D}_{sct}, \quad t\in\mathrm{T}, \tag{8.31}$$

from which $p$ and $v_k$ for observation points in $\mathbb{D}_{sct}$ and time instants in T can, in principle, be determined. The Green's states $G^q = G^q(\boldsymbol{x}|\boldsymbol{x}',t)$, $\Gamma_k^q = \Gamma_k^q(\boldsymbol{x}|\boldsymbol{x}',t)$, $G_l^f = G_l^f(\boldsymbol{x}|\boldsymbol{x}',t)$ and $\Gamma_{l,k}^f = \Gamma_{l,k}^f(\boldsymbol{x}|\boldsymbol{x}',t)$ are the time-domain counterparts of the $s$-domain Green's states $\hat{G}^q$, $\hat{\Gamma}_k^q$, $\hat{G}_l^f$ and $\hat{\Gamma}_{l,k}^f$. Further, $C_t\{\cdot,\cdot\} = C_t\{\cdot,\cdot\}(\boldsymbol{x}',t)$ denotes the temporal convolution defined by Eq. (1.19).

For the special case of a homogeneous background medium, the pertaining system of integral equations of Eqs. (8.30) and (8.31) reduces to

$$p(\boldsymbol{x},t) - \int_{\boldsymbol{x}'\in\mathbb{D}_{sct}} \left[\rho(\kappa-\kappa^s)\frac{\partial_t^2 p(\boldsymbol{x}',t-\frac{|\boldsymbol{x}-\boldsymbol{x}'|}{c})}{4\pi|\boldsymbol{x}-\boldsymbol{x}'|}\right.$$

$$\left. - (\rho-\rho^s)\partial_k\frac{\partial_t v_k(\boldsymbol{x}',t-\frac{|\boldsymbol{x}-\boldsymbol{x}'|}{c})}{4\pi|\boldsymbol{x}-\boldsymbol{x}'|}\right]\mathrm{dV}$$

$$= p^{inc}(\boldsymbol{x},t), \quad \boldsymbol{x}\in\mathbb{D}_{sct}, \quad t\in\mathrm{T}, \tag{8.32}$$

and

$$\left[\frac{1}{3}+\frac{2}{3}\frac{\rho^s(\boldsymbol{x})}{\rho}\right] v_l(\boldsymbol{x},t) - \int_{\boldsymbol{x}'\in\mathbb{D}_{sct}} \left[-(\kappa-\kappa^s)\partial_l\frac{\partial_t p(\boldsymbol{x}',t-\frac{|\boldsymbol{x}-\boldsymbol{x}'|}{c})}{4\pi|\boldsymbol{x}-\boldsymbol{x}'|}\right.$$

$$\left. + \frac{\rho-\rho^s}{\rho}\partial_l\partial_k\frac{v_k(\boldsymbol{x}',t-\frac{|\boldsymbol{x}-\boldsymbol{x}'|}{c})}{4\pi|\boldsymbol{x}-\boldsymbol{x}'|}\right]\mathrm{dV}$$

$$= v_l^{inc}(\boldsymbol{x},t), \quad \boldsymbol{x}\in\mathbb{D}_{sct}, \quad t\in\mathrm{T}. \tag{8.33}$$

When there is no contrast in the volume density of mass ($\rho^s = \rho$) we observe that we are left with one integral equation for the acoustic pressure only, viz.,

$$p(\boldsymbol{x},t) + \int_{\boldsymbol{x}' \in \mathbb{D}_{sct}} [(c^s)^{-2} - c^{-2}] \frac{\partial_t^2 p(\boldsymbol{x}', t - \frac{|\boldsymbol{x}-\boldsymbol{x}'|}{c})}{4\pi|\boldsymbol{x}-\boldsymbol{x}'|} \mathrm{dV} = p^{inc}(\boldsymbol{x},t) \, ,$$

$$\boldsymbol{x} \in \mathbb{D}_{sct}, \quad t \in \mathrm{T} \, , \tag{8.34}$$

where $c = (\rho\kappa)^{-\frac{1}{2}}$ is the wave speed of the embedding and $c^s = (\rho\kappa^s)^{-\frac{1}{2}}$ is the wave speed in a point $\boldsymbol{x}$ of the scatterer.

*The Born approximation in the time-domain*

For small values of $\kappa - \kappa^s$ and $\rho - \rho^s$, the system of integral equations can be solved iteratively by using the Neumann expansion of Section 2.5. The first term of this expansion yields

$$p(\boldsymbol{x},t) \cong p^{inc}(\boldsymbol{x},t) \, , \tag{8.35}$$

$$v_l(\boldsymbol{x},t) \cong v_l^{inc}(\boldsymbol{x},t) \, . \tag{8.36}$$

This is the so-called first Born approximation in the time domain.

For the special case of a homogeneous background medium and no contrast in the volume density of mass, DE HOOP (1991) has shown that the Neumann expansion of integral equation (8.34) converges for all $t \in \mathrm{T}$, provided that $|(c^s)^{-2} - c^{-2}| < c^{-2}$, independently of the size of the scatterer. This follows directly from the $s$-domain results for real $s$.

## 8.2. The boundary-integral equation formulation

We investigate the direct or forward scattering of acoustic waves by a contrasting fluid domain of bounded extent, present in a fluid embedding of infinite extent. Let $\mathbb{D}_{sct}$ be the bounded domain occupied by the fluid scatterer and let $\rho^s = \rho^s(\boldsymbol{x})$ be its volume density of mass and $\kappa^s = \kappa^s(\boldsymbol{x})$ its compressibility. The enclosing boundary of the scatterer is denoted as $\partial\mathbb{D}_{sct}$; the normal vector $\nu_k$ on $\partial\mathbb{D}_{sct}$ is directed away from $\mathbb{D}_{sct}$. The domain exterior to $\partial\mathbb{D}_{sct}$ is denoted by $\mathbb{D}'_{sct}$. In $\mathbb{D}'_{sct}$, the embedding, an inhomogeneous, isotropic fluid is present; its volume density of mass is denoted by $\rho = \rho(\boldsymbol{x})$ and its compressibility by $\kappa = \kappa(\boldsymbol{x})$ (Fig. 8.1). We assume that both the Green's states (cf. Chapter 7) of the embedding and the Green's states of the medium of the scatterer can be determined. We start our analysis in the $s$-domain. The results in the time domain are obtained by applying the standard rules for inversion from the $s$-domain.

### 8.2.1. Boundary-integral representations in the $s$-domain

The total acoustic wavefield in the embedding $\{\hat{p}, \hat{v}_k\}$ is decomposed into the incident wavefield $\{\hat{p}^{inc}, \hat{v}_k^{inc}\}$ and the scattered wavefield $\{\hat{p}^{sct}, \hat{v}_k^{sct}\}$. The incident wavefield is the wavefield that would be present in the entire configuration if the domain $\mathbb{D}_{sct}$ showed no contrast with the embedding. The Green's states of the embedding are denoted as $\hat{G}^q$, $\hat{\Gamma}_k^q$, $\hat{G}_l^f$ and $\hat{\Gamma}_{l,k}^f$, while the Green's states of the of the object medium are denoted as $\hat{G}^{q^s}$, $\hat{\Gamma}_k^{q^s}$, $\hat{G}_l^{f^s}$ and $\hat{\Gamma}_{l,k}^{f^s}$. The total wavefield is generated by sources that are located outside the scattering domain. Since these sources remain present even if the scattering domain is thought to be absent, they also serve as sources for the incident wavefield. The incident wavefield quantities can be calculated with the representations obtained in Chapter 7.

Outside the scattering object, we define the scattered wavefield being the difference between the total wavefield and the incident wavefield. Hence,

$$\{\hat{p}^{sct}, \hat{v}_k^{sct}\} = \{\hat{p} - \hat{p}^{inc}, \hat{v}_k - \hat{v}_k^{inc}\}\,. \tag{8.37}$$

The scattered wavefield is sourcefree in $\mathbb{D}'_{sct}$ and its acoustic wavefield quantities satisfy the equations

$$\partial_k \hat{p}^{sct} + s\rho \hat{v}_k^{sct} = 0, \quad \boldsymbol{x} \in \mathbb{D}'_{sct}\,, \tag{8.38}$$

$$\partial_k \hat{v}_k^{sct} + s\kappa \hat{p}^{sct} = 0, \quad \boldsymbol{x} \in \mathbb{D}'_{sct}\,. \tag{8.39}$$

If both $\hat{p}^{sct}$ and $\nu_k \hat{v}_k^{sct}$ on $\partial\mathbb{D}_{sct}$ were known, Eqs. (8.38) and (8.39) would constitute an acoustic radiation problem as discussed in Section 7.2 (cf. Eq. (7.63)). Hence, we may write

$$\hat{p}^{sct}(\boldsymbol{x}^R, s) = \int_{\boldsymbol{x}\in\partial\mathbb{D}_{sct}} [\hat{G}^q\, \hat{v}_k^{sct}(\boldsymbol{x}, s) + \hat{\Gamma}_k^q\, \hat{p}^{sct}(\boldsymbol{x}, s)]\nu_k \mathrm{dA} \qquad \text{when } \boldsymbol{x}^R \in \mathbb{D}'_{sct}\,, \tag{8.40}$$

in which the Green's states $\hat{G}^q = \hat{G}^q(\boldsymbol{x}^R|\boldsymbol{x},s)$ and $\hat{\Gamma}^q_k = \hat{\Gamma}^q_k(\boldsymbol{x}^R|\boldsymbol{x},s)$ are discussed in Chapter 7.

In the special case that the medium in the domain exterior to the scatterer is homogeneous, we have

$$\hat{G}^q(\boldsymbol{x}^R|\boldsymbol{x},s) = s\rho\hat{G}(\boldsymbol{x}^R - \boldsymbol{x},s)\,, \tag{8.41}$$

and

$$\hat{\Gamma}^q_k(\boldsymbol{x}^R|\boldsymbol{x},s) = -\partial^R_k\hat{G}(\boldsymbol{x}^R - \boldsymbol{x},s)\,, \tag{8.42}$$

where

$$\hat{G}(\boldsymbol{x}^R - \boldsymbol{x},s) = \frac{\exp(-\hat{\gamma}|\boldsymbol{x}^R - \boldsymbol{x}|)}{4\pi|\boldsymbol{x}^R - \boldsymbol{x}|} \quad \text{with } \hat{\gamma} = s(\kappa\rho)^{\frac{1}{2}} = \frac{s}{c}\,. \tag{8.43}$$

In view of the boundary conditions that the total wavefield quantities $\hat{p}$ and $\nu_k\hat{v}_k$ are continuous through $\partial\mathbb{D}_{sct}$, we prefer an integral representation with the total wave quantities in the integral of the right-hand side of Eq. (8.40). To this end, we consider the relations with respect to the incident wavefield quantities. The incident wavefield is sourcefree in $\mathbb{D}_{sct}$ and its acoustic wavefield quantities satisfy the equations

$$\partial_k\hat{p}^{inc} + s\rho\hat{v}^{inc}_k = 0, \quad \boldsymbol{x} \in \mathbb{D}_{sct}\,, \tag{8.44}$$

$$\partial_k\hat{v}^{inc}_k + s\kappa\hat{p}^{inc} = 0, \quad \boldsymbol{x} \in \mathbb{D}_{sct}\,. \tag{8.45}$$

Note that the incident wavefield is present in $\mathbb{D}_{sct}$ when we assume that the scatterer shows no contrast with respect to its embedding. Therefore, the material quantities $\rho$ and $\kappa$ of the embedding occur in Eqs. (8.44) and (8.45). We now use Eq. (7.83) by replacing $\{\hat{p},\hat{v}_k\}$ by $\{\hat{p}^{inc},\hat{v}^{inc}_k\}$, $\{\mathbb{D},\partial\mathbb{D},\mathbb{D}'\}$ by $\{\mathbb{D}_{sct},\partial\mathbb{D}_{sct},\mathbb{D}'_{sct}\}$, and considering the observation point $\boldsymbol{x}^R$ in $\mathbb{D}'_{sct}$. We directly obtain

$$0 = \int_{\boldsymbol{x}\in\partial\mathbb{D}_{sct}} [\hat{G}^q\,\hat{v}^{inc}_k(\boldsymbol{x},s) + \hat{\Gamma}^q_k\,\hat{p}^{inc}(\boldsymbol{x},s)]\nu_k\mathrm{dA} \qquad \text{when } \boldsymbol{x}^R \in \mathbb{D}'_{sct}\,. \tag{8.46}$$

Adding the results of Eqs. (8.40) and (8.46), we arrive at

$$\hat{p}^{sct}(\boldsymbol{x}^R,s) = \int_{\boldsymbol{x}\in\partial\mathbb{D}_{sct}} [\hat{G}^q\,\hat{v}_k(\boldsymbol{x},s) + \hat{\Gamma}^q_k\,\hat{p}(\boldsymbol{x},s)]\nu_k\mathrm{dA} \qquad \text{when } \boldsymbol{x}^R \in \mathbb{D}'_{sct}\,. \tag{8.47}$$

Equation (8.47) is the desired representation of the scattered wavefield outside the scatterer. Together with the incident wavefield it determines the total wavefield outside the scatterer.

In order to obtain integral representations of the wavefield in the interior of the contrasting domain, we start with the observation, that this interior wavefield satisfy the equations

$$\partial_k \hat{p} + s\rho^s \hat{v}_k = 0, \quad \boldsymbol{x} \in \mathbb{D}_{sct} , \tag{8.48}$$

$$\partial_k \hat{v}_k + s\kappa^s \hat{p} = 0, \quad \boldsymbol{x} \in \mathbb{D}_{sct} . \tag{8.49}$$

We now employ Eq. (7.82) by replacing $\{\rho, \kappa\}$ by $\{\rho^s, \kappa^s\}$, $\{\mathbb{D}, \partial\mathbb{D}, \mathbb{D}'\}$ by $\{\mathbb{D}_{sct}, \partial\mathbb{D}_{sct}, \mathbb{D}'_{sct}\}$, and considering the observation point $\boldsymbol{x}^R$ in $\mathbb{D}_{sct}$. We directly obtain

$$\hat{p}(\boldsymbol{x}^R, s) = -\int_{\boldsymbol{x}\in\partial\mathbb{D}_{sct}} [\hat{G}^{q^s}\, \hat{v}_k(\boldsymbol{x}, s) + \hat{\Gamma}_k^{q^s}\, \hat{p}(\boldsymbol{x}, s)]\nu_k \mathrm{dA}$$
$$\text{when } \boldsymbol{x}^R \in \mathbb{D}_{sct} . \tag{8.50}$$

The Green's states in the latter equation are the acoustic wavefield quantities of a point source of volume injection, when the medium inside $\mathbb{D}_{sct}$ is characterized by the material quantities $\rho^s$ and $\kappa^s$ in stead of $\rho$ and $\kappa$.

In the special case that the medium in the domain interior to the scatterer is homogeneous, we have

$$\hat{G}^{q^s}(\boldsymbol{x}^R|\boldsymbol{x}, s) = s\rho^s \hat{G}^s(\boldsymbol{x}^R - \boldsymbol{x}, s) , \tag{8.51}$$

and

$$\hat{\Gamma}_k^{q^s}(\boldsymbol{x}^R|\boldsymbol{x}, s) = -\partial_k^R \hat{G}^s(\boldsymbol{x}^R - \boldsymbol{x}, s) , \tag{8.52}$$

where

$$\hat{G}^s(\boldsymbol{x}^R - \boldsymbol{x}, s) = \frac{\exp(-\hat{\gamma}^s|\boldsymbol{x}^R - \boldsymbol{x}|)}{4\pi|\boldsymbol{x}^R - \boldsymbol{x}|} \quad \text{with } \hat{\gamma}^s = s(\kappa^s\rho^s)^{\frac{1}{2}} = \frac{s}{c^s} . \tag{8.53}$$

Equation (8.50) is the desired integral representation for the wavefield in the interior of the contrasting domain. Once the total wavefield quantities $\hat{p}$ and $\nu_k \hat{v}_k$ on the boundary surface of the contrasting domain are known, the integral representations of Eqs. (8.47) and (8.50) enable us to calculate the acoustic wavefield quantities in the whole space, both in the interior and in the exterior of the scatterer.

### 8.2.2. Boundary-integral equations in the $s$-domain

The boundary-integral equations are obtained by letting in Eqs. (8.40) and (8.46) the point of observation approach the boundary $\partial\mathbb{D}_{sct}$ of the scatterer. Then we may write

$$\frac{1}{2}\hat{p}^{sct}(\boldsymbol{x},s) = \int_{\boldsymbol{x}'\in\partial\mathbb{D}_{sct}} [\hat{G}^q\,\hat{v}_k^{sct}(\boldsymbol{x}',s) + \hat{\Gamma}_k^q\,\hat{p}^{sct}(\boldsymbol{x}',s)]\nu_k'\mathrm{dA} \quad \text{when } \boldsymbol{x}\in\partial\mathbb{D}_{sct}\,, \tag{8.54}$$

$$-\frac{1}{2}\hat{p}^{inc}(\boldsymbol{x},s) = \int_{\boldsymbol{x}'\in\partial\mathbb{D}_{sct}} [\hat{G}^q\,\hat{v}_k^{inc}(\boldsymbol{x}',s) + \hat{\Gamma}_k^q\,\hat{p}^{inc}(\boldsymbol{x}',s)]\nu_k'\mathrm{dA} \quad \text{when } \boldsymbol{x}\in\partial\mathbb{D}_{sct}\,, \tag{8.55}$$

where $\nu_k'$ denotes the normal vector at a point $\boldsymbol{x}'$ of the boundary surface $\partial\mathbb{D}_{sct}$ pointing away from $\mathbb{D}_{sct}$. Adding Eqs. (8.54) and (8.55), while using Eq. (8.37), we arrive at

$$\frac{1}{2}\hat{p}(\boldsymbol{x},s) - \int_{\boldsymbol{x}'\in\partial\mathbb{D}_{sct}} [\hat{G}^q(\boldsymbol{x}|\boldsymbol{x}',s)\hat{v}_k(\boldsymbol{x}',s) + \hat{\Gamma}_k^q(\boldsymbol{x}|\boldsymbol{x}',s)\hat{p}(\boldsymbol{x}',s)]\nu_k'\mathrm{dA} = \hat{p}^{inc}(\boldsymbol{x},s), \quad \boldsymbol{x}\in\partial\mathbb{D}_{sct}\,. \tag{8.56}$$

Similarly, when we let the point of observation in Eq. (8.50) approach the boundary $\partial\mathbb{D}_{sct}$, then we may write

$$\frac{1}{2}\hat{p}(\boldsymbol{x},s) + \int_{\boldsymbol{x}'\in\partial\mathbb{D}_{sct}} [\hat{G}^{q^s}(\boldsymbol{x}|\boldsymbol{x}',s)\hat{v}_k(\boldsymbol{x}',s) + \hat{\Gamma}_k^{q^s}(\boldsymbol{x}|\boldsymbol{x}',s)\hat{p}(\boldsymbol{x}',s)]\nu_k'\mathrm{dA} = 0, \quad \boldsymbol{x}\in\partial\mathbb{D}_{sct}\,. \tag{8.57}$$

It is noted that the integrals in the left-hand sides of Eqs. (8.56) and (8.57) have to be interpreted as their principle values, i.e., the integrals are, when necessary, calculated by a limiting procedure that excludes the singularity at $\boldsymbol{x} = \boldsymbol{x}'$ in a symmetrical manner. Equations (8.56) and (8.57) constitute a system of two integral equations with two unknown quantities, viz., $\hat{p}(\boldsymbol{x},s)$ and $\nu_k\hat{v}_k(\boldsymbol{x},s)$ on $\partial\mathbb{D}_{sct}$. From this system of integral equations the quantities $\hat{p}$ and $\nu_k\hat{v}_k$ on $\partial\mathbb{D}_{sct}$ can, in principle, be determined. In practice, the integral equations associated with the direct scattering problem have to be solved with the aid of numerical methods. In special cases, analytical approximations to the values of the acoustic wavefield quantities at the boundary surface of the scatterer can be made. The iterative solution of integral equations has been discussed in Chapter 2. Once the solution of the integral equations has been obtained, the scattered-wavefield quantities and the interior-wavefield quantities follow by evaluating the right-hand sides of Eqs. (8.47) and (8.50), respectively.

### 8.2.3. Boundary-integral representations in the time domain

By applying the standard rules for inversion from the $s$-domain to the time domain, the time-domain representations are directly obtained from Eqs. (8.47) and (8.50) as

$$\int_{\boldsymbol{x}\in\partial\mathbb{D}_{sct}} [C_t\{G^q, \nu_k v_k\} + C_t\{\Gamma_k^q, \nu_k p\}]\mathrm{dA} = \chi_T(t)p^{sct}(\boldsymbol{x}^R, t) \quad \text{when } \boldsymbol{x}^R \in \mathbb{D}'_{sct} , \tag{8.58}$$

$$\int_{\boldsymbol{x}\in\partial\mathbb{D}_{sct}} [C_t\{G^{q^s}, \nu_k v_k\} + C_t\{\Gamma_k^{q^s}, \nu_k p\}]\mathrm{dA} = -\chi_T(t)p(\boldsymbol{x}^R, t) \quad \text{when } \boldsymbol{x}^R \in \mathbb{D}_{sct} , \tag{8.59}$$

in which the Green's states $G^q = G^q(\boldsymbol{x}^R|\boldsymbol{x}, t)$, $\Gamma_k^q = \Gamma_k^q(\boldsymbol{x}^R|\boldsymbol{x}, t)$, $G^{q^s} = G^{q^s}(\boldsymbol{x}^R|\boldsymbol{x}, t)$ and $\Gamma_k^{q^s} = \Gamma_k^{q^s}(\boldsymbol{x}^R|\boldsymbol{x}, t)$ are the time-domain counterparts of the $s$-domain Green's states $\hat{G}^q$, $\hat{\Gamma}_k^q$, $\hat{G}^{q^s}$ and $\hat{\Gamma}_k^{q^s}$. Further, $C_t\{\cdot,\cdot\} = C_t\{\cdot,\cdot\}(\boldsymbol{x}, t)$ denotes the temporal convolution defined by Eq. (1.19).

The time-domain representations of Eqs. (8.58) - (8.59) can also be derived by remaining in the time domain. Performing a similar analysis using the time-domain reciprocity theorem of convolution type of Section 5.2 we obtain the same results.

For the special case of homogeneous media the pertaining system of Eqs. (8.58) and (8.59) reduces to

$$\int_{\boldsymbol{x}\in\partial\mathbb{D}_{sct}} \left[\rho\frac{\partial_t v_k(\boldsymbol{x}, t - \frac{|\boldsymbol{x}^R-\boldsymbol{x}|}{c})}{4\pi|\boldsymbol{x}^R - \boldsymbol{x}|} - \partial_k^R \frac{p(\boldsymbol{x}, t - \frac{|\boldsymbol{x}^R-\boldsymbol{x}|}{c})}{4\pi|\boldsymbol{x}^R - \boldsymbol{x}|}\right] \nu_k \mathrm{dA} \tag{8.60}$$

$$= \chi_{\mathrm{T}}(t)p^{sct}(\boldsymbol{x}^R, t) \quad \text{when } \boldsymbol{x}^R \in \mathbb{D}'_{sct} ,$$

and

$$\int_{\boldsymbol{x}\in\partial\mathbb{D}_{sct}} \left[ \rho^s \frac{\partial_t v_k(\boldsymbol{x}, t - \frac{|\boldsymbol{x}^R - \boldsymbol{x}|}{c^s})}{4\pi|\boldsymbol{x}^R - \boldsymbol{x}|} - \partial_k^R \frac{p(\boldsymbol{x}, t - \frac{|\boldsymbol{x}^R - \boldsymbol{x}|}{c^s})}{4\pi|\boldsymbol{x}^R - \boldsymbol{x}|} \right] \nu_k \mathrm{dA} \tag{8.61}$$

$$= -\chi_{\mathrm{T}}(t) p^{sct}(\boldsymbol{x}^R, t) \quad \text{when} \quad \boldsymbol{x}^R \in \mathbb{D}_{sct} \,,$$

where $c^s$ is the wave speed in the medium of $\mathbb{D}_{sct}$.

### 8.2.4. Boundary-integral equations in the time domain

In a similar way as in the previous section, the time-domain counterparts of the integral equations (8.56) and (8.57) are obtained by applying the standard rules for inversion from the $s$-domain to the time domain:

$$\frac{1}{2} p(\boldsymbol{x}, t) - \int_{\boldsymbol{x}'\in\partial\mathbb{D}_{sct}} [C_t\{G^q, \nu_k' v_k\} + C_t\{\Gamma_k^q, \nu_k' p\}] \mathrm{dA}$$

$$= p^{inc}(\boldsymbol{x}, t), \quad \boldsymbol{x} \in \partial\mathbb{D}_{sct}, \quad t \in \mathrm{T} \,, \tag{8.62}$$

$$\frac{1}{2} p(\boldsymbol{x}, t) + \int_{\boldsymbol{x}'\in\partial\mathbb{D}_{sct}} [C_t\{G^{q^s}, \nu_k' v_k\} + C_t\{\Gamma_k^{q^s}, \nu_k' p\}] \mathrm{dA}$$

$$= 0, \quad \boldsymbol{x} \in \partial\mathbb{D}_{sct}, \quad t \in \mathrm{T} \,, \tag{8.63}$$

in which the Green's states $G^q = G^q(\boldsymbol{x}|\boldsymbol{x}', t)$, $\Gamma_k^q = \Gamma_k^q(\boldsymbol{x}|\boldsymbol{x}', t)$, $G^{q^s} = G^{q^s}(\boldsymbol{x}|\boldsymbol{x}', t)$ and $\Gamma_k^{q^s} = \Gamma_k^{q^s}(\boldsymbol{x}|\boldsymbol{x}', t)$ are the time-domain counterparts of the $s$-domain Green's states $\hat{G}^q$, $\hat{\Gamma}_k^q$, $\hat{G}^{q^s}$ and $\hat{\Gamma}_k^{q^s}$. Further, $C_t\{\cdot,\cdot\} = C_t\{\cdot,\cdot\}(\boldsymbol{x}', t)$ denotes the temporal convolution defined by Eq. (1.19).

For the special case of homogeneous media the pertaining system of Eqs. (8.62) and (8.63) reduces to

$$\frac{1}{2} p(\boldsymbol{x}, t) - \int_{\boldsymbol{x}'\in\partial\mathbb{D}_{sct}} \left[ \rho \frac{\partial_t v_k(\boldsymbol{x}', t - \frac{|\boldsymbol{x} - \boldsymbol{x}'|}{c})}{4\pi|\boldsymbol{x} - \boldsymbol{x}'|} - \partial_k \frac{p(\boldsymbol{x}', t - \frac{|\boldsymbol{x} - \boldsymbol{x}'|}{c})}{4\pi|\boldsymbol{x} - \boldsymbol{x}'|} \right] \nu_k' \mathrm{dA}$$

$$= p^{inc}(\boldsymbol{x}, t), \quad \boldsymbol{x} \in \partial\mathbb{D}_{sct}, \quad t \in \mathrm{T} \,, \tag{8.64}$$

and

$$\frac{1}{2}p(\boldsymbol{x},t) + \int_{\boldsymbol{x}'\in\partial\mathbb{D}_{sct}} \left[ \rho^s \frac{\partial_t v_k(\boldsymbol{x}', t - \frac{|\boldsymbol{x}-\boldsymbol{x}'|}{c^s})}{4\pi|\boldsymbol{x}-\boldsymbol{x}'|} - \partial_k \frac{p(\boldsymbol{x}', t - \frac{|\boldsymbol{x}-\boldsymbol{x}'|}{c^s})}{4\pi|\boldsymbol{x}-\boldsymbol{x}'|} \right] \nu'_k \mathrm{dA}$$

$$= 0, \quad \boldsymbol{x} \in \partial\mathbb{D}_{sct} \quad t \in \mathrm{T}, \tag{8.65}$$

where $c^s$ is the wave speed in the medium of $\mathbb{D}_{sct}$.

### 8.2.5. The case of an impenetrable scatterer

We now investigate the case that the scatterer is impenetrable for acoustic wave motion and we consider the two cases dealt with in Section 3.1.2.

*The special case of a pressure-free scatterer (void)*

The presence of a void is accounted for by the explicit boundary condition (cf. Eq (3.11))

$$\lim_{\varepsilon\downarrow 0} \hat{p}(\boldsymbol{x} + \varepsilon\boldsymbol{\nu}, s) = 0, \quad \boldsymbol{x} \in \partial\mathbb{D}_{sct} \, . \tag{8.66}$$

After insertion of this boundary condition into the integral representation of Eq. (8.47), we obtain

$$\hat{p}^{sct}(\boldsymbol{x}^R, s) = \int_{\boldsymbol{x}\in\partial\mathbb{D}_{sct}} \hat{G}^q(\boldsymbol{x}^R|\boldsymbol{x}, s)\nu_k \hat{v}_k(\boldsymbol{x}, s)\mathrm{dA}$$

$$\text{when} \quad \boldsymbol{x}^R \in \mathbb{D}'_{sct} \, . \tag{8.67}$$

With the aid of Eq. (8.67) the acoustic pressure can be calculated as soon as $\nu_k \hat{v}_k(\boldsymbol{x}, s)$ at the boundary $\partial\mathbb{D}_{sct}$ is known. The latter quantity can be determined from the integral equation

$$\int_{\boldsymbol{x}'\in\partial\mathbb{D}_{sct}} \hat{G}^q(\boldsymbol{x}|\boldsymbol{x}', s)\nu'_k \hat{v}_k(\boldsymbol{x}', s)\mathrm{dA} = -\hat{p}^{inc}(\boldsymbol{x}, s) \, ,$$

$$\boldsymbol{x} \in \partial\mathbb{D}_{sct} \, , \tag{8.68}$$

that directly follows from Eq. (8.56) upon inserting the boundary condition of Eq. (8.66). This is an integral equation of the first kind.

An integral equation of the second kind can be derived as follows. Using

$$\hat{v}_l^{sct}(\boldsymbol{x}^R, s) = \frac{-1}{s\rho(\boldsymbol{x}^R)}\partial_l^R \hat{p}^{sct}(\boldsymbol{x}^R, s) \quad \text{when } \boldsymbol{x}^R \in \mathbb{D}'_{sct} , \tag{8.69}$$

and $\hat{v}_l^{sct} = \hat{v}_l - \hat{v}_l^{inc}$, we obtain from Eq. (8.67) the representation

$$\hat{v}_l(\boldsymbol{x}^R, s) = \hat{v}_l^{inc}(\boldsymbol{x}^R, s) - \frac{1}{s\rho(\boldsymbol{x}^R)}\partial_l^R \int_{\boldsymbol{x}\in\partial\mathbb{D}_{sct}} \hat{G}^q(\boldsymbol{x}^R|\boldsymbol{x}, s)\nu_k \hat{v}_k(\boldsymbol{x}, s)\mathrm{dA}$$

$$\text{when } \boldsymbol{x}^R \in \mathbb{D}'_{sct} . \tag{8.70}$$

Now let the point of observation approach the boundary $\partial\mathbb{D}_{sct}$ and multiply both sides of Eq. (8.70) through by $\nu_l$. Performing the limiting procedure of Section 7.2.2, we arrive at the integral equation of the second kind

$$\frac{1}{2}\nu_k \hat{v}_k(\boldsymbol{x}, s) + \frac{1}{s\rho(\boldsymbol{x})} \int_{\boldsymbol{x}'\in\partial\mathbb{D}_{sct}} \nu_l \partial_l \hat{G}^q(\boldsymbol{x}|\boldsymbol{x}', s)\nu'_k \hat{v}_k(\boldsymbol{x}', s)\mathrm{dA}$$

$$= \nu_k \hat{v}_k^{inc}(\boldsymbol{x}, s) , \quad \boldsymbol{x} \in \partial\mathbb{D}_{sct} , \tag{8.71}$$

from which $\nu_k \hat{v}_k$ at the boundary $\partial\mathbb{D}_{sct}$ can be determined.

The time-domain counterparts of the latter results are obtained in a similar way as before by either inversion of the $s$-domain results or by direct application of the time-domain reciprocity theorem of convolution type.

*The special case of an immovable rigid scatterer*

The presence of an immovable rigid scatterer is accounted for by the explicit boundary condition (cf. Eq. (3.12))

$$\lim_{\varepsilon \downarrow 0} \nu_k \hat{v}_k(\boldsymbol{x} + \varepsilon \boldsymbol{\nu}, s) = 0, \quad \boldsymbol{x} \in \partial\mathbb{D}_{sct} . \tag{8.72}$$

After insertion of this boundary condition into the integral representation of Eq. (8.47), we obtain

$$\hat{p}^{sct}(\boldsymbol{x}^R, s) = \int_{\boldsymbol{x}\in\partial\mathbb{D}_{sct}} \hat{\Gamma}_k^q(\boldsymbol{x}^R|\boldsymbol{x}, s)\nu_k \hat{p}(\boldsymbol{x}, s)\mathrm{dA}$$

$$\text{when } \boldsymbol{x}^R \in \mathbb{D}'_{sct} . \tag{8.73}$$

With the aid of Eq. (8.73), the acoustic pressure can be calculated as soon as $\hat{p}(\boldsymbol{x}, s)$ at the boundary $\partial\mathbb{D}_{sct}$ is known. The latter can be determined from the integral equation

$$\frac{1}{2}\hat{p}(\boldsymbol{x}, s) - \int_{\boldsymbol{x}'\in\partial\mathbb{D}_{sct}} \hat{\Gamma}^q_k(\boldsymbol{x}|\boldsymbol{x}', s)\nu'_k\hat{p}(\boldsymbol{x}', s)\mathrm{dA} = \hat{p}^{inc}(\boldsymbol{x}, s) \, , \qquad \boldsymbol{x} \in \partial\mathbb{D}'_{sct} \, , \tag{8.74}$$

that directly follows from Eq. (8.56) upon inserting the boundary condition of Eq. (8.72). This is an integral equation of the second kind.

An integral equation of the first kind can be derived as follows. Using

$$\hat{v}^{sct}_l(\boldsymbol{x}^R, s) = \frac{-1}{s\rho(\boldsymbol{x}^R)}\partial^R_l\hat{p}^{sct}(\boldsymbol{x}^R, s) \quad \text{when } \boldsymbol{x}^R \in \mathbb{D}'_{sct} \, , \tag{8.75}$$

we obtain from Eq. (8.73) the representation

$$\hat{v}^{sct}_l(\boldsymbol{x}^R, s) = -\frac{1}{s\rho(\boldsymbol{x}^R)}\partial^R_l\int_{\boldsymbol{x}\in\partial\mathbb{D}_{sct}} \hat{\Gamma}^q_k(\boldsymbol{x}^R|\boldsymbol{x}, s)\nu_k\hat{p}(\boldsymbol{x}, s)\mathrm{dA} \qquad \text{when } \boldsymbol{x}^R \in \mathbb{D}'_{sct} \, . \tag{8.76}$$

Now let the point of observation approach the boundary $\partial\mathbb{D}_{sct}$ and multiply both sides of Eq. (8.76) through by $\nu_l$. Performing the limiting procedure of Section 7.2.2, we arrive at the integral equation of the first kind

$$\lim_{\varepsilon\downarrow 0}\frac{1}{s\rho(\boldsymbol{x})}\partial_\varepsilon\int_{\boldsymbol{x}'\in\partial\mathbb{D}_{sct}} \hat{\Gamma}^q_k(\boldsymbol{x}+\varepsilon\boldsymbol{\nu}|\boldsymbol{x}', s)\nu'_k\hat{p}(\boldsymbol{x}', s)\mathrm{dA} = \nu_k\hat{v}^{inc}_k(\boldsymbol{x}, s), \quad \boldsymbol{x} \in \partial\mathbb{D}_{sct} \, , \tag{8.77}$$

from which $\hat{p}(\boldsymbol{x}, s)$ at the boundary $\partial\mathbb{D}_{sct}$ can be determined. In this latter equation, special care has to be taken in the limit that $\varepsilon$ tends to zero.

The time-domain counterparts of the latter results are obtained in a similar way as before by either inversion of the $s$-domain results or by direct application of the time-domain reciprocity theorem of convolution type.

# Chapter 9

# Scattering by a Disk

As an example of the problem of scattering by a contrasting domain, we consider the problem of an acoustic wave scattered by an infinitely thin disk. The disk is either perfectly compliant or perfectly rigid and immovable. We first consider the case that the disk is located in a homogeneous embedding. Secondly, we consider the case that the disk is located in an inhomogeneous embedding: a homogeneous halfspace. For both cases, we derive an integral equation with an operator of convolution type, which allows us to employ the standard Fourier-transformation techniques in the computations.

## 9.1. Scattering by a planar object of vanishing thickness

Let a planar object of vanishing thickness occupy the domain

$$\mathbb{D}_{sct} = \{\boldsymbol{x} \in \mathbb{R}^3;\ (x_1, x_2) \in \Lambda_{sct},\ x_3 = h\}\,, \tag{9.1}$$

where $\Lambda_{sct}$ is a bounded domain in the $(x_1, x_2)$-plane (see Fig. 9.1). The physical properties of the object are characterized by boundary conditions to be satisfied upon approaching either side of the object. The embedding domain is denoted as $\mathbb{D}'_{sct}$. Noting that in Chapter 8 the enclosing boundary of the scatterer is denoted as $\partial\mathbb{D}_{sct}$, we observe that in our case it consists of the two sides of the planar object, viz.,

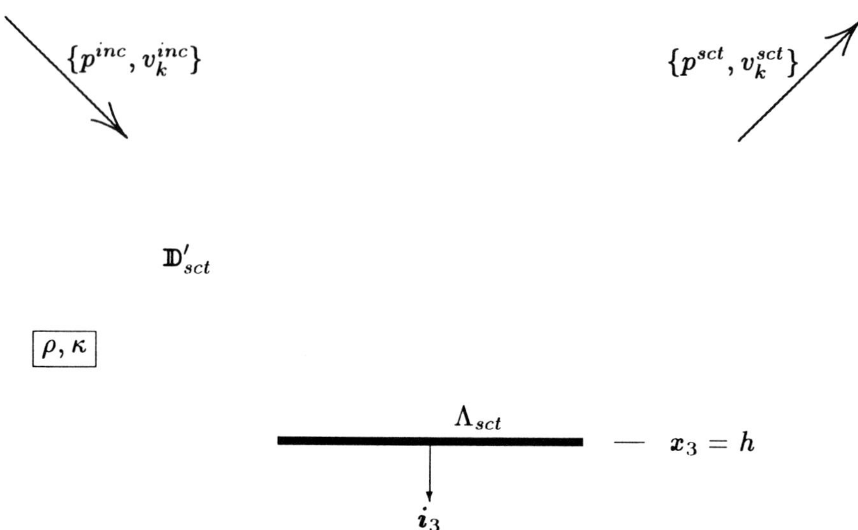


Figure 9.1. Planar disk in an unbounded domain $D'_{sct}$.

$$\partial \mathbb{D}_{sct} = \{\boldsymbol{x} \in \mathbb{R}^3;\ (x_1, x_2) \in \Lambda_{sct},\ x_3 = \lim_{\varepsilon \downarrow 0}(h \pm \varepsilon)\}\,. \tag{9.2}$$

When the point of observation $\boldsymbol{x}^R$ is located outside $\partial \mathbb{D}_{sct}$, we may replace the expression for the scattered wavefield of Eq. (8.47) by

$$\hat{p}^{sct}(\boldsymbol{x}^R, s) = \int_{(x_1,x_2)\in\Lambda_{sct}} [\hat{G}^q\, \partial\hat{q}(x_1, x_2, s) + \hat{\Gamma}_3^q\, \partial\hat{f}_3(x_1, x_2, s)]\mathrm{dA}$$
$$\text{when } \boldsymbol{x}^R \in \mathbb{D}'_{sct}\,. \tag{9.3}$$

The Green's states $\hat{G}^q = \hat{G}^q(\boldsymbol{x}^R|x_1, x_2, h, s)$ and $\hat{\Gamma}_3^q = \hat{\Gamma}_3^q(\boldsymbol{x}^R|x_1, x_2, h, s)$ are discussed in Chapter 7. Further, in Eq. (9.3) we have defined an equivalent surface density of injected volume time rate

$$\partial\hat{q} = \lim_{\varepsilon\downarrow 0}\, [\hat{v}_3(x_1, x_2, h+\varepsilon, s) - \hat{v}_3(x_1, x_2, h-\varepsilon, s)]\big|_{(x_1,x_2)\in\Lambda_{sct}}\,, \tag{9.4}$$

and the equivalent surface-force density

$$\partial\hat{f}_3 = \lim_{\varepsilon\downarrow 0}[\hat{p}(x_1, x_2, h+\varepsilon, s) - \hat{p}(x_1, x_2, h-\varepsilon, s)]\big|_{(x_1,x_2)\in\Lambda_{sct}} . \tag{9.5}$$

The introduction of these surface sources is consistent with Eq. (7.9) where expressions in terms of volume sources are presented. As soon as we know the values of $\partial\hat{q}$ and $\partial\hat{f}_3$ we can calculate the acoustic quantities in the receiver location $\boldsymbol{x}^R \in \mathbb{D}'_{sct}$, using Eq. (9.3).

*The case of a perfectly compliant lamina*

If the scattering object of vanishing thickness is a perfectly compliant lamina, the acoustic pressure vanishes on either side of it:

$$\lim_{\varepsilon\downarrow 0} \hat{p}(x_1, x_2, h \pm \varepsilon, s) = 0, \quad (x_1, x_2) \in \Lambda_{sct} ; \tag{9.6}$$

this implies that $\partial\hat{f}_3 = 0$ in Eq. (9.3). Specifically, this representation reduces to

$$\hat{p}^{sct}(\boldsymbol{x}^R, s) = \int_{(x_1,x_2)\in\Lambda_{sct}} \hat{G}^q(\boldsymbol{x}^R|x_1, x_2, h, s)\partial\hat{q}(x_1, x_2, s)\mathrm{dA}$$
$$\text{when } \boldsymbol{x}^R \in \mathbb{D}'_{sct} . \tag{9.7}$$

With the aid of Eq. (9.7), the acoustic pressure can be calculated as soon as $\partial\hat{q}(x_1, x_2, s)$ on $\Lambda_{sct}$ is known. The source function $\partial\hat{q}$ can be obtained by letting the point of observation $\boldsymbol{x}^R$ approach $\Lambda_{sct}$. Employing the boundary condition

$$\hat{p}^{sct}(\boldsymbol{x}^R, s) = -\hat{p}^{inc}(\boldsymbol{x}^R, s) \quad \text{when } \boldsymbol{x}^R = \lim_{\varepsilon\downarrow 0}(x_1, x_2, h \pm \varepsilon), \quad (x_1, x_2) \in \Lambda_{sct} , \tag{9.8}$$

we arrive at the integral equation of the first kind (cf. Eq. (8.68))

$$\lim_{\varepsilon\downarrow 0} \int_{(x'_1,x'_2)\in\Lambda_{sct}} \hat{G}^q(x_1, x_2, h+\varepsilon|x'_1, x'_2, h, s)\partial\hat{q}(x'_1, x'_2, s)\mathrm{dA}$$
$$= -\hat{p}^{inc}(x_1, x_2, h, s), \quad (x_1, x_2) \in \Lambda_{sct} . \tag{9.9}$$

This integral equation for $\partial\hat{q}$ has to be solved numerically. Aspects of iterative solutions are discussed in Chapter 2.

*The case of an immovable rigid disk*

If the scattering object of vanishing thickness is an immovable perfectly rigid disk, the acoustic particle velocity vanishes on either side of it:

$$\lim_{\varepsilon\downarrow 0}\hat{v}_3(\boldsymbol{x}_1,\boldsymbol{x}_2,h\pm\varepsilon,s)=0,\quad(\boldsymbol{x}_1,\boldsymbol{x}_2)\in\Lambda_{sct}\ ; \tag{9.10}$$

this implies that $\partial\hat{q}=0$ in Eq. (9.3). Specifically, this representation reduces to

$$\hat{p}^{sct}(\boldsymbol{x}^R,s)=\int_{(x_1,x_2)\in\Lambda_{sct}}\hat{\Gamma}_3^q(\boldsymbol{x}^R|\boldsymbol{x}_1,\boldsymbol{x}_2,h,s)\partial\hat{f}_3(\boldsymbol{x}_1,\boldsymbol{x}_2,s)\mathrm{dA}\quad\text{when}\ \ \boldsymbol{x}^R\in\mathbb{D}'_{sct}\ . \tag{9.11}$$

With the aid of Eq. (9.11), the acoustic pressure can be calculated as soon as $\partial\hat{f}_3(\boldsymbol{x}_1,\boldsymbol{x}_2,s)$ at $\Lambda_{sct}$ is known. In order to derive an integral equation for this source function we first determine (cf. Eq. (7.70))

$$\hat{v}_3^{sct}(\boldsymbol{x}^R,s)=\frac{-1}{s\rho(\boldsymbol{x}^R)}\partial_3^R\hat{p}^{sct}(\boldsymbol{x}^R,s)\quad\text{when}\ \ \boldsymbol{x}^R\in\mathbb{D}'_{sct}\ , \tag{9.12}$$

to obtain the representation

$$\hat{v}_3^{sct}(\boldsymbol{x}^R,s)=\frac{-1}{s\rho(\boldsymbol{x}^R)}\partial_3^R\int_{(x_1,x_2)\in\Lambda_{sct}}\hat{\Gamma}_3^q(\boldsymbol{x}^R|\boldsymbol{x}_1,\boldsymbol{x}_2,h,s)\partial\hat{f}_3(\boldsymbol{x}_1,\boldsymbol{x}_2,s)\mathrm{dA}\quad\text{when}\ \ \boldsymbol{x}^R\in\mathbb{D}'_{sct}\ . \tag{9.13}$$

Now let the point of observation approach the boundary $\partial\mathbb{D}_{sct}$. Employing the boundary condition

$$\hat{v}_3^{sct}(\boldsymbol{x}^R,s)=-\hat{v}_3^{inc}(\boldsymbol{x}^R,s)\quad\text{when}\ \ \boldsymbol{x}^R=\lim_{\varepsilon\downarrow 0}(\boldsymbol{x}_1,\boldsymbol{x}_2,h\pm\varepsilon),\ \ (\boldsymbol{x}_1,\boldsymbol{x}_2)\in\Lambda_{sct}\ , \tag{9.14}$$

we arrive at the integral equation of the first kind (cf. Eq. (8.77))

$$\frac{1}{s\rho(\boldsymbol{x}_1,\boldsymbol{x}_2,h)}\lim_{\varepsilon\downarrow 0}\partial_\varepsilon\int_{(x_1',x_2')\in\Lambda_{sct}}\hat{\Gamma}_3^q(\boldsymbol{x}_1,\boldsymbol{x}_2,h+\varepsilon|\boldsymbol{x}_1',\boldsymbol{x}_2',h,s)\partial\hat{f}_3(\boldsymbol{x}_1',\boldsymbol{x}_2',s)\mathrm{dA}$$
$$=\hat{v}_3^{inc}(\boldsymbol{x}_1,\boldsymbol{x}_2,h,s),\quad(\boldsymbol{x}_1,\boldsymbol{x}_2)\in\Lambda_{sct}\ , \tag{9.15}$$

from which $\partial\hat{f}_3(\boldsymbol{x}_1,\boldsymbol{x}_2,s)$ at the boundary $\partial\mathbb{D}_{sct}$ can be determined. In this latter equation, special care has to be taken in the limit that $\varepsilon$ tends to zero.

This integral equation has to be solved numerically. Aspects of iterative solutions are discussed in Chapter 2.

## 9.2. Disk in a homogeneous embedding

In the special case that the disk is located in a homogeneous embedding the Green's states $\hat{G}^q(\boldsymbol{x}^R|\boldsymbol{x},s)$ and $\hat{\Gamma}_3^q(\boldsymbol{x}^R|\boldsymbol{x},s)$ are given by (cf. Chapter 8)

$$\hat{G}^q(\boldsymbol{x}^R|\boldsymbol{x},s) = s\rho\hat{G}(\boldsymbol{x}^R-\boldsymbol{x},s)\,, \tag{9.16}$$

and

$$\hat{\Gamma}_3^q(\boldsymbol{x}^R|\boldsymbol{x},s) = -\partial_3^R\hat{G}(\boldsymbol{x}^R-\boldsymbol{x},s)\,, \tag{9.17}$$

where

$$\hat{G}(\boldsymbol{x}^R-\boldsymbol{x},s) = \frac{\exp(-\hat{\gamma}|\boldsymbol{x}^R-\boldsymbol{x}|)}{4\pi|\boldsymbol{x}^R-\boldsymbol{x}|} \quad \text{with } \hat{\gamma} = s(\kappa\rho)^{\frac{1}{2}} = \frac{s}{c}\,. \tag{9.18}$$

The integral equation (9.9) can now be written as

$$s\rho\lim_{\varepsilon\downarrow 0}\int_{(x_1',x_2')\in\Lambda_{sct}}\hat{G}(x_1-x_2',x_2-x_2',\varepsilon,s)\partial\hat{q}(x_1',x_2',s)\mathrm{dA}$$
$$= -\hat{p}^{inc}(x_1,x_2,h,s)\,,\quad (x_1,x_2)\in\Lambda_{sct}\,, \tag{9.19}$$

for the case of a compliant lamina,

while the integral equation (9.15) is rewritten as

$$\frac{1}{s\rho}\lim_{\varepsilon\downarrow 0}\partial_\varepsilon^2\int_{(x_1',x_2')\in\Lambda_{sct}}\hat{G}(x_1-x_2',x_2-x_2',\varepsilon,s)\partial\hat{f}_3(x_1',x_2',s)\mathrm{dA}$$
$$= -\hat{v}_3^{inc}(x_1,x_2,h,s)\,,\quad (x_1,x_2)\in\Lambda_{sct}\,, \tag{9.20}$$

for the case of a rigid disk.

In view of the discussion how to solve these integral equations, they are written in our operator notation of Chapter 2 as

$$L\hat{u} = \hat{f}\,,\qquad (x_1,x_2)\in\Lambda_{sct}\,, \tag{9.21}$$

where

$$L\hat{u} = s\rho \lim_{\varepsilon \downarrow 0} \int_{(x_1', x_2') \in \Lambda_{sct}} \hat{G}(x_1 - x_1', x_2 - x_2', \varepsilon, s) \hat{u}(x_1', x_2', s) \mathrm{dA} \,,$$

$$\begin{aligned} \hat{f}(x_1, x_2, s) &= -\hat{p}^{inc}(x_1, x_2, h, s)\,, \\ \hat{u}(x_1, x_2, s) &= \partial \hat{q}(x_1, x_2, s)\,, \end{aligned} \tag{9.22}$$

for the case of a compliant lamina,

and

$$L\hat{u} = \frac{1}{s\rho} \lim_{\varepsilon \downarrow 0} \partial_\varepsilon^2 \int_{(x_1', x_2') \in \Lambda_{sct}} \hat{G}(x_1 - x_1', x_2 - x_2', \varepsilon, s) \hat{u}(x_1', x_2', s) \mathrm{dA} \,,$$

$$\begin{aligned} \hat{f}(x_1, x_2, s) &= -\hat{v}_3^{inc}(x_1, x_2, h, s)\,, \\ \hat{u}(x_1, x_2, s) &= \partial \hat{f}_3(x_1, x_2, s)\,, \end{aligned} \tag{9.23}$$

for the case of a rigid disk.

Obviously, they are integral equations of the convolution type and can be solved with the help of the iterative schemes of Chapter 2, where the operator can be determined with the Fourier transformations, as it has been discussed in Section 2.7.

*Computation of the operator*

Introducing the characteristic function

$$\chi_{\Lambda_{sct}} = \{1, \frac{1}{2}, 0\} \quad \text{when} \quad (x_1, x_2) \in \{\Lambda_{sct}, \partial\Lambda_{sct}, \Lambda'_{sct}\}\,, \tag{9.24}$$

where the boundary $\partial\Lambda_{sct}$ denotes the edge of the disk and $\Lambda'_{sct}$ denotes the complement of $\Lambda_{sct} \cup \partial\Lambda_{sct}$ in the horizontal plane $\{(x_1, x_2) \in \mathbb{R}^2; x_3 = h\}$. With this characteristic function we may write the operator expressions as

$$L\hat{u} = s\rho \lim_{\varepsilon \downarrow 0} F^{-1}\Big\{\overline{G}(js\alpha_1, js\alpha_2, \varepsilon, s)\, F\{\chi_{\Lambda_{sct}} \hat{u}\}\Big\}$$

for the case of a compliant lamina, (9.25)

$$L\hat{u} = \frac{1}{s\rho} \lim_{\varepsilon \downarrow 0} \partial_\varepsilon^2 F^{-1}\Big\{\overline{G}(js\alpha_1, js\alpha_2, \varepsilon, s)\, F\{\chi_{\Lambda_{sct}} \hat{u}\}\Big\}$$

for the case of a rigid disk. (9.26)

where the two-dimensional Fourier transformation is denoted by the operator $F$ and its inverse by $F^{-1}$, i.e. (cf. Eqs. (1.46) and (1.47))

$$\overline{u} = F\{\hat{u}\} = \int_{(x_1,x_2)\in\mathbb{R}^2} \exp(jsα_1x_1 + jsα_2x_2)\hat{u}(x_1,x_2,x_3,s)\mathrm{dA}\,, \tag{9.27}$$

$$F^{-1}\{\overline{u}\} = \frac{1}{(2\pi)^2}\int_{(sα_1,sα_2)\in\mathbb{R}^2} \exp(-jsα_1x_1 - jsα_2x_2)\overline{u}(jsα_1,jsα_2,x_3,s)\mathrm{dA}. \tag{9.28}$$

In Eqs. (9.25) and (9.26), $\overline{G}(jsα_1,jsα_2,x_3,s)$ denotes the Fourier transform of $\hat{G}(x_1,x_2,x_3,s)$ with respect to the horizontal coordinates. The result follows from the combination of Eqs. (4.58) and (4.59) as

$$\overline{G}(jsα_1,jsα_2,x_3,s) = \frac{\exp(-s\Gamma|x_3|)}{2s\Gamma}\,, \tag{9.29}$$

where

$$\Gamma = \left(\frac{1}{c^2} + α_1^2 + α_2^2\right)^{\frac{1}{2}}\,, \quad \mathrm{Re}(\Gamma) > 0\,. \tag{9.30}$$

Hence, we finally arrive at

$$L\hat{u} = F^{-1}\{\overline{g}\,F\{\chi_{\Lambda_{sct}}\hat{u}\}\}\,, \tag{9.31}$$

where

$$\overline{g} = s\rho\lim_{\varepsilon\downarrow 0}\frac{\exp(-s\Gamma\varepsilon)}{2s\Gamma} = \frac{\rho}{2\Gamma} \quad \text{for the case of a compliant lamina}\,, \tag{9.32}$$

$$\overline{g} = \frac{1}{s\rho}\lim_{\varepsilon\downarrow 0}\partial_\varepsilon^2\frac{\exp(-s\Gamma\varepsilon)}{2s\Gamma} = \frac{\Gamma}{2\rho} \quad \text{for the case of a rigid disk}\,. \tag{9.33}$$

In the derivation to arrive at Eqs. (9.31) and (9.33) we have interchanged the differentiations with respect to $\varepsilon$ and the integrations of the inverse Fourier transform. This is is admissible as long as

$$\lim_{|(sα_1,sα_2)|\to\infty} |(sα_1,sα_2)|\,F\{\chi_{\Lambda_{sct}}\hat{u}\} = 0 \quad \text{as } (sα_1,sα_2)\in\mathbb{R}^2\,, \tag{9.34}$$

where $|(sα_1,sα_2)| = [(sα_1)^2 + (sα_2)^2]^{\frac{1}{2}}$. This condition is satisfied if $\hat{u} = \partial\hat{f}_3$ vanishes at the edge $\partial\Lambda_{sct}$ of the disk domain $\Lambda_{sct}$. This is the so-called edge condition of the wavefield problem of a rigid disk (JONES, 1952).

*Computation of the adjoint operator*

We are now able to use the various iterative schemes discussed in Chapter 2. In order to be able to use these schemes, we need the adjoint $L^*$ of the operator $L$. This operator is now obtained as

$$L^*\hat{v} = F^{-1}\{\overline{g}^\star F\{\chi_{\Lambda_{sct}}\hat{v}\}\} . \tag{9.35}$$

*Computation of the preconditioning operators*

In Section 2.7 it is found that for the type of convolution operators at hand, a preconditioning operator may be found as

$$P\hat{v} = F^{-1}\Big\{(\overline{g})^{-1} F\{\chi_{\Lambda_{sct}}\hat{v}\}\Big\} , \tag{9.36}$$

while its adjoint is given by

$$P^*\hat{u} = F^{-1}\Big\{(\overline{g}^\star)^{-1} F\{\chi_{\Lambda_{sct}}\hat{u}\}\Big\} . \tag{9.37}$$

*Computation of the scattered wavefield*

Once the operator equation (9.21) is solved, which means that $\hat{u} = \partial\hat{q}$ for the case of a compliant lamina and $\hat{u} = \partial\hat{f}_3$ for the case of a rigid disk are known, we can determine the scattered acoustic pressure from the integral representations (9.7) and (9.11) as

$$\hat{p}^{sct}(\boldsymbol{x}^R, s) = s\rho \int_{(x_1,x_2)\in\Lambda_{sct}} \hat{G}(x_1^R - x_1, x_2^R - x_2, x_3^R - h, s)\partial\hat{q}(x_1, x_2, s)\mathrm{dA} ,$$

$$\text{for the case of a compliant lamina,} \tag{9.38}$$

and

$$\hat{p}^{sct}(\boldsymbol{x}^R, s) = -\int_{(x_1,x_2)\in\Lambda_{sct}} \partial_3^R \hat{G}(x_1^R - x_1, x_2^R - x_2, x_3^R - h, s)\partial\hat{f}_3(x_1, x_2, s)\mathrm{dA} ,$$

$$\text{for the case of a rigid disk,} \tag{9.39}$$

when $\boldsymbol{x}^R \in \mathbb{D}'_{sct}$. Using the Fourier transformations defined in Eqs. (9.27) - (9.28) and the Fourier transform of the Green's function given in Eq. (9.29), we observe that the scattered acoustic pressure can be computed as

$$\hat{p}^{sct}(x_1, x_2, x_3, s) = F^{-1}\left\{\frac{\rho}{2\Gamma}\exp(-s\Gamma|x_3 - h|)\, F\{\chi_{\Lambda_{sct}}\partial\hat{q}\}\right\}$$

$$\text{for the case of a compliant lamina}, \qquad (9.40)$$

and

$$\hat{p}^{sct}(x_1, x_2, x_3, s) = F^{-1}\left\{\frac{1}{2}\,\mathrm{sign}(x_3 - h)\exp(-s\Gamma|x_3 - h|)\, F\{\chi_{\Lambda_{sct}}\partial\hat{f}_3\}\right\}$$

$$\text{for the case of a rigid disk}. \qquad (9.41)$$

## 9.3. Analytic solution for a pressure-free plane

We now assume that the support of the disk $\Lambda_{sct}$ becomes unbounded in the $x_1$- and $x_2$-directions and we replace $\Lambda_{sct}$ by $\mathbb{R}^2$. Assuming that the background is homogeneous, we observe that the preconditioning operator becomes the exact inverse of the operator $L$. The solution becomes

$$\hat{u} = L^{-1}\hat{f} = F^{-1}\left\{(\overline{g})^{-1}F\{\hat{f}\}\right\}, \qquad (9.42)$$

while the preconditioning operator becomes

$$P\hat{f} = F^{-1}\left\{(\overline{g})^{-1}F\{\hat{f}\}\right\} = L^{-1}\hat{f}\,. \qquad (9.43)$$

In particular, we consider the case of a perfectly compliant lamina. When $\Lambda_{sct}$ becomes unbounded in the horizontal directions, we end up with a pressure-free plane at $x_3 = h$. For simplicity, we take $h = 0$. We consider the source and receiver positions below this pressure-free plane $x_3 = 0$ (see Fig. 9.2). Following Eq. (9.42), the solution in the two-dimensional Fourier domain is given by

$$\partial\overline{q}(js\alpha_1, js\alpha_2, s) = -\frac{2\Gamma}{\rho}\overline{p}^{inc}(js\alpha_1, js\alpha_2, 0, s)\,. \qquad (9.44)$$

Furthermore, the Fourier transform of the scattered wavefield of Eq. (9.40) may be written as

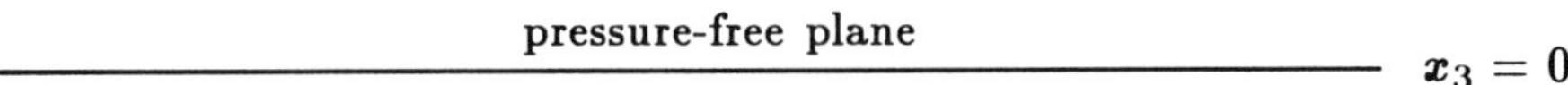


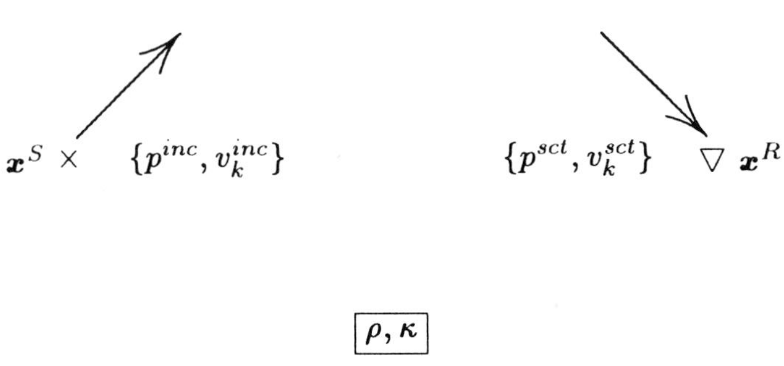


Figure 9.2. A point source in a homogeneous halfspace.

$$\begin{aligned}\bar{p}^{sct}(js\alpha_1, js\alpha_2, x_3, s) &= \frac{\rho}{2\Gamma}\exp(-s\Gamma x_3)\,\partial\bar{q}(js\alpha_1, js\alpha_2, s) \\ &= -\exp(-s\Gamma x_3)\,\bar{p}^{inc}(js\alpha_1, js\alpha_2, 0, s)\,. \qquad (9.45)\end{aligned}$$

The two-dimensional inverse Fourier transformation of Eq. (9.45) yields the desired expression for the scattered pressure wavefield in the halfspace below the pressure-free plane at $x_3 = 0$.

*Monopole source*

As first example, we consider a monopole transducer at $x^S$ generating the incident wavefield $\hat{p}^{inc}$. From Eq. (4.82) it follows that the acoustic pressure of this incident wavefield may be written as

$$\hat{p}^{inc}(x, s) = s\rho\hat{q}^S\hat{G}(x - x^S, s)\,, \qquad (9.46)$$

where $\hat{q}^S$ is the $s$-domain time rate of volume injection. The two-dimensional

Fourier transform of this wavefield at $x_3 = 0$ is given by

$$\bar{p}^{inc}(js\alpha_1, js\alpha_2, 0, s) = \frac{\rho}{2\Gamma}\hat{q}^S \exp(js\alpha_1 x_1^S + js\alpha_2 x_2^S - s\Gamma x_3^S)\,. \tag{9.47}$$

Substituting this result in Eq. (9.45) we obtain

$$\bar{p}^{sct}(js\alpha_1, js\alpha_2, x_3, s) = \frac{-\rho}{2\Gamma}\hat{q}^S \exp[js\alpha_1 x_1^S + js\alpha_2 x_2^S - s\Gamma(x_3 + x_3^S)]\,. \tag{9.48}$$

Inversion of Eq. (9.48) to the spatial domain can now be done by inspection. The result is

$$\hat{p}^{sct}(\boldsymbol{x}, s) = -s\rho\hat{q}^S \hat{G}(x_1 - x_1^S, x_2 - x_2^S, x_3 + x_3^S, s)\,. \tag{9.49}$$

We observe that the scattered wavefield seems to be generated by a monopole source located at the virtual point imaged by the reflector at $x_3 = 0$, the so-called ghost reflection.

*Dipole source*

As second example, we consider a dipole transducer at $\boldsymbol{x}^S$ generating the incident wavefield $\hat{p}^{inc}$. From Eq. (4.91) it follows that the acoustic pressure of this incident wavefield may be written as

$$\hat{p}^{inc}(\boldsymbol{x}, s) = \hat{f}_k^S \partial_k^S \hat{G}(\boldsymbol{x} - \boldsymbol{x}^S, s)\,, \tag{9.50}$$

where $\hat{f}_k^S$ is the $s$-domain time rate of volume force and $\partial_k^S$ denotes the spatial derivative with respect to $x_k^S$. The two-dimensional Fourier transform of this wavefield at $x_3 = 0$ is given by

$$\bar{p}^{inc}(js\alpha_1, js\alpha_2, 0, s) = \frac{1}{2s\Gamma}\hat{f}_k^S \partial_k^S \exp(js\alpha_1 x_1^S + js\alpha_2 x_2^S - s\Gamma x_3^S)\,. \tag{9.51}$$

Substituting this result in Eq. (9.45) we obtain

$$\bar{p}^{sct}(js\alpha_1, js\alpha_2, x_3, s) = \frac{-1}{2s\Gamma}\hat{f}_k^S \partial_k^S \exp[js\alpha_1 x_1^S + js\alpha_2 x_2^S - s\Gamma(x_3 + x_3^S)]. \tag{9.52}$$

Inversion of Eq. (9.52) can now be done by inspection. The result is

$$\hat{p}^{sct}(\boldsymbol{x}, s) = -\hat{f}_k^S \partial_k^S \hat{G}(x_1 - x_1^S, x_2 - x_2^S, x_3 + x_3^S, s)\,. \tag{9.53}$$

We observe that the scattered wavefield seems to be generated by a dipole source located at the virtual point imaged by the reflector at $x_3 = 0$, the so-called ghost reflection.

## 9.4. Disk in a homogeneous halfspace

As the most simple inhomogeneous embedding, we now turn our attention to the halfspace as background (see Fig. 9.3). We assume that the disk is located at $x_3 = h$ in a homogeneous halfspace $0 < x_3 < \infty$, where at $x_3 = 0$ the total pressure wavefield vanishes:

$$\lim_{x_3 \downarrow 0} \hat{p}(\boldsymbol{x}, s) = 0 \,. \tag{9.54}$$

First we compute the incident wavefield in the scattering configuration. This is the wavefield that would be present in the entire configuration if the domain $\Lambda_{sct}$ showed no contrast with its embedding. This wavefield follows directly from the results of Section 9.3. The superposition of the representations of Eqs. (9.46) and (9.49) yields the total wavefield generated by a monopole source in the halfspace in absence of the disk scatterer, while the superposition of the representations of Eqs. (9.50) and (9.53) yields the total wavefield generated by a dipole source in the halfspace in absence of the disk scatterer; we directly conclude that the acoustic pressure of the present incident wavefield $\{\hat{p}^{inc,H}, \hat{v}_k^{inc,H}\}$ in our inhomogeneous embedding is arrived as

$$\hat{p}^{inc,H}(\boldsymbol{x}, s) = s\rho\hat{q}^S \hat{G}^H(\boldsymbol{x}|\boldsymbol{x}^S, s) \quad \text{for a monopole source} \,, \tag{9.55}$$

and

$$\hat{p}^{inc,H}(\boldsymbol{x}, s) = \hat{f}_k^S \partial_k^S \hat{G}^H(\boldsymbol{x}|\boldsymbol{x}^S, s) \quad \text{for a dipole source} \,, \tag{9.56}$$

where

$$\hat{G}^H(\boldsymbol{x}|\boldsymbol{x}^S, s) = \frac{\exp(-\hat{\gamma}|\boldsymbol{x} - \boldsymbol{x}^S|)}{4\pi|\boldsymbol{x} - \boldsymbol{x}^S|} - \frac{\exp(-\hat{\gamma}|\boldsymbol{x} - \boldsymbol{x}^{S^I}|)}{4\pi|\boldsymbol{x} - \boldsymbol{x}^{S^I}|} \,, \tag{9.57}$$

in which

$$\boldsymbol{x}^{S^I} = (x_1^S, x_2^S, -x_3^S) \tag{9.58}$$

denotes the image point of $\boldsymbol{x}^S$ with respect to the reflecting surface at $x_3 = 0$.

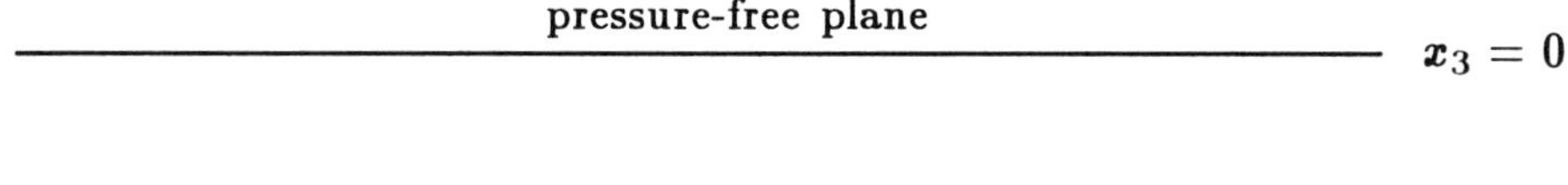


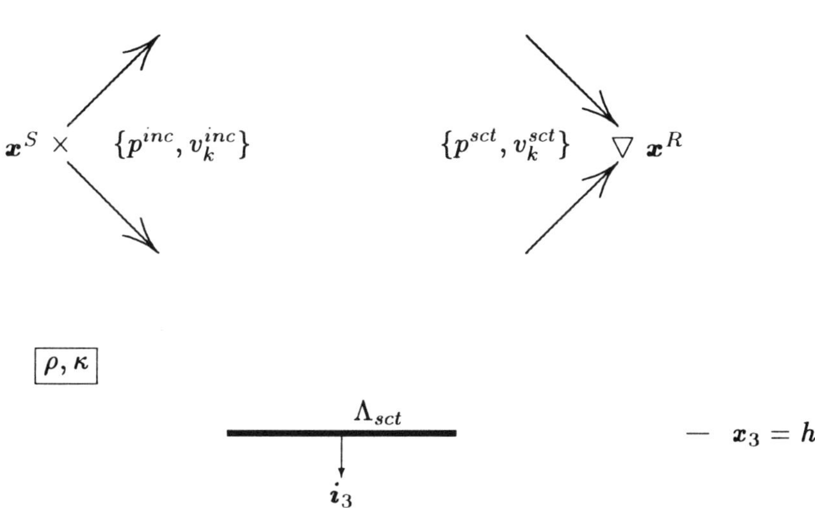


Figure 9.3. Planar disk in a homogeneous halfspace.

In order to be able to use the wavefield representation of Eq. (9.3), we need the Green's states of the inhomogeneous background, i.e., the Green's states of the homogeneous halfspace. The latter follow direct from the results of the monopole source in Eq. (9.55), viz.,

$$\hat{G}^q(\boldsymbol{x}^R|\boldsymbol{x}, s) = s\rho\hat{G}^H(\boldsymbol{x}^R|\boldsymbol{x}, s) , \tag{9.59}$$

and from Eq. (7.70) we directly obtain

$$\hat{\Gamma}_3^q(\boldsymbol{x}^R|\boldsymbol{x}, s) = -\partial_3^R\hat{G}^H(\boldsymbol{x}^R|\boldsymbol{x}, s) . \tag{9.60}$$

The integral equation (9.9) can now be written as

$$s\rho \lim_{\varepsilon \downarrow 0} \int_{(x_1', x_2') \in \Lambda_{sct}} \hat{G}^H(x_1, x_2, h+\varepsilon | x_1', x_2', h, s) \partial \hat{q}(x_1', x_2', s) \mathrm{dA}$$

$$= -\hat{p}^{inc,H}(x_1, x_2, h, s), \quad (x_1, x_2) \in \Lambda_{sct}, \tag{9.61}$$

for the case of a compliant lamina,

while the integral equation (9.15) is rewritten as

$$\frac{1}{s\rho} \lim_{\varepsilon \downarrow 0} \partial_\varepsilon^2 \int_{(x_1', x_2') \in \Lambda_{sct}} \hat{G}^H(x_1, x_2, h+\varepsilon | x_1', x_2', h, s) \partial \hat{f}_3(x_1', x_2', s) \mathrm{dA}$$

$$= -\hat{v}_3^{inc,H}(x_1, x_2, h, s), \quad (x_1, x_2) \in \Lambda_{sct}, \tag{9.62}$$

for the case of a rigid disk.

In view of the discussion how to solve these integral equations, they are written in our operator notation of Chapter 2 as

$$L\hat{u} = \hat{f}, \quad (x_1, x_2) \in \Lambda_{sct}, \tag{9.63}$$

where

$$L\hat{u} = s\rho \lim_{\varepsilon \downarrow 0} \int_{(x_1', x_2') \in \Lambda_{sct}} \hat{G}^h(x_1 - x_1', x_2 - x_2', \varepsilon, s) \hat{u}(x_1', x_2', s) \mathrm{dA},$$

$$\begin{aligned} \hat{f}(x_1, x_2, s) &= -\hat{p}^{inc,H}(x_1, x_2, h, s), \\ \hat{u}(x_1, x_2, s) &= \partial \hat{q}(x_1, x_2, s), \end{aligned} \tag{9.64}$$

for the case of a compliant lamina,

and

$$L\hat{u} = \frac{1}{s\rho} \lim_{\varepsilon \downarrow 0} \partial_\varepsilon^2 \int_{(x_1', x_2') \in \Lambda_{sct}} \hat{G}^h(x_1 - x_1', x_2 - x_2', \varepsilon, s) \hat{u}(x_1', x_2', s) \mathrm{dA},$$

$$\tag{9.65}$$

$$\begin{aligned} \hat{f}(x_1, x_2, s) &= -\hat{v}_3^{inc,H}(x_1, x_2, h, s), \\ \hat{u}(x_1, x_2, s) &= \partial \hat{f}_3(x_1, x_2, s), \end{aligned}$$

for the case of a rigid disk, (9.66)

while, in both cases,

$$\hat{G}^h(x_1, x_2, \varepsilon, s) = \hat{G}(x_1, x_2, \varepsilon, s) - \hat{G}(x_1, x_2, \varepsilon + 2h, s) . \tag{9.67}$$

Obviously, they are integral equations of the convolution type and can be solved with the help of the iterative schemes of Chapter 2, where the operator can be determined with the Fourier transforms as it has been discussed in Section 2.7.

*Computation of the operator*

Introducing the characteristic function

$$\chi_{\Lambda_{sct}} = \{1, \frac{1}{2}, 0\} \quad \text{when} \quad (x_1, x_2) \in \{\Lambda_{sct}, \partial\Lambda_{sct}, \Lambda'_{sct}\} , \tag{9.68}$$

where the boundary $\partial\Lambda_{sct}$ denotes the edge of the disk and $\Lambda'_{sct}$ denotes the complement of $\Lambda_{sct} \cup \partial\Lambda_{sct}$ in the horizontal plane $\{(x_1, x_2) \in \mathbb{R}_2; x_3 = h\}$. With this characteristic function we may write the operator expressions as

$$L\hat{u} = s\rho \lim_{\varepsilon \downarrow 0} F^{-1}\left\{\overline{G}^h(js\alpha_1, js\alpha_2, \varepsilon, s)\, F\{\chi_{\Lambda_{sct}} \hat{u}\}\right\}$$

$$\text{for the case of a compliant lamina,} \tag{9.69}$$

$$L\hat{u} = \frac{1}{s\rho} \lim_{\varepsilon \downarrow 0} \partial_\varepsilon^2 F^{-1}\left\{\overline{G}^h(js\alpha_1, js\alpha_2, \varepsilon, s)\, F\{\chi_{\Lambda_{sct}} \hat{u}\}\right\}$$

$$\text{for the case of a rigid disk ,} \tag{9.70}$$

where the two-dimensional Fourier transformation and its inverse are defined in Eqs. (9.27) and (9.28), and $\overline{G}^h(js\alpha_1, js\alpha_2, \varepsilon, s)$ denotes the Fourier transform of $\hat{G}^h(x_1, x_2, \varepsilon, s)$ with respect to the horizontal coordinates and it is given by (cf. Eq. (9.29))

$$\overline{G}^h(js\alpha_1, js\alpha_2, \varepsilon, s) = \frac{\exp(-s\Gamma|\varepsilon|)}{2s\Gamma} - \frac{\exp(-s\Gamma|\varepsilon + 2h|)}{2s\Gamma} , \tag{9.71}$$

where $\Gamma$ is defined in Eq. (9.30). Hence, we finally arrive at

$$L\hat{u} = F^{-1}\left\{\overline{g}^h\, F\{\chi_{\Lambda_{sct}} \hat{u}\}\right\} , \tag{9.72}$$

where

$$\begin{aligned}\overline{g}^h &= s\rho \lim_{\varepsilon\downarrow 0} \frac{\exp(-s\Gamma\varepsilon)}{2s\Gamma} - \frac{\rho}{2\Gamma}\exp(-2s\Gamma h) \\ &= \frac{\rho}{2\Gamma}[1-\exp(-2s\Gamma h)] \quad \text{for the case of a compliant lamina,} \qquad (9.73)\end{aligned}$$

$$\begin{aligned}\overline{g}^h &= \frac{1}{s\rho} \lim_{\varepsilon\downarrow 0} \partial_\varepsilon^2 \frac{\exp(-s\Gamma\varepsilon)}{2s\Gamma} - \frac{\Gamma}{2\rho}\exp(-2s\Gamma h) \\ &= \frac{\Gamma}{2\rho}[1-\exp(-2s\Gamma h)] \quad \text{for the case of a rigid disk.} \qquad (9.74)\end{aligned}$$

In the derivation to arrive at Eqs. (9.72) and (9.74) we have interchanged the differentiations with respect to $\varepsilon$ and the integrations of the inverse Fourier transform. This is is admissible as long as

$$\lim_{|(s\alpha_1, s\alpha_2)|\to\infty} |(s\alpha_1, s\alpha_2)|\, F\{\chi_{\Lambda_{sct}}\hat{u}\} = 0 \quad \text{as } (s\alpha_1, s\alpha_2) \in \mathbb{R}^2\,, \qquad (9.75)$$

where $|(s\alpha_1, s\alpha_2)| = [(s\alpha_1)^2 + (s\alpha_2)^2]^{\frac{1}{2}}$. This condition is satisfied if $\hat{u}(x_1, x_2) = \partial\hat{f}_3(x_1, x_2, 0, s)$ vanishes at the edge $\partial\Lambda_{sct}$ of the disk domain $\Lambda_{sct}$. This is the so-called edge condition of the wavefield problem of a rigid disk.

*Computation of the adjoint operator*

We are now able to use the various iterative schemes discussed in Chapter 2. In order to be able to use these schemes, we need the adjoint $L^*$ of the operator $L$. This operator is now obtained as

$$L^*\hat{v} = F^{-1}\Big\{\overline{g}^{h*}\, F\{\chi_{\Lambda_{sct}}\hat{v}\}\Big\}\,. \qquad (9.76)$$

*Computation of the preconditioning operators*

In Section 2.7 it is found that, for the type of convolution operators at hand, a preconditioning operator may be found as

$$P\hat{v} = F^{-1}\Big\{(\overline{g}^h)^{-1}\, F\{\chi_{\Lambda_{sct}}\hat{v}\}\Big\}\,, \qquad (9.77)$$

while its adjoint is given by

$$P^*\hat{u} = F^{-1}\Big\{(\overline{g}^{h*})^{-1}\, F\{\chi_{\Lambda_{sct}}\hat{u}\}\Big\}\,. \qquad (9.78)$$

*Computation of the scattered wavefield*

Once the operator equation (9.63) is solved, which means that $\hat{u} = \partial\hat{q}$ for the case of a compliant lamina and $\hat{u} = \partial\hat{f}_3$ for the case of a rigid disk are known, we can determine the scattered acoustic pressure from the integral representations (9.7) and (9.11) as

$$\hat{p}^{sct}(\boldsymbol{x}^R, s) = s\rho \int_{(x_1,x_2)\in\Lambda_{sct}} \hat{G}^h(x_1^R - x_1, x_2^R - x_2, x_3^R - h, s)\partial\hat{q}(x_1, x_2, s)\mathrm{dA}\,,$$

$$\text{for the case of a compliant lamina}, \tag{9.79}$$

$$\hat{p}^{sct}(\boldsymbol{x}^R, s) = -\int_{(x_1,x_2)\in\Lambda_{sct}} \partial_3^R\hat{G}^h(x_1^R - x_1, x_2^R - x_2, x_3^R - h, s)\partial\hat{f}_3(x_1, x_2, s)\mathrm{dA}\,,$$

$$\text{for the case of a rigid disk}, \tag{9.80}$$

when $\boldsymbol{x}^R \in \mathbb{D}'_{sct}$. Using the Fourier transformations defined in Eqs. (9.27) - (9.28) and the Fourier transform of the Green's function given in Eq. (9.71), we observe that the scattered acoustic pressure can be computed as

$$\hat{p}^{sct}(x_1, x_2, x_3, s)$$

$$= F^{-1}\left\{ \frac{\rho}{2\Gamma} \left\{\exp[-s\Gamma|x_3 - h|] - \exp[-s\Gamma(x_3 + h)]\right\} F\{\chi_{\Lambda_{sct}}\partial\hat{q}\} \right\}$$

$$\text{for the case of a compliant lamina}, \tag{9.81}$$

and

$$\hat{p}^{sct}(x_1, x_2, x_3, s)$$

$$= F^{-1}\left\{ \frac{1}{2} \left\{\mathrm{sign}(x_3 - h)\exp[-s\Gamma|x_3 - h|] - \exp[-s\Gamma(x_3 + h)]\right\} F\{\chi_{\Lambda_{sct}}\partial\hat{f}_3\} \right\}$$

$$\text{for the case of a rigid disk}. \tag{9.82}$$

After the discussion of the scattering by a disk in a homogeneous embedding and in a homogeneous halfspace, respectively, we shall now consider the consequences of a two-dimensional version of the pertaining scattering problem.

## 9.5. Two-dimensional scattering by a strip

In this section we consider the two-dimensional scattering problem, in which the disk domain $\Lambda_{sct}$ becomes a strip domain, presented by

$$\Lambda_{sct} = \{\boldsymbol{x} \in \mathbb{R}^3;\ -a < x_1 < a,\ -\infty < x_2 < \infty,\ x_3 = h\}\ , \tag{9.83}$$

where $a$ is half the width of the strip. Further, in this problem, we assume that the sources generate a two-dimensional acoustic wavefield independent of the $x_2$-direction. Hence, we only deal with the two-dimensional position vector $\boldsymbol{x}_T = (x_1, x_3)$.

*The two-dimensional Green's function*

In order to discuss the changes we have to make for a two-dimensional acoustic wavefield problem, we consider the two-dimensional wavefield generated by a monopole line source. The expression for this wavefield is obtained by integrating the expression for the monopole point source of Eq. (9.46) with respect to $x_2$

$$\hat{p}(x_1, x_3, s) = s\rho\hat{q}^S \int_{x_2 \in \mathbb{R}} \hat{G}(x_1 - x_1^S, x_2, x_3 - x_3^S)\mathrm{d}x_2\ , \tag{9.84}$$

where (cf. Eq. (4.60))

$$\hat{G}(x_1, x_2, x_3, s) = \frac{1}{(2\pi)^2} \int_{(s\alpha_1, s\alpha_2) \in \mathbb{R}^2} \frac{\exp(-js\alpha_1 x_1 - js\alpha_2 x_2 - s\Gamma|x_3|)}{2s\Gamma}\mathrm{dA} \tag{9.85}$$

and

$$\Gamma = \left(\frac{1}{c^2} + \alpha_1^2 + \alpha_2^2\right)^{\frac{1}{2}}\ , \quad \mathrm{Re}(\Gamma) > 0\ . \tag{9.86}$$

Interchanging the order of integrations we arrive at

$$\hat{p}(x_1, x_3, s) = s\rho\hat{q}^S \hat{\mathcal{G}}(x_1 - x_1^S, x_3 - x_3^S)\ , \tag{9.87}$$

in which the two-dimensional Green's function $\hat{\mathcal{G}}$ is given by

$$\hat{\mathcal{G}}(x_1, x_3, s) = \frac{1}{2\pi} \int_{s\alpha_1 \in \mathbb{R}} \frac{\exp(-js\alpha_1 x_1 - s\Upsilon|x_3|)}{2s\Upsilon}\mathrm{d}(s\alpha_1) \tag{9.88}$$

and

$$\Upsilon = \left(\frac{1}{c^2} + \alpha_1^2\right)^{\frac{1}{2}}, \quad \mathrm{Re}(\Upsilon) > 0\,. \tag{9.89}$$

As far as the spatial dependence is concerned, we now show that $\hat{\mathcal{G}}$ only depends on the two-dimensional distance from the origin,

$$|\boldsymbol{x}_T| = (x_1^2 + x_3^2)^{\frac{1}{2}}\,. \tag{9.90}$$

Writing

$$x_1 = |\boldsymbol{x}_T|\cos(|\phi|)\,, \quad x_3 = |\boldsymbol{x}_T|\sin(|\phi|)\,, \quad -\pi \le \phi < \pi\,, \tag{9.91}$$

and replacing the real integration variable $s\alpha_1$ in Eq. (9.88) by the complex variable $\Phi$ through

$$s\alpha_1 = -j\frac{s}{c}\cos(|\phi| - j\Phi)\,, \tag{9.92}$$

we arrive at

$$\hat{\mathcal{G}}(x_1, x_3, s) = \frac{1}{4\pi}\int_{\Phi=\Phi(s\alpha_1)|_{s\alpha_1\in\mathbb{R}}} \exp\left[-\frac{s}{c}|\boldsymbol{x}_T|\cosh(\Phi)\right]\mathrm{d}\Phi\,. \tag{9.93}$$

The next step is to perform the integration in the complex $(s\alpha_1)$-plane along such a path that $\Phi$ becomes real. From Eq. (9.92) we observe that this path is a hyperbola (see Fig. 9.4). We therefore extend the integrand of Eq. (9.93) in the complex $(s\alpha_1)$-plane. Since we want to keep the integrand single-valued, we introduce branch cuts from the branch point $s\alpha_1 = js/c$ to $\mathrm{Re}(s\alpha_1) \to -\infty$ and from the branch point $s\alpha_1 = -js/c$ to $\mathrm{Re}(s\alpha_1) \to \infty$ and keep $\mathrm{Re}(\Upsilon) \ge 0$ in the entire cut $(s\alpha_1)$-plane. On virtue of Cauchy's theorem and Jordan's lemma, the integration path along the real $(s\alpha_1)$-axis is now deformed into the hyperbola ($\Phi \in \mathbb{R}$). After this deformation, we have

$$\hat{\mathcal{G}}(x_1, x_3, s) = \frac{1}{4\pi}\int_{-\infty}^{\infty} \exp\left[-\frac{s}{c}|\boldsymbol{x}_T|\cosh(\Phi)\right]\mathrm{d}\Phi\,. \tag{9.94}$$

Introducing the modified Bessel function of the second kind and zero order, (ABRAMOWITZ and STEGUN, 1968, p. 376)

$$K_0(z) = \int_0^{\infty} \exp[-z\cosh(\Phi)]\mathrm{d}\Phi\,, \tag{9.95}$$

we may write Eq. (9.94) as

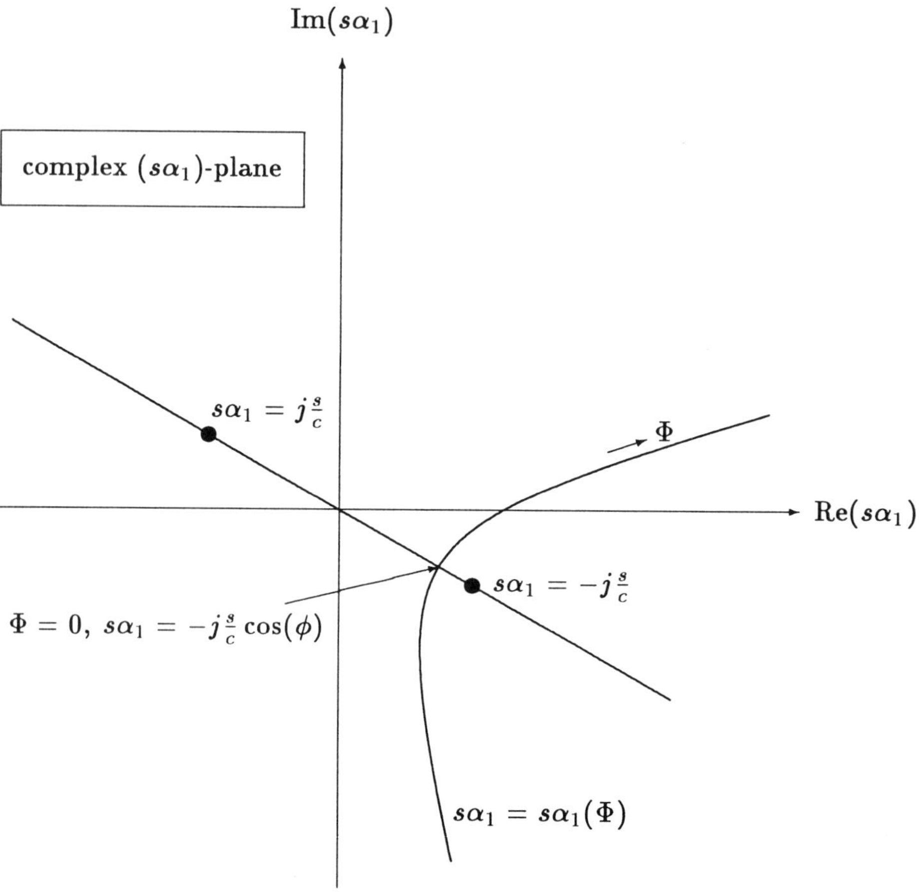


Figure 9.4. Complex ($s\alpha_1$)-plane with modified integration path $s\alpha_1(\Phi)$.

$$\hat{\mathcal{G}}(\boldsymbol{x}_1, \boldsymbol{x}_3, s) = \frac{1}{2\pi} K_0(\frac{s}{c}|\boldsymbol{x}_T|)\,. \tag{9.96}$$

Note that in the limiting case of $s \to j\omega$, we arrive at

$$\hat{\mathcal{G}}(\boldsymbol{x}_1, \boldsymbol{x}_3, s) = \frac{-j}{4} H_0^{(2)}(\frac{\omega}{c}|\boldsymbol{x}_T|)\,, \tag{9.97}$$

in which $H_0^{(2)}$ is the Hankel function of the second kind and zero order.

Before continuing our analysis, we present the time-domain result of the two-dimensional Green's function. We introduce the new integration variable $\tau$,

$$\tau = T_0 \cosh(\Phi), \quad T_0 = \frac{|\boldsymbol{x}_T|}{c}, \tag{9.98}$$

in Eq. (9.94). Then, we have

$$\hat{\mathcal{G}}(x_1, x_3, s) = \frac{1}{2\pi} \int_{T_0}^{\infty} \frac{\exp(-s\tau)}{(\tau^2 - T_0^2)^{\frac{1}{2}}} \mathrm{d}\tau . \tag{9.99}$$

The expression of the two-dimensional Green's function in the time domain is directly found by inspection (assuming $s$ is real), viz.,

$$\mathcal{G}(x_1, x_3, s) = \frac{\chi_{\mathbb{R}^+}(t - T_0)}{2\pi (t^2 - T_0^2)^{\frac{1}{2}}}, \tag{9.100}$$

where $\chi_{\mathbb{R}^+}(t)$ is the characteristic function related to the domain $\mathbb{R}^+ = \{t \in \mathbb{R}, t > 0\}$. This result can also be derived directly from Eq. (9.88), using the Cagniard-de Hoop method (DE HOOP, 1960).

Comparing the three-dimensional Green's function of Eq. (9.85) and the two-dimensional Green's function of Eq. (9.88), we conclude that we can adapt our three-dimensional analysis of the previous chapters to a two-dimensional one by simply replacing the two-dimensional Fourier transforms with respect to the horizontal coordinates by the one-dimensional Fourier transform with respect to $x_1$.

We define the one-dimensional Fourier transformation with respect to the horizontal coordinate $x_1$, again denoted by the operator $F$ and its inverse by $F^{-1}$, viz.,

$$\overline{u} = F\{\hat{u}\} = \int_{x_1 \in \mathbb{R}} \exp(js\alpha_1 x_1) \hat{u}(x_1, x_3, s) \mathrm{d}x_1 , \tag{9.101}$$

$$F^{-1}\{\overline{u}\} = \frac{1}{2\pi} \int_{s\alpha_1 \in \mathbb{R}} \exp(-js\alpha_1 x_1) \overline{u}(js\alpha_1, x_3, s) \mathrm{d}(s\alpha_1) . \tag{9.102}$$

Subsequently, we discuss the modifications we have to make for both the problem of a strip in a homogeneous embedding and the problem of a strip in a homogeneous half-space. In order to be able to use the various iterative schemes discussed in Chapter 2 for the solution of the integral equation of the scattering problem at hand, we observe that the operator expressions needed in the various schemes, directly follow from Eqs. (9.31), (9.35)-(9.37), (9.72), (9.76)-(9.78).

For the configurations of the strip in a homogeneous embedding and the strip in a homogeneous halfspace, the various operator expressions are given below.

*Strip in a homogeneous embedding*

The operator expressions are now given by

$$L\hat{u} = F^{-1}\{\overline{g}\, F\{\chi_{\Lambda_{sct}}\hat{u}\}\} \;, \tag{9.103}$$

$$L^{*}\hat{v} = F^{-1}\{\overline{g}^{\star}\, F\{\chi_{\Lambda_{sct}}\hat{v}\}\} \;, \tag{9.104}$$

$$P\hat{v} = F^{-1}\Big\{(\overline{g})^{-1}\, F\{\chi_{\Lambda_{sct}}\hat{v}\}\Big\} \;, \tag{9.105}$$

$$P^{*}\hat{u} = F^{-1}\Big\{(\overline{g}^{\star})^{-1}\, F\{\chi_{\Lambda_{sct}}\hat{u}\}\Big\} \;. \tag{9.106}$$

where

$$\overline{g} = \frac{\rho}{2\Upsilon} \quad \text{for the case of a compliant strip,} \tag{9.107}$$

$$\overline{g} = \frac{\Upsilon}{2\rho} \quad \text{for the case of a rigid strip .} \tag{9.108}$$

Once the operator equation

$$L\hat{u} = \hat{f} \tag{9.109}$$

is solved, the two-dimensional wavefield scattered by the strip can be computed from (cf. Eqs. (9.40) - (9.41))

$$\hat{p}^{sct}(x_1, x_3, s) = F^{-1}\left\{\frac{\rho}{2\Upsilon}\exp(-s\Upsilon|x_3 - h|)\, F\{\chi_{\Lambda_{sct}}\hat{u}\}\right\}$$

$$\text{for the case of a compliant strip,} \tag{9.110}$$

$$\hat{p}^{sct}(x_1, x_3, s) = F^{-1}\left\{\frac{1}{2}\,\mathrm{sign}(x_3 - h)\exp(-s\Upsilon|x_3 - h|)\, F\{\chi_{\Lambda_{sct}}\hat{u}\}\right\}$$

$$\text{for the case of a rigid strip.} \tag{9.111}$$

*Strip in a homogeneous halfspace*

The operator expressions are now given by

$$L\hat{u} = F^{-1}\Big\{\overline{g}^h \, F\{\chi_{\Lambda_{sct}}\hat{u}\}\Big\} \, , \tag{9.112}$$

$$L^*\hat{v} = F^{-1}\Big\{\overline{g}^{h^*} \, F\{\chi_{\Lambda_{sct}}\hat{v}\}\Big\} \, , \tag{9.113}$$

$$P\hat{v} = F^{-1}\Big\{(\overline{g}^h)^{-1} \, F\{\chi_{\Lambda_{sct}}\hat{v}\}\Big\} \, , \tag{9.114}$$

$$P^*\hat{u} = F^{-1}\Big\{(\overline{g}^{h^*})^{-1} \, F\{\chi_{\Lambda_{sct}}\hat{u}\}\Big\} \, . \tag{9.115}$$

where

$$\overline{g}^h = \frac{\rho}{2\Upsilon}[1 - \exp(-2s\Upsilon h)] \quad \text{for the case of a compliant strip,} \tag{9.116}$$

$$\overline{g}^h = \frac{\Upsilon}{2\rho}[1 - \exp(-2s\Upsilon h)] \quad \text{for the case of a rigid strip} \, . \tag{9.117}$$

Once the operator equation

$$L\hat{u} = \hat{f} \tag{9.118}$$

is solved, the two-dimensional wavefield scattered by the strip can be computed from (cf. Eqs. (9.81) - (9.82))

$$\hat{p}^{sct}(x_1, x_2, x_3, s) = F^{-1}\left\{\frac{\rho}{2\Upsilon}\{\exp[-s\Upsilon|x_3-h|] - \exp[-s\Upsilon(x_3+h)]\} \, F\{\chi_{\Lambda_{sct}}\partial\hat{q}\}\right\} \tag{9.119}$$

for the case of a compliant strip ,

$$\hat{p}^{sct}(x_1, x_2, x_3, s) = F^{-1}\left\{\frac{1}{2}\{\mathrm{sign}(x_3-h)\exp[-s\Upsilon|x_3-h|] - \exp[-s\Upsilon(x_3+h)]\} \, F\{\chi_{\Lambda_{sct}}\partial\hat{f}_3\}\right\}$$

for the case of a rigid strip . (9.120)

All these various operator expressions are used in the numerical computations discussed in the remainder of this chapter.

### 9.5.1. Numerical performance of the iterative schemes

We investigate the performance of the various iterative schemes in the frequency domain by taking $s \rightarrow j\omega$. In our numerical computations we take

$$s = j\omega(1 - 0.01j), \quad \omega = kc\,, \tag{9.121}$$

where $k$ is the angular wavenumber. Then, we may write for the vertical slowness

$$\Upsilon = \left(\frac{1}{c^2} - \frac{(s\alpha_1)^2}{\omega^2(1-0.01j)^2}\right)^{\frac{1}{2}}, \quad \mathrm{Re}(\Upsilon) > 0\,. \tag{9.122}$$

Here, $s\alpha_1$ is the (real) Fourier-transform parameter of the Fourier transformation defined by Eqs. (9.101) - (9.102).

For simplicity we take the incident acoustic wavefield to be a plane wave with unit amplitude and zero phase at the strip (see Fig. 9.5), viz.,

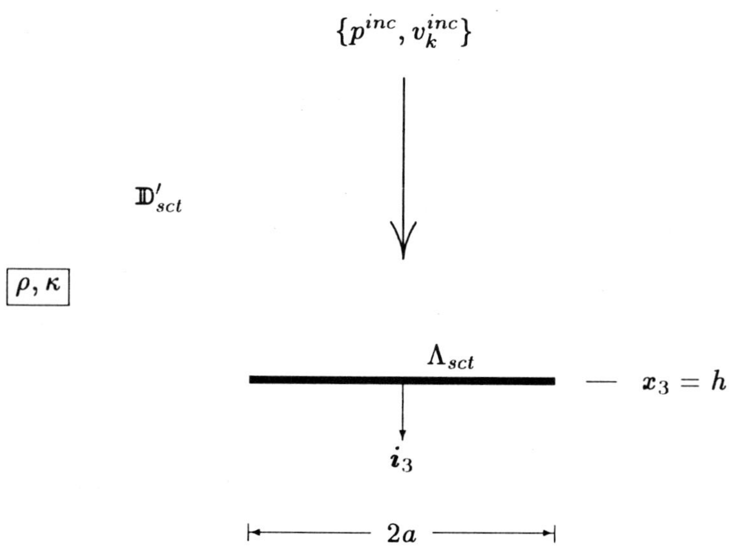


Figure 9.5. Plane-wave scattering by a strip.

$$\hat{p}^{inc}(x_1, x_3, s) = \exp[-\frac{s}{c}(x_3 - h)] , \tag{9.123}$$

while we only consider the case of the strip to be perfectly compliant. In this case $\hat{f} = -\hat{p}^{inc}$.

All operator expressions are computed by a 4096-points FFT routine. All integrals occurring in the inner products of the different iterative schemes are computed numerically with the aid of a simple summation of the function values at the sample points. The number of sample points on the strip (of width $2a$) amounts to 17, 35, 81 and 181, when $ka$ =0.2, 1, 5 and 25, respectively. The latter numbers are determined experimentally and are chosen such that numerical discretization errors are less than the error made in the resulting approximation of our pertaining wavefield values at the strip. As soon as the number of iterations grows larger, the danger of loss of significant figures turns up. For this reason, all computations have been carried out in double precision, while the residual in the operator equation has each time been determined by substituting the obtained approximate solution in this equation and not by using the recursive relation for the successive residuals that for a number of cases is available. In those cases where a loss of significant figures was expected, a check has been carried out against the corresponding computation in single precision. In the conjugate-gradient scheme an additional check is provided by the orthogonality relations that have to be satisfied. Once a discrepancy in these occurs, the orthogonality relations are enforced by falling back on the recursive scheme defined by Eqs. (2.42) - (2.44).

*Recursive solution with $T = I$*

We first consider the iterative solution of the recursive scheme summarized in Eqs. (2.42) - (2.44), in which we take $T = I$. In Fig. 9.6 we present the numerical results for the root-mean-square error $\widehat{\text{ERR}}$ (cf. Eq. (2.10)) as a function of the number of iterations.

*Preconditioned recursive solution with $T = P$*

Subsequently, we consider the recursive scheme of Eqs. (2.42) - (2.44), in which we take $T = P$. In Fig. 9.7 we present the numerical results for the root-mean-square error $\widehat{\text{ERR}}$ as a function of the number of iterations.

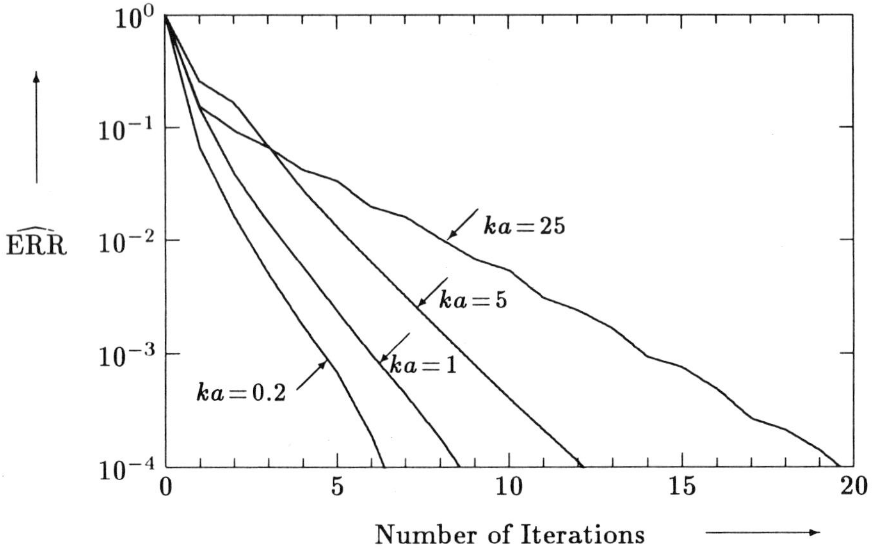


Figure 9.6. Results of the recursive scheme with $T = I$.

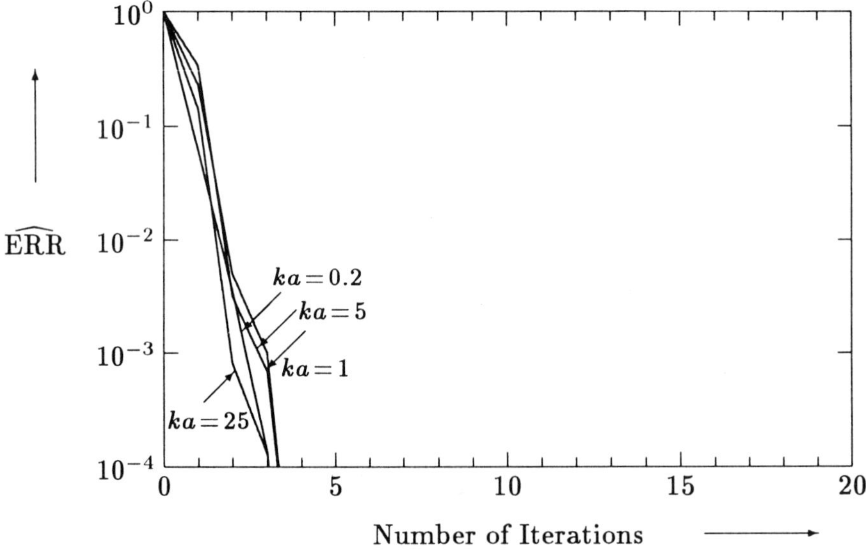


Figure 9.7. Results of the preconditioned recursive scheme with $T = P$.

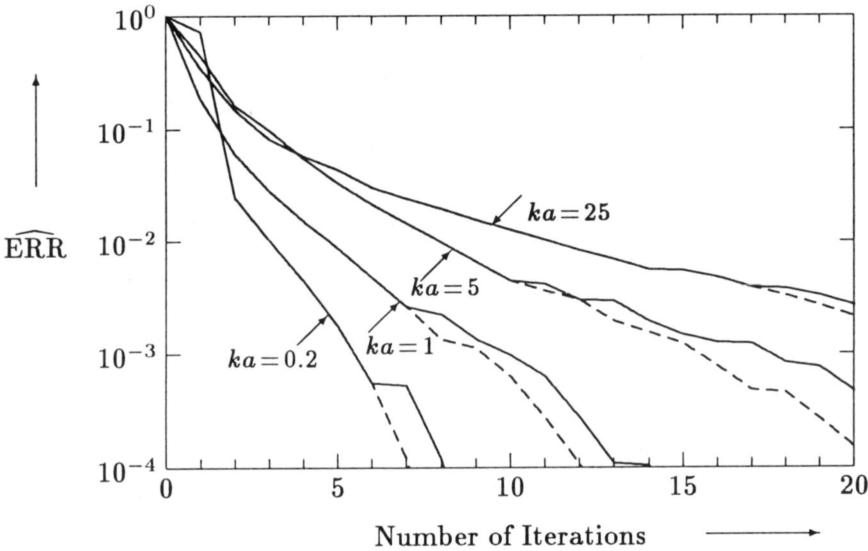


Figure 9.8. Results of the conjugate-gradient scheme ($T = L^\star$).

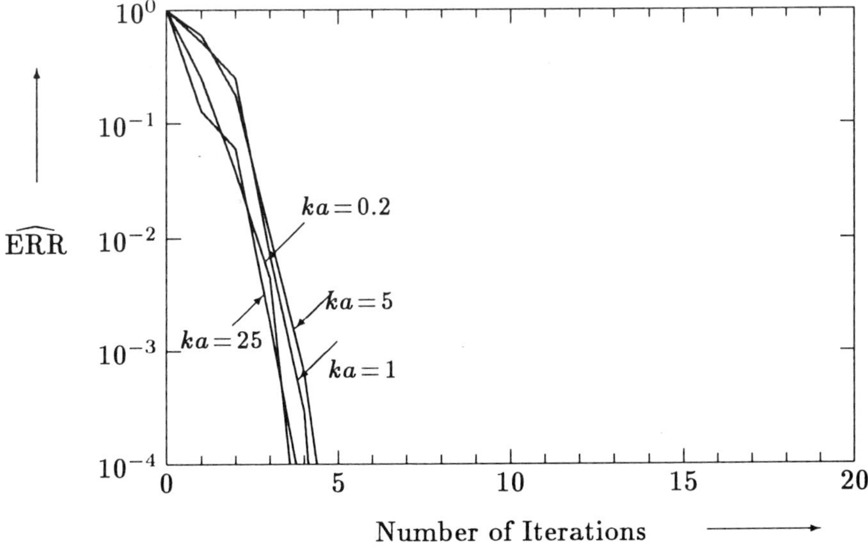


Figure 9.9. Results of the preconditioned conjugate-gradient scheme ($T = PP^\star L^\star$).

Comparing the results with those of the non-preconditioned recursive scheme shown in Fig. 9.6, we then notice that it is superior to the non-preconditioned scheme, even when we take into account that the computation time of one iteration is now nearly doubled. For all values of $ka$ considered the preconditioned scheme converges within a very few iterations ($\widehat{\mathrm{ERR}} < 0.0001$). Obviously, we have constructed a very efficient preconditioner for the present strip problem.

*Conjugate-gradient method:* $T = L^{\star}$

We now consider the recursive scheme of Eqs. (2.42) - (2.44), in which we take $T = L^{\star}$. Since $LT = LL^{\star}$ is a selfadjoint operator we can use a conjugate-gradient scheme (cf. Sections 2.4 and 2.6); in particular, we can employ the scheme of Eqs. (2.60) - (2.62). In Fig. 9.8 we present the numerical results for the root-mean-square error $\widehat{\mathrm{ERR}}$ as a function of the number of iterations. Comparing the results with those of the recursive scheme of Fig. 9.6, we observe that the rate of convergence has been decreased in taking $T = L^{\star}$ in stead of $T = I$. The advantage of the conjugate-gradient scheme is that the orthogonalization of the expansion functions is automatically enforced and storage of these expansion functions of all previous iterations is superfluous. However, after a number of iterations in the conjugate-gradient scheme, loss of significant figures leads to a non-satisfaction of the orthogonality conditions. Then, the convergence slows down for a few iterations. If, however, we enforce the orthogonality by falling back on the recursive scheme of Eqs. (2.42) - (2.44), the convergence is maintained. In Fig. 9.8 we observe this phenomenon. The dashed lines represent the results when the orthogonalization is enforced by using the recursive scheme with full orthogonalization.

*Preconditioned conjugate-gradient method:* $T = PP^{\star}L^{\star}$

Subsequently, we consider the recursive scheme of Eqs. (2.42) - (2.44), in which we take $T = PP^{\star}L^{\star}$. Since $LT = LPP^{\star}L^{\star}$ is a selfadjoint operator we can use a conjugate-gradient scheme (cf. Sections 2.4 and 2.6); in particular, we employ the scheme of Eqs. (2.60) - (2.62). In Fig. 9.9 we present the numerical results for the root-mean-square error $\widehat{\mathrm{ERR}}$ as a function of the number of iterations. Comparing the results with those of the non-preconditioned conjugate-gradient scheme of Fig. 9.8, we observe that the convergence has been considerably increased, although the results of the preconditioned non-symmetrized recursive scheme of Fig. 9.7 exhibit a better convergence. However, in the latter scheme the computer storage and computer time required for each iteration increase with an increasing number of iterations. Therefore, we prefer the preconditioned conjugate-gradient method for the construction of the synthetic data of the wavefield scattered by a strip.

### 9.5.2. Synthetic data of scattering by a strip

To study the scattering properties of the strip we consider the situation where the incident wavefield originates from a monopole line source, while the wavefield is measured by a line receiver. This implies that in our numerical analysis we restrict ourselves to the two-dimensional scattering problem, since both the excitation and the configuration are independent of $x_2$.

In the case of a homogeneous medium, the incident wavefield generated by a monopole line source at $(x_1^S, x_3^S)$ is given by (cf. Eqs. (9.87) and (9.88))

$$\hat{p}^{inc}(x_1, x_3, s) = \hat{W}(s)F^{-1}\left\{\frac{\exp(js\alpha_1 x_1^S - s\Upsilon|x_3 - x_3^S|)}{2s\Upsilon}\right\}, \tag{9.124}$$

where the the inverse Fourier transformation $F^{-1}$ is defined by Eq. (9.102). $\hat{W}$ is known as the Laplace transform of the source wavelet, which is related to the $s$-domain time rate of volume injection $\hat{q}^S$ through

$$\hat{W}(s) = s\rho\hat{q}^S(s). \tag{9.125}$$

In the case of a homogeneous halfspace, the incident wavefield generated by a monopole line source at $(x_1^S, x_3^S)$ is given by

$$\hat{p}^{inc,H}(x_1, x_3, s)$$
$$= \hat{W}(s)F^{-1}\left\{\frac{\exp(js\alpha_1 x_1^S)}{2s\Upsilon}\left\{\exp[-s\Upsilon|x_3 - x_3^S|] - \exp[-s\Upsilon(x_3 + x_3^S)]\right\}\right\}. \tag{9.126}$$

*The numerical procedure*

In our numerical computations we take

$$s = 4 + j\omega \,. \tag{9.127}$$

Then, we may write for the vertical slowness

$$\Upsilon = \left( \frac{1}{c^2} + \frac{(s\alpha_1)^2}{(4 + j\omega)^2} \right)^{\frac{1}{2}} \,, \quad \mathrm{Re}(\Upsilon) > 0 \,. \tag{9.128}$$

Here, $s\alpha_1$ is the (real) Fourier-transform parameter of the Fourier transformation defined by Eqs. (9.101) - (9.102). Note that we have taken a real part that is independent of $\omega$. This means that the temporal forward and inverse Fourier transformations can be employed. Apart from a factor $\exp(4t)$, these transforms are standard Fourier transforms and are computed with FFT routines.

The shape of the input source wavelet $W(t)$ in the time domain with a sample rate of 2 ms is shown in Fig. 9.10.

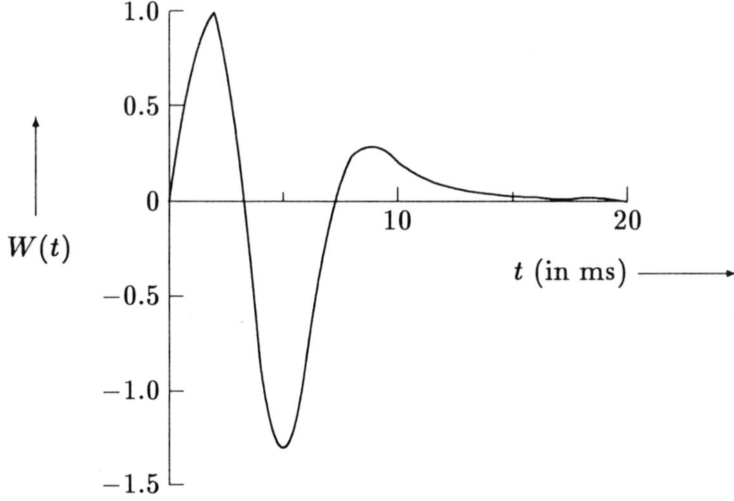


Figure 9.10. Input source wavelet $W(t)$ in $10^6$ Pa m.

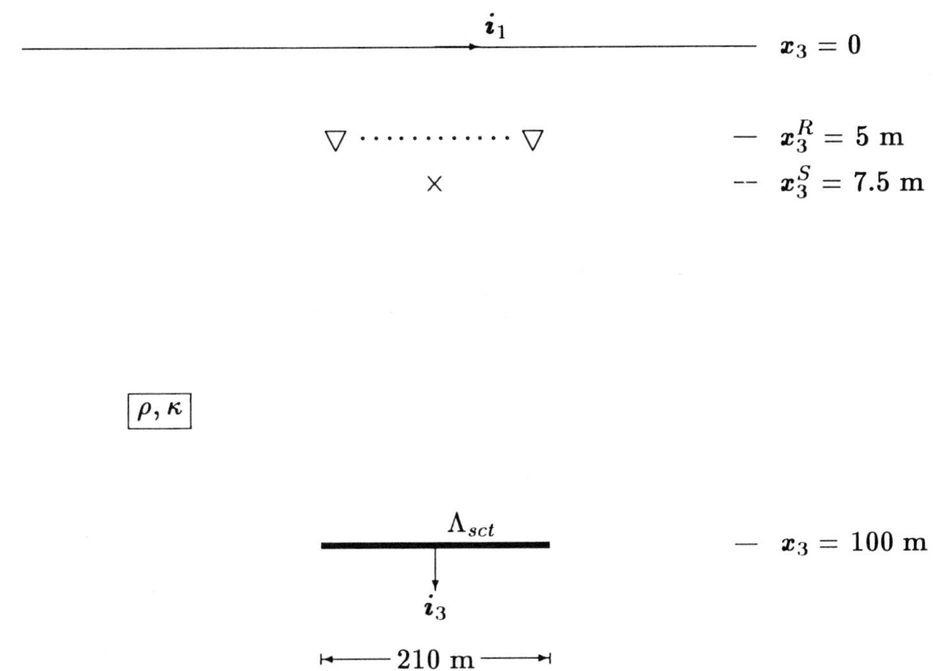


Figure 9.11. The two-dimensional scattering configuration.

As a scattering configuration we consider the situation shown in Fig. 9.11. The receiver level is 5 m below the reference level $x_3 = 0$, and the source level is 7.5 m below $x_3 = 0$. The center of the strip is located at $x_1 = 0$ and $x_3 = 100$ m, while the width of the strip amounts to 210 m. The mass density of the fluid embedding is 1024 kg/m$^3$, the compressibility is $4.6 \times 10^{-10}$ Pa$^{-1}$, while the acoustic wave speed amounts to 1457 m/s.

To arrive at a numerical solution of our scattering problem we employ the preconditioned conjugate-gradient method with $T = PP^\star L^\star$. In particular, we employ the scheme of Eqs. (2.60) - (2.62). All operator expressions are computed by a 1024-points FFT routine with a spatial sample interval of 3.5 m. The 1024 sample points are chosen such that the outer-left position is at $x_1 = -1788.5$ m, while the outer-right position is at $x_1 = 1788.5$ m.

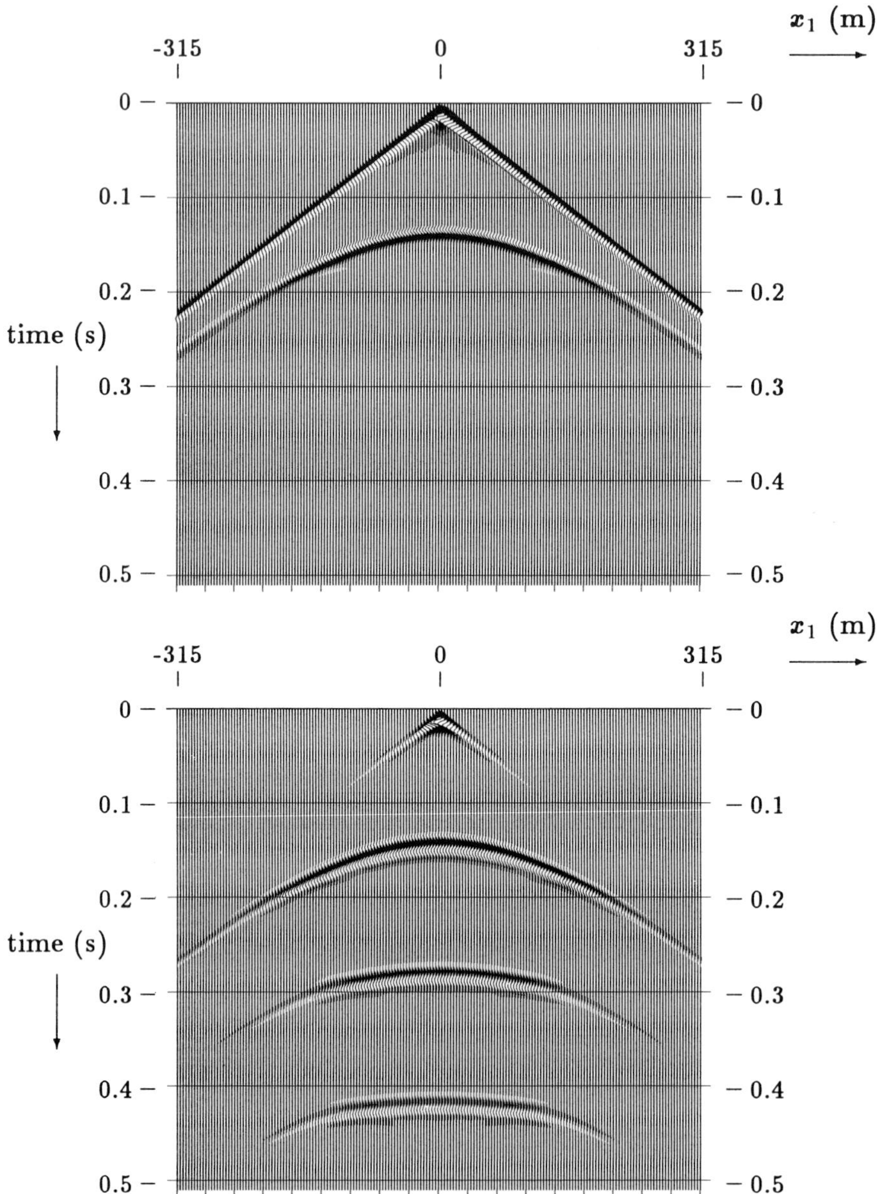


Figure 9.12. The total wavefield for a compliant strip in a homogeneous embedding (top) and in a homogeneous halfspace (bottom), respectively. The source position is above the center of the strip ($x_1^S = 0$).

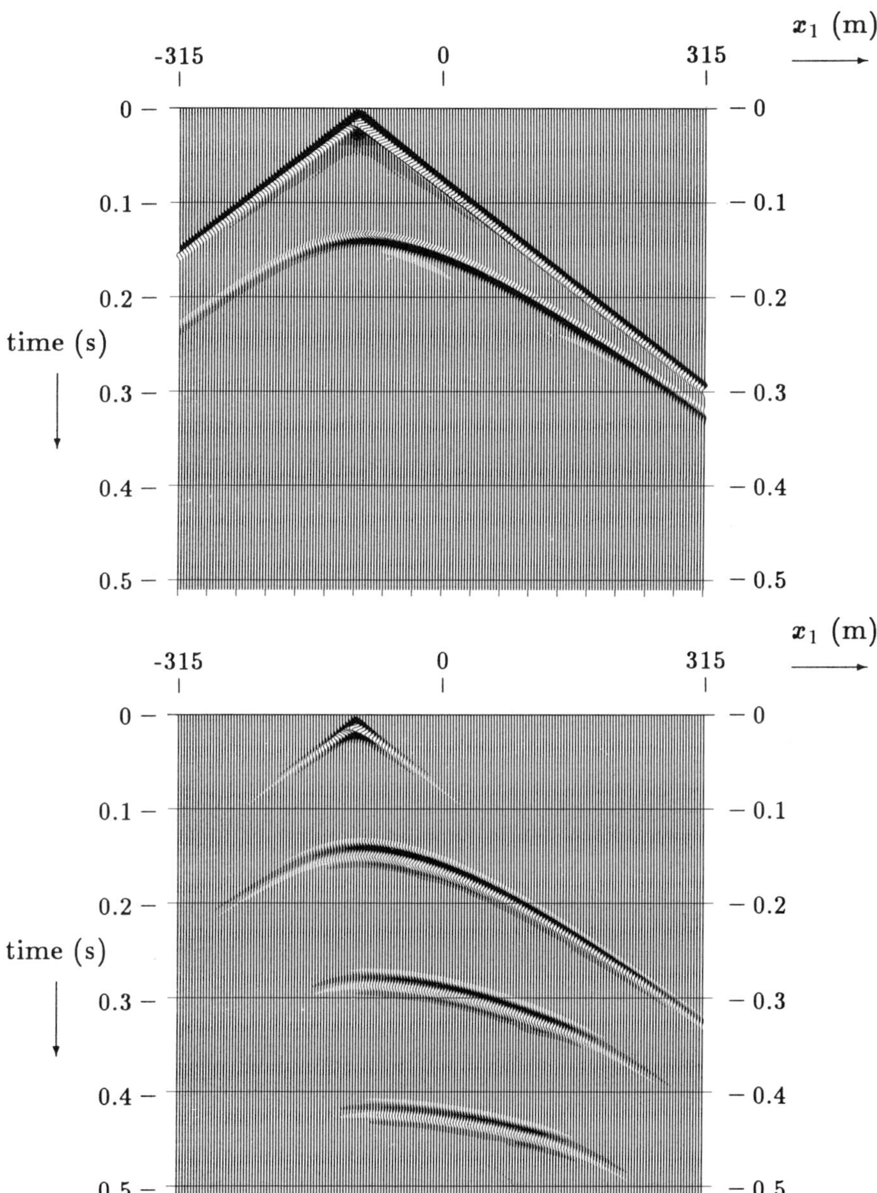


Figure 9.13. The total wavefield for a compliant strip in a homogeneous embedding (top) and in a homogeneous halfspace (bottom), respectively. The source position is above the left edge of the strip ($x_1^S = -105$ m).

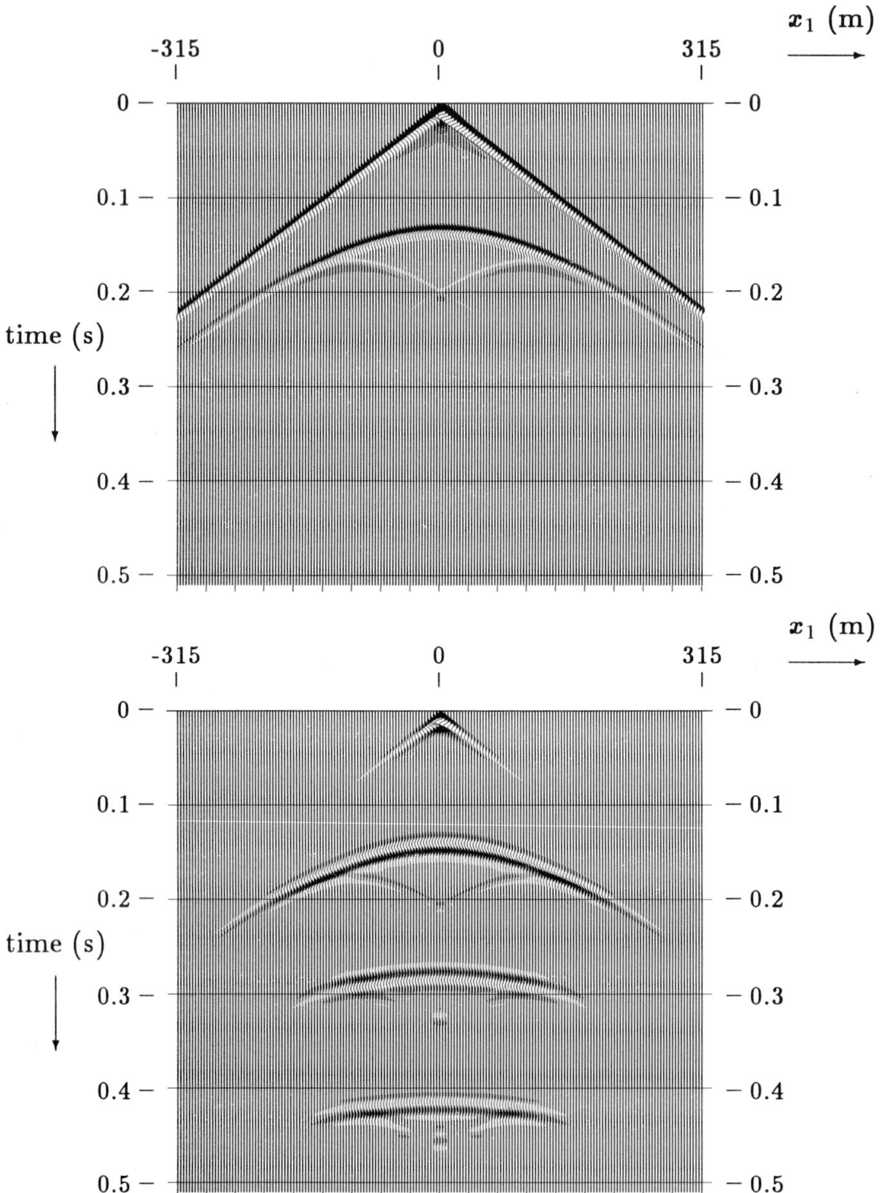


Figure 9.14. The total wavefield for a rigid strip in a homogeneous embedding (top) and in a homogeneous halfspace (bottom), respectively. The source position is above the center of the strip ($x_1^S = 0$).

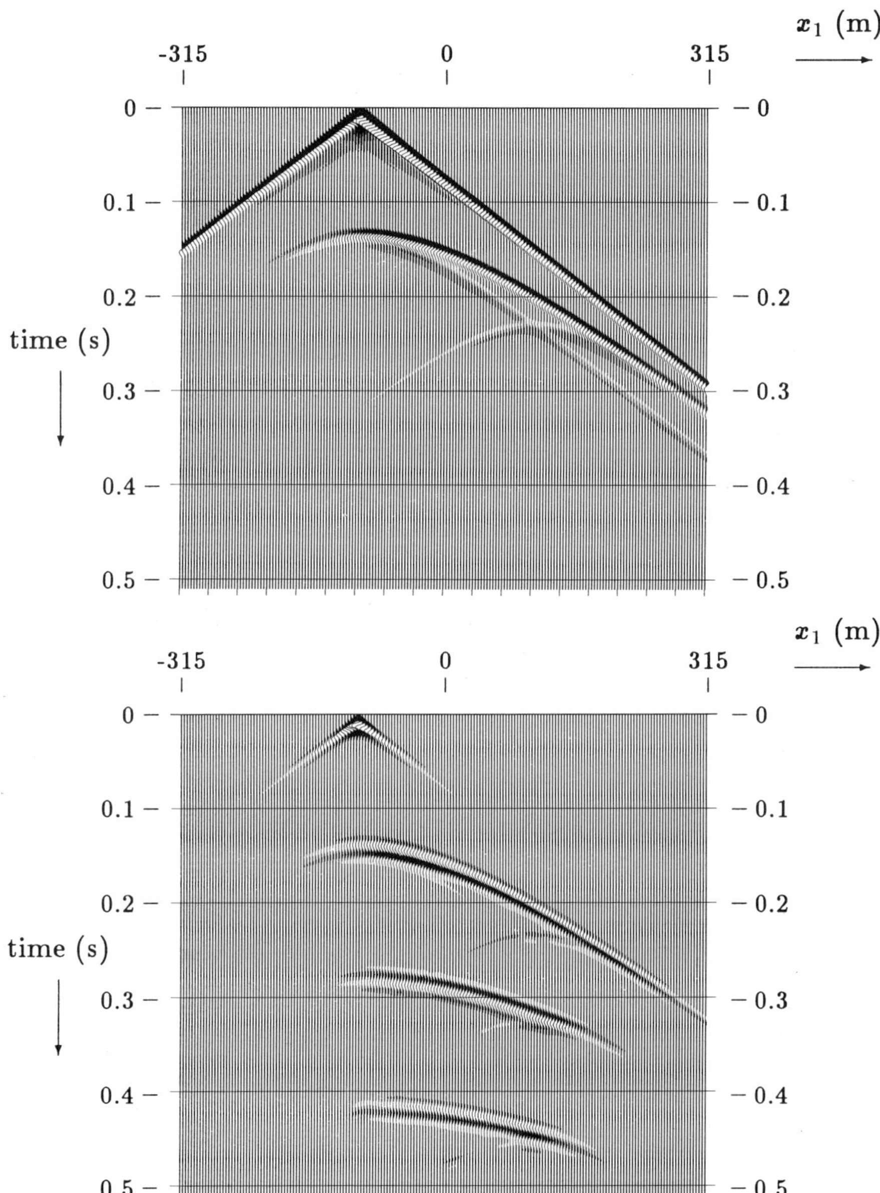


Figure 9.15. The total wavefield for a rigid strip in a homogeneous embedding (top) and in a homogeneous halfspace (bottom), respectively. The source position is above the left edge of the strip ($x_1^S = -105$ m).

All integrals occurring in the inner products of the iterative scheme are computed numerically with the aid of a simple summation of the function values at the sample points. It is noted that the number of spatial sample points at the strip is kept at 61, independent of the angular frequency.

The temporal Fourier transformation needed for the transform to the frequency domain and the inverse transformation to the time domain is computed by a 512-points FFT routine with a temporal sample interval of 2 ms. Hence, the interval 0 - 0.51 s is the time interval in which we model and present the causal wave motion.

To simulate a two-dimensional seismic experiment we have computed the synthetic seismograms for 255 different source positions, starting at $x_1^S = -444.5$ m and ending at $x_1^S = 444.5$ m with an increment of 3.5 m. We have used 255 receiver positions per source, arranged in a symmetrical split-spread fashion, starting at $x_1^R = x_1^S - 444.5$ m and ending at $x_1^R = x_1^S + 444.5$ m, where $x_1^S$ is the horizontal source position. The Fourier transforms with respect to the horizontal source and receiver coordinates are computed by a 512-points FFT routine.

*The compliant strip*

The operator expressions of Eqs. (9.103) - (9.106) are used for the strip in the homogeneous embedding and the operator expressions of Eqs. (9.112) - (9.115) are used for the strip in the homogeneous halfspace.

After the surface sources on the compliant strip are determined, the scattered wavefield in the configuration is computed using Eq. (9.110) in the case of the homogeneous embedding, while Eq. (9.119) is used for the homogeneous halfspace. The incident wavefield from the monopole source follows directly from Eqs. (9.124) and (9.126) for the homogeneous embedding and the homogeneous halfspace, respectively. Then, linear superposition of the incident wavefield and the scattered wavefield yields the result for the total wavefield.

Fig. 9.12 shows the total wavefield for the lateral source position at $x_1^S = 0$ m. Fig. 9.13 shows the total wavefield for the lateral source position at $x_1^S = -105$ m. It is obvious that the main contribution to the wavefield scattered by the compliant strip is caused by its center. The total pressure wavefield vanishes at the edges of the compliant strip; therefore, there are no edge-diffraction curves visible.

*The rigid strip*

The operator expressions of Eqs. (9.103) - (9.106) are used for the strip in the homogeneous embedding and the operator expressions of Eqs. (9.112) - (9.115) are used for the strip in the homogeneous halfspace.

After the surface sources on the rigid strip are determined, the scattered wavefield in the configuration is computed using Eq. (9.111) in the case of the homogeneous embedding, while Eq. (9.120) is used for the homogeneous halfspace. The incident wavefield from the monopole source follows directly from Eqs. (9.124) and (9.126) for the homogeneous embedding and the homogeneous halfspace, respectively. Then, linear superposition of the incident wavefield and the scattered wavefield yields the result for the total wavefield.

Fig. 9.14 shows the total wavefield for the lateral source position at $x_1^S =$ 0 m. Fig. 9.15 shows the total wavefield for the lateral source position at $x_1^S = -105$ m. Since the total pressure exhibits a singular behavior at the edges of rigid strip, the edge-diffraction curves corresponding to edge diffraction are clearly visible.

This concludes the analysis of the scattering by a disk. The numerical results obtained in this chapter are collected in a dataset and will be used as input for some seismic-processing tools developed in the remainder of this book.

# Chapter 10

# Wavefield Decomposition

In this chapter we show that in a horizontal plane in a homogeneous subdomain the acoustic wavefield may be written as a superposition of downgoing and upgoing wave constituents. In the analysis the $s$-domain field reciprocity of Section 5.1 and the $s$-domain power reciprocity of Section 5.3 play a fundamental role.

We consider two interfaces $\partial\mathbb{D}_0$ and $\partial\mathbb{D}_1$. We assume that the medium in the domain $\mathbb{D}$ between these interfaces is homogeneous with constitutive parameters $\rho$ and $\kappa$. We further assume that the interface $\partial\mathbb{D}_0$ and $\partial\mathbb{D}_1$ do not overlap, i.e., $x_{3,min}^{(1)} > x_{3,max}^{(0)}$, where $x_{3,max}^{(0)}$ denotes the maximum value of $x_3$ on the interface $\partial\mathbb{D}_0$, while $x_{3,min}^{(1)}$ denotes the minimum value of $x_3$ on the interface $\partial\mathbb{D}_1$. This means that there always exists a horizontal plane at $x_3^R$ such that $x_{3,max}^{(0)} < x_3^R < x_{3,min}^{(1)}$ (see Fig. 10.1).

## 10.1. Decomposition based on field reciprocity

We apply the reciprocity theorem of Section 5.1 to the domain $\mathbb{D}$ inside the interfaces $\partial\mathbb{D}_0$ and $\partial\mathbb{D}_1$ (see Fig. 10.1). The normal $\nu_k$ to the interfaces is directed towards the domain $\mathbb{D}$. State $A$ is taken to be the actual wavefield that is generated by sources confined to a bounded domain in $\mathbb{D}'$. The

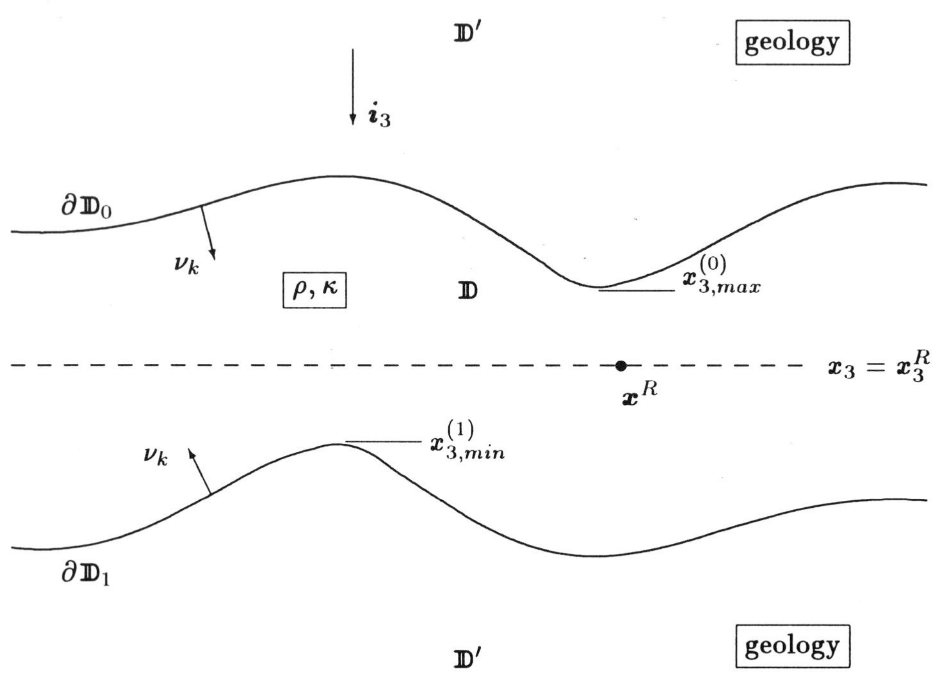


Figure 10.1. A homogeneous subdomain $\mathbb{D}$ bounded by the interfaces $\partial\mathbb{D}_0$ and $\partial\mathbb{D}_1$.

wavefield of State $B$ is generated by a point source of volume injection (cf. Table 10.1). Using the Green's states of Eq. (7.62) we arrive at

$$\hat{p}(\boldsymbol{x}^R, s)$$

$$= \int_{\boldsymbol{x}\in(\partial\mathbb{D}_0\cup\partial\mathbb{D}_1)} [\hat{G}^q(\boldsymbol{x}^R|\boldsymbol{x}, s)\hat{v}_k(\boldsymbol{x}, s) + \hat{\Gamma}_k^q(\boldsymbol{x}^R|\boldsymbol{x}, s)\hat{p}(\boldsymbol{x}, s)]\nu_k \mathrm{dA}$$

$$\text{when } \boldsymbol{x}^R \in \mathbb{D}\,. \tag{10.1}$$

From this representation we observe that the wavefield at $\boldsymbol{x}^R$ consists of contributions of surfaces sources located at $\partial\mathbb{D}_0$ and $\partial\mathbb{D}_1$.

Table 10.1. States in the field reciprocity theorem

| | State $A$ (actual state) | State $B$ (volume-injection Green's state) |
|---|---|---|
| Field state | $\{\hat{p}, \hat{v}_k\}(\boldsymbol{x}, s)$ | $\{\hat{p}^q, \hat{v}_k^q\}(\boldsymbol{x}\vert\boldsymbol{x}^R, s)$ |
| Material state | $\{\rho, \kappa\}$ | $\{\rho, \kappa\}$ |
| Source state | $\{0, 0\}$ | $\{\hat{q}^B(s)\delta(\boldsymbol{x}-\boldsymbol{x}^R), 0\}$ |
| Domain $\mathbb{D}$ (see Fig. 10.1) | | |

The Green's states (for a homogeneous background) are given by

$$\hat{G}^q(\boldsymbol{x}^R|\boldsymbol{x}, s) = s\rho\hat{G}(\boldsymbol{x}^R - \boldsymbol{x}, s) \tag{10.2}$$

and

$$\hat{\Gamma}_k^q(\boldsymbol{x}^R|\boldsymbol{x}, s) = -\partial_k^R\hat{G}(\boldsymbol{x}^R - \boldsymbol{x}, s)\,, \tag{10.3}$$

where

$$\hat{G}(\boldsymbol{x}, s) = \frac{\exp(-\frac{s}{c}|\boldsymbol{x}|)}{4\pi|\boldsymbol{x}|}, \quad \text{with } c = (\kappa\rho)^{-\frac{1}{2}}\,. \tag{10.4}$$

In the derivation of Eq. (10.1) we have taken into account that contributions of the bounding surfaces at $(x_1^2 + x_2^2) \to \infty$ vanish, since the integrand of Eq. (10.1) is of Order $((x_1^2 + x_2^2)^{-1})$ as $(x_1^2 + x_2^2) \to \infty$. The latter asymptotic behavior follows directly from the Green's function representation of Eq. (10.4) and the far-field approximations of Eq. (4.36).

It is most convenient to carry the decomposition of the wavefield in the domain of the Fourier transform with respect to the horizontal coordinates. We therefore use the Fourier representation of the Green's function, given by Eq. (9.29),

$$\overline{G}(js\alpha_1, js\alpha_2, x_3, s) = \frac{\exp(-s\Gamma|x_3|)}{2s\Gamma} , \tag{10.5}$$

where

$$\Gamma = \left(\frac{1}{c^2} + \alpha_1^2 + \alpha_2^2\right)^{\frac{1}{2}} , \quad \mathrm{Re}(\Gamma) > 0 . \tag{10.6}$$

Transforming Eq. (10.1) to the Fourier domain, using the representations of Eqs. (10.2), (10.3) and (10.5), noting that $|x_3^R - x_3| = x_3^R - x_3$ when $\boldsymbol{x} \in \partial\mathbb{D}_0$ and $|x_3^R - x_3| = x_3 - x_3^R$ when $\boldsymbol{x} \in \partial\mathbb{D}_1$, and interchanging the order of integrations, we arrive at the decomposition into the downgoing and upgoing wavefields

$$\hat{p}(x_1, x_2, x_3^R, s) = \hat{p}^{down}(x_1, x_2, x_3^R, s) + \hat{p}^{up}(x_1, x_2, x_3^R, s) , \tag{10.7}$$

where the spectral counterparts (see Eq. (9.27) for the definition of the spatial Fourier transform with respect to the horizontal coordinates) are given by

$$\overline{p}^{down}(js\alpha_1, js\alpha_2, x_3^R, s) = \overline{P}^{down}(js\alpha_1, js\alpha_2, s)\exp(-s\Gamma x_3^R) \tag{10.8}$$

and

$$\overline{p}^{up}(js\alpha_1, js\alpha_2, x_3^R, s) = \overline{P}^{up}(js\alpha_1, js\alpha_2, s)\exp(s\Gamma x_3^R) . \tag{10.9}$$

The amplitude $\overline{P}^{down}$ of the downgoing wavefield of Eq. (10.8) consists of contributions of surface sources at $\partial\mathbb{D}_0$, while the amplitude $\overline{P}^{up}$ of the upgoing wavefield of Eq. (10.9) consists of contributions of surface sources at $\partial\mathbb{D}_1$. These amplitudes are expressed as

$$\overline{P}^{down}(js\alpha_1, js\alpha_2, s)$$

$$= \frac{1}{2s\Gamma}\int_{\boldsymbol{x}\in\partial\mathbb{D}_0} [\hat{v}_k(\boldsymbol{x}, s)\, s\rho \exp(js\alpha_1 x_1 + js\alpha_2 x_2 + s\Gamma x_3)$$

$$+ \hat{p}(\boldsymbol{x}, s)\, \partial_k \exp(js\alpha_1 x_1 + js\alpha_2 x_2 + s\Gamma x_3)]\nu_k \mathrm{dA} , \tag{10.10}$$

$$\overline{P}^{up}(js\alpha_1, js\alpha_2, s)$$

$$= \frac{1}{2s\Gamma}\int_{\boldsymbol{x}\in\partial\mathbb{D}_1} [\hat{v}_k(\boldsymbol{x}, s)\, s\rho \exp(js\alpha_1 x_1 + js\alpha_2 x_2 - s\Gamma x_3)$$

$$+ \hat{p}(\boldsymbol{x}, s)\, \partial_k \exp(js\alpha_1 x_1 + js\alpha_2 x_2 - s\Gamma x_3)]\nu_k \mathrm{dA} . \tag{10.11}$$

*Condition for downgoing waves*

The particle velocity associated with the downgoing wavefield follows directly from Eqs. (10.8) and (7.70). The vertical component of the particle velocity and the acoustic pressure of the downgoing wavefield are related to each other as

$$\rho\overline{v}_3^{down}(js\alpha_1, js\alpha_2, x_3^R, s) - \Gamma\overline{p}^{down}(js\alpha_1, js\alpha_2, x_3^R, s) = 0\,. \tag{10.12}$$

*Condition for upgoing waves*

The particle velocity associated with the upgoing wavefield follows directly from Eqs. (10.9) and (7.70). The vertical component of the particle velocity and the acoustic pressure of the upgoing wavefield are related to each other as

$$\rho\overline{v}_3^{up}(js\alpha_1, js\alpha_2, x_3^R, s) + \Gamma\overline{p}^{up}(js\alpha_1, js\alpha_2, x_3^R, s) = 0\,. \tag{10.13}$$

Equations (10.7) - (10.13) show the decomposition in downgoing and upgoing wavefields in a homogeneous subdomain of infinite extent in the horizontal directions. The downgoing wavefield $\hat{p}^{down}(\boldsymbol{x}, s)$ is obtained from the integral representation of Eq. (10.1) with surface contributions from $\partial\mathbb{D}_0$ only. This downgoing wavefield is used as a forward extrapolator in the redatuming of seismic data (see Section 10.3.1).

When $\partial\mathbb{D}_0$ and $\partial\mathbb{D}_1$ are plane interfaces, Eqs. (10.10) and (10.11) can be recognized as spatial Fourier transforms. The results are given below.

*Plane interface $\partial\mathbb{D}_0$*

In the case that $\partial\mathbb{D}_0$ is a plane interface at $x_3 = x_3^{(0)}$, the expression for the amplitude $\overline{P}^{down}$ reduces to

$$\overline{P}^{down}(js\alpha_1, js\alpha_2, s)$$
$$= \frac{\exp(s\Gamma x_3^{(0)})}{2\Gamma}\left[\rho\overline{v}_3(js\alpha_1, js\alpha_2, x_3^{(0)}, s) + \Gamma\overline{p}(js\alpha_1, js\alpha_2, x_3^{(0)}, s)\right]. \tag{10.14}$$

Combining Eqs. (10.8) and (10.14), the Fourier transform of the downgoing wavefield becomes

$$\bar{p}^{down}(js\alpha_1, js\alpha_2, x_3^R, s)$$

$$= \frac{\exp[-s\Gamma(x_3^R - x_3^{(0)})]}{2\Gamma}\left[\rho\bar{v}_3(js\alpha_1, js\alpha_2, x_3^{(0)}, s) + \Gamma\bar{p}(js\alpha_1, js\alpha_2, x_3^{(0)}, s)\right]. \tag{10.15}$$

From Eq. (10.13) it is obvious that the upgoing waves do not contribute in the right-hand side of Eq. (10.15).

*Plane interface* $\partial\mathbb{D}_1$

In the case that $\partial\mathbb{D}_1$ is a plane interface at $x_3 = x_3^{(1)}$, the expression for the amplitude $\overline{P}^{up}$ reduces to

$$\overline{P}^{up}(js\alpha_1, js\alpha_2, s)$$

$$= \frac{\exp(-s\Gamma x_3^{(1)})}{-2\Gamma}\left[\rho\bar{v}_3(js\alpha_1, js\alpha_2, x_3^{(1)}, s) - \Gamma\bar{p}(js\alpha_1, js\alpha_2, x_3^{(1)}, s)\right]. \tag{10.16}$$

Combining Eqs. (10.9) and (10.16), the Fourier transform of the upgoing wavefield becomes

$$\bar{p}^{up}(js\alpha_1, js\alpha_2, x_3^R, s)$$

$$= \frac{\exp[s\Gamma(x_3^R - x_3^{(1)})]}{-2\Gamma}\left[\rho\bar{v}_3(js\alpha_1, js\alpha_2, x_3^{(1)}, s) - \Gamma\bar{p}(js\alpha_1, js\alpha_2, x_3^{(1)}, s)\right]. \tag{10.17}$$

From Eq. (10.12) it is obvious that the downgoing waves do not contribute in the right-hand side of Eq. (10.17).

## 10.2. Decomposition based on power reciprocity

We apply the reciprocity theorem of Section 5.3 to the domain $\mathbb{D}$ inside the interfaces $\partial\mathbb{D}_0$ and $\partial\mathbb{D}_1$ (see Fig. 10.1). The normal $\nu_k$ to the interfaces is directed towards the domain $\mathbb{D}$. State $A$ is taken to be the actual wavefield that is generated by sources confined to a bounded domain in $\mathbb{D}'$. State $B$ is the anti-causal wavefield generated by a point source of volume injection (cf. Table 10.2). Using the Green's states of Eq. (7.62), in which we replace $s$ by $-s$, we arrive at

$$\hat{p}(\boldsymbol{x}^R, s)$$

$$= \int_{\boldsymbol{x}\in(\partial\mathbb{D}_0\cup\partial\mathbb{D}_1)} [-\hat{G}^q(\boldsymbol{x}^R|\boldsymbol{x}, -s)\hat{v}_k(\boldsymbol{x}, s) + \hat{\Gamma}_k^q(\boldsymbol{x}^R|\boldsymbol{x}, -s)\hat{p}(\boldsymbol{x}, s)]\nu_k \mathrm{dA}$$

$$\text{when } \boldsymbol{x}^R \in \mathbb{D}\,. \tag{10.18}$$

From this representation we observe that the wavefield at $\boldsymbol{x}^R$ consists of contributions of surfaces sources at $\partial\mathbb{D}_0$ and $\partial\mathbb{D}_1$.

The anti-causal Green's states (for a homogeneous background) are given by

$$\hat{G}^q(\boldsymbol{x}^R|\boldsymbol{x}, -s) = -s\rho\hat{G}(\boldsymbol{x}^R - \boldsymbol{x}, -s) \tag{10.19}$$

and

$$\hat{\Gamma}_k^q(\boldsymbol{x}^R|\boldsymbol{x}, -s) = -\partial_k^R\hat{G}(\boldsymbol{x}^R - \boldsymbol{x}, -s)\,, \tag{10.20}$$

Table 10.2. States in the power reciprocity theorem

| | State $A$ (actual state) | State $B$ (volume-injection Green's state) |
|---|---|---|
| Field state | $\{\hat{p}, \hat{v}_k\}(\boldsymbol{x}, s)$ | $\{\hat{p}^q, \hat{v}_k^q\}(\boldsymbol{x}\|\boldsymbol{x}^R, -s)$ |
| Material state | $\{\rho, \kappa\}$ | $\{\rho, \kappa\}$ |
| Source state | $\{0, 0\}$ | $\{\hat{q}^B(-s)\delta(\boldsymbol{x} - \boldsymbol{x}^R), 0\}$ |
| Domain $\mathbb{D}$ (see Fig. 10.1) | | |

where the anti-causal Green's function is given by

$$\hat{G}(\boldsymbol{x},-s)=\frac{\exp(\frac{s}{c}|\boldsymbol{x}|)}{4\pi|\boldsymbol{x}|}, \quad \text{with } c=(\kappa\rho)^{-\frac{1}{2}}\,. \tag{10.21}$$

In the derivation of Eq. (10.18) we have taken into account that contributions of the bounding surfaces at $(x_1^2+x_2^2)\to\infty$ vanish, since the integrand of Eq. (10.18) is of Order $((x_1^2+x_2^2)^{-1})$ as $(x_1^2+x_2^2)\to\infty$. The latter asymptotic behavior follows directly from the Green's function representation of Eq. (10.21) and the far-field approximations of Eq. (4.36). The arguments in the exponential function of the Green's function and the far-field expression of the actual wavefield have opposite signs.

We may also apply the reciprocity theorem of Section 5.3 to the domain $\mathbb{D}$ inside the interfaces $\partial\mathbb{D}_0$ and $\partial\mathbb{D}_1$ (see Fig. 10.1), but now State $B$ is taken to be the anti-causal counterpart of the actual wavefield, while State $A$ is the causal wavefield generated by a point source of volume injection (cf. Table 10.3). We then arrive at

Table 10.3. States in the power reciprocity theorem

| | State $A$ (volume-injection Green's state) | State $B$ (actual state) |
|---|---|---|
| Field state | $\{\hat{p}^q,\hat{v}_k^q\}(\boldsymbol{x}\|\boldsymbol{x}^R,s)$ | $\{\hat{p},\hat{v}_k\}(\boldsymbol{x},-s)$ |
| Material state | $\{\rho,\kappa\}$ | $\{\rho,\kappa\}$ |
| Source state | $\{\hat{q}^B(s)\delta(\boldsymbol{x}-\boldsymbol{x}^R),0\}$ | $\{0,0\}$ |
| Domain $\mathbb{D}$ (see Fig. 10.1) | | |

$$\hat{p}(\boldsymbol{x}^R,-s)$$
$$= \int_{\boldsymbol{x}\in(\partial\mathbb{D}_0\cup\partial\mathbb{D}_1)} [-\hat{G}^q(\boldsymbol{x}^R|\boldsymbol{x},s)\hat{v}_k(\boldsymbol{x},-s) + \hat{\Gamma}_k^q(\boldsymbol{x}^R|\boldsymbol{x},s)\hat{p}(\boldsymbol{x},-s)]\nu_k \mathrm{dA}\,, \tag{10.22}$$

when $\boldsymbol{x}^R \in \mathbb{D}$. Comparing Eqs. (10.18) and (10.22), it is obvious that the causal actual wavefield can be obtained from Eq. (10.22) by simply replacing $-s$ by $s$.

To carry out the decomposition of the actual wavefield, we write the Green's function as a plane-wave representation, which is obtained from Eq. (4.60). This representation is used in Eqs. (10.2), (10.3) and the results are substituted in the right-hand side of Eq. (10.22). Changing the order of integrations, we then have

$$\hat{p}(\boldsymbol{x}^R,-s) = \frac{1}{(2\pi)^2}\int_{(s\alpha_1,s\alpha_2)\in\mathbb{R}^2} \frac{\exp(-js\alpha_1 x_1^R - js\alpha_2 x_2^R)}{2s\Gamma}\mathrm{dA}$$
$$\int_{\boldsymbol{x}\in(\partial\mathbb{D}_0\cup\partial\mathbb{D}_1)} [-\hat{v}_k(\boldsymbol{x},-s)s\rho\exp(js\alpha_1 x_1 + js\alpha_2 x_2 - s\Gamma|\boldsymbol{x}^R - \boldsymbol{x}|)$$
$$+\hat{p}(\boldsymbol{x},-s)\partial_k \exp(js\alpha_1 x_1 + js\alpha_2 x_2 - s\Gamma|\boldsymbol{x}^R - \boldsymbol{x}|)]\nu_k \mathrm{dA}\,, \tag{10.23}$$

where

$$\Gamma = \left(\frac{1}{c^2} + \alpha_1^2 + \alpha_2^2\right)^{\frac{1}{2}}, \quad \mathrm{Re}(\Gamma) > 0\,. \tag{10.24}$$

We now switch back to the causal wavefield by replacing $-s$ by $s$. Noting that $|x_3^R - x_3| = x_3^R - x_3$ when $\boldsymbol{x} \in \partial\mathbb{D}_0$ and $|x_3^R - x_3| = x_3 - x_3^R$ when $\boldsymbol{x} \in \partial\mathbb{D}_1$, we arrive at the decomposition into the upgoing and downgoing wavefields

$$\hat{p}(x_1,x_2,x_3^R,s) = \hat{p}^{up}(x_1,x_2,x_3^R,s) + \hat{p}^{down}(x_1,x_2,x_3^R,s)\,, \tag{10.25}$$

where the spectral counterparts (see Eq. (9.27) for the definition of the spatial Fourier transform with respect to the horizontal coordinates) are given by

$$\overline{p}^{up}(js\alpha_1,js\alpha_2,x_3^R,s) = \overline{P}^{up}(js\alpha_1,js\alpha_2,s)\exp(s\Gamma x_3^R) \tag{10.26}$$

and

$$\overline{p}^{down}(js\alpha_1,js\alpha_2,x_3^R,s) = \overline{P}^{down}(js\alpha_1,js\alpha_2,s)\exp(-s\Gamma x_3^R)\,. \tag{10.27}$$

The amplitude $\overline{P}^{up}$ of the upgoing wavefield of Eq. (10.26) consists of contributions of surface sources at $\partial\mathbb{D}_0$, while the amplitude $\overline{P}^{down}$ of the downgoing wavefield of Eq. (10.27) consists of contributions of surface sources at $\partial\mathbb{D}_1$. These amplitudes are expressed as

$$\overline{P}^{up}(js\alpha_1, js\alpha_2, s)$$

$$= \frac{-1}{2s\Gamma}\int_{\boldsymbol{x}\in\partial\mathbb{D}_0}[\hat{v}_k(\boldsymbol{x},s)\,s\rho\exp(js\alpha_1x_1 + js\alpha_2x_2 - s\Gamma x_3)$$

$$+\hat{p}(\boldsymbol{x},s)\,\partial_k\exp(js\alpha_1x_1 + js\alpha_2x_2 - s\Gamma x_3)]\nu_k\mathrm{dA}\,, \tag{10.28}$$

$$\overline{P}^{down}(js\alpha_1, js\alpha_2, s)$$

$$= \frac{-1}{2s\Gamma}\int_{\boldsymbol{x}\in\partial\mathbb{D}_1}[\hat{v}_k(\boldsymbol{x},s)\,s\rho\exp(js\alpha_1x_1 + js\alpha_2x_2 + s\Gamma x_3)$$

$$+\hat{p}(\boldsymbol{x},s)\,\partial_k\exp(js\alpha_1x_1 + js\alpha_2x_2 + s\Gamma x_3)]\nu_k\mathrm{dA}\,. \tag{10.29}$$

*Condition for upgoing waves*

The particle velocity associated with the upgoing wavefield follows directly from Eqs. (10.26) and (7.70). The vertical component of the particle velocity and the acoustic pressure of the upgoing wavefield are related to each other as

$$\rho\overline{v}_3^{up}(js\alpha_1, js\alpha_2, x_3^R, s) + \Gamma\overline{p}^{up}(js\alpha_1, js\alpha_2, x_3^R, s) = 0\,. \tag{10.30}$$

*Condition for downgoing waves*

The particle velocity associated with the downgoing wavefield follows directly from Eqs. (10.27) and (7.70). The vertical component of the particle velocity and the acoustic pressure of the downgoing wavefield are related to each other as

$$\rho\overline{v}_3^{down}(js\alpha_1, js\alpha_2, x_3^R, s) - \Gamma\overline{p}^{down}(js\alpha_1, js\alpha_2, x_3^R, s) = 0\,. \tag{10.31}$$

Equations (10.25) - (10.31) show the decomposition in upgoing and downgoing wavefields in a homogeneous subdomain of infinite extent in the horizontal directions. The upgoing wavefield $\hat{p}^{up}(\boldsymbol{x}, s)$ is obtained from the integral representation of Eq. (10.18) with surface contributions from $\partial\mathbb{D}_0$ only. This upgoing wavefield is used as an inverse extrapolator in the redatuming of seismic data (see Section 10.3.2).

When $\partial\mathbb{D}_0$ and $\partial\mathbb{D}_1$ are plane interfaces, Eqs. (10.28) and (10.29) can be recognized as spatial Fourier transforms. The results are given below.

*Plane interface $\partial\mathbb{D}_0$*

In the case that $\partial\mathbb{D}_0$ is a plane interface at $x_3 = x_3^{(0)}$, the expression for the amplitude $\overline{P}^{up}$ reduces to

$$\overline{P}^{up}(js\alpha_1, js\alpha_2, s)$$

$$= \frac{\exp(-s\Gamma x_3^{(0)})}{-2\Gamma}\left[\rho\overline{v}_3(js\alpha_1, js\alpha_2, x_3^{(0)}, s) - \Gamma\overline{p}(js\alpha_1, js\alpha_2, x_3^{(0)}, s)\right]. \tag{10.32}$$

Combining Eqs. (10.26) and (10.32), the Fourier transform of the upgoing wavefield becomes

$$\overline{p}^{up}(js\alpha_1, js\alpha_2, x_3^R, s)$$

$$= \frac{\exp[s\Gamma(x_3^R - x_3^{(0)})]}{-2\Gamma}\left[\rho\overline{v}_3(js\alpha_1, js\alpha_2, x_3^{(0)}, s) - \Gamma\overline{p}(js\alpha_1, js\alpha_2, x_3^{(0)}, s)\right]. \tag{10.33}$$

From Eq. (10.31) it is obvious that the downgoing waves do not contribute in the right-hand side of Eq. (10.33).

*Plane interface $\partial\mathbb{D}_1$*

In the case that $\partial\mathbb{D}_1$ is a plane interface at $x_3 = x_3^{(1)}$, the expression for the amplitude $\overline{P}^{down}$ reduces to

$$\overline{P}^{down}(js\alpha_1, js\alpha_2, s)$$

$$= \frac{\exp(s\Gamma x_3^{(1)})}{2\Gamma}\left[\rho\overline{v}_3(js\alpha_1, js\alpha_2, x_3^{(1)}, s) + \Gamma\overline{p}(js\alpha_1, js\alpha_2, x_3^{(1)}, s)\right]. \tag{10.34}$$

Combining Eqs. (10.27) and (10.34), the Fourier transform of the downgoing wavefield becomes

$$\bar{p}^{down}(js\alpha_1, js\alpha_2, x_3^R, s)$$

$$= \frac{\exp[-s\Gamma(x_3^R - x_3^{(1)})]}{2\Gamma}\Big[\rho\bar{v}_3(js\alpha_1, js\alpha_2, x_3^{(1)}, s) + \Gamma\bar{p}(js\alpha_1, js\alpha_2, x_3^{(1)}, s)\Big]. \tag{10.35}$$

From Eq. (10.30) it is obvious that the upgoing waves do not contribute in the right-hand side of Eq. (10.35).

## 10.3. Redatuming of seismic data

In the procedure of redatuming of seismic data, we want to meet two objectives, viz., we want to extrapolate wavefield quantities downwards (i.e. in the positive $x_3$-direction), the so-called forward extrapolation, and we want to extrapolate wavefield quantities upwards (i.e. in the negative $x_3$-direction), the so-called inverse extrapolation.

### 10.3.1. Forward extrapolation

In the forward extrapolation of seismic data we want the wavefield to be extrapolated downwards in the homogeneous domain $\mathbb{D}$ (see Fig. 10.1), using the wavefield at $\partial\mathbb{D}_0$. From the results of Section 10.1 it follows that the downgoing wavefield is obtained as

$$\hat{p}^{down}(\boldsymbol{x}^R, s) = \int_{\boldsymbol{x}\in\partial\mathbb{D}_0} [\hat{G}^q(\boldsymbol{x}^R|\boldsymbol{x}, s)\hat{v}_k(\boldsymbol{x}, s) + \hat{\Gamma}_k^q(\boldsymbol{x}^R|\boldsymbol{x}, s)\hat{p}(\boldsymbol{x}, s)]\nu_k \mathrm{dA}$$

$$\text{when } x_3^R > x_{3,max}^{(0)}. \tag{10.36}$$

Taking the limit that $s \to j\omega$, we obtain its frequency-domain counterpart. Then, the integral operator of Eq. (10.36) is known as the forward extrapolator (WAPENAAR and BERKHOUT, 1989, chapter V).

For completeness, we present the frequency-domain counterpart of the Green's states:

$$\hat{G}^q(\boldsymbol{x}^R|\boldsymbol{x},j\omega) = j\omega\rho\hat{G}(\boldsymbol{x}^R-\boldsymbol{x},j\omega) \tag{10.37}$$

and

$$\hat{\Gamma}_k^q(\boldsymbol{x}^R|\boldsymbol{x},j\omega) = -\partial_k^R\hat{G}(\boldsymbol{x}^R-\boldsymbol{x},j\omega)\,, \tag{10.38}$$

where the causal Green's function is given by

$$\hat{G}(\boldsymbol{x}^R-\boldsymbol{x},j\omega) = \frac{\exp(-j\frac{\omega}{c}|\boldsymbol{x}^R-\boldsymbol{x}|)}{4\pi|\boldsymbol{x}^R-\boldsymbol{x}|}\,. \tag{10.39}$$

With the introduction of $p_1 = j\alpha_1$ and $p_2 = j\alpha_2$, the spatial Fourier representation of the Green's function (cf. Eq. (10.5)) is given by

$$\overline{G}(j\omega p_1, j\omega p_2, x_3, j\omega) = \frac{\exp(-j\omega\Gamma|x_3|)}{2j\omega\Gamma}\,, \tag{10.40}$$

where

$$\Gamma = \left(\frac{1}{c^2} - p_1^2 - p_2^2\right)^{\frac{1}{2}}\,, \tag{10.41}$$

with either $\Gamma$ is real and positive if $p_1^2 + p_2^2 < c^{-2}$ or $\Gamma$ is imaginary and negative if $p_1^2 + p_2^2 > c^{-2}$. The choice of the square root follows directly from the definition of $\Gamma$ in the complex $s$-domain ($\mathrm{Re}(s) > 0$). The limiting procedure of $s \to j\omega$ determines the signs of the real and imaginary part. The plane-wave constituent of Eq. (10.40) is of either propagating ($\Gamma$ is real) or exponentially decaying ($\Gamma$ is negative imaginary).

By applying the standard rules for inversion from the $s$-domain to the time domain, the time-domain representation of the downgoing wavefield is directly obtained from Eq. (10.36) as

$$p^{down}(\boldsymbol{x}^R,t) = \int_{\boldsymbol{x}\in\partial\mathbb{D}_0}\left[\rho\frac{\partial_t v_k(\boldsymbol{x},t-\frac{|\boldsymbol{x}^R-\boldsymbol{x}|}{c})}{4\pi|\boldsymbol{x}^R-\boldsymbol{x}|} - \partial_k^R\frac{p(\boldsymbol{x},t-\frac{|\boldsymbol{x}^R-\boldsymbol{x}|}{c})}{4\pi|\boldsymbol{x}^R-\boldsymbol{x}|}\right]\nu_k\mathrm{dA}$$

$$\text{when } x_3^R > x_{3,max}^{(0)}\,. \tag{10.42}$$

Observe that in the time-domain result the downgoing wavefield is obtained from the time-retarded contribution of the monopole and dipole sources distributed over the interface $\partial\mathbb{D}_0$.

### 10.3.2. Inverse extrapolation

In the inverse extrapolation of seismic data we want the wavefield to be extrapolated upwards in the homogeneous domain $\mathbb{D}$ (see Fig. 10.1), using the wavefield at $\partial\mathbb{D}_0$. From the results of Section 10.2 it follows that the upgoing wavefield is obtained as

$$\hat{p}^{up}(\boldsymbol{x}^R, s)$$
$$= \int_{\boldsymbol{x}\in\partial\mathbb{D}_0} [-\hat{G}^q(\boldsymbol{x}^R|\boldsymbol{x}, -s)\hat{v}_k(\boldsymbol{x}, s) + \hat{\Gamma}_k^q(\boldsymbol{x}^R|\boldsymbol{x}, -s)\hat{p}(\boldsymbol{x}, s)]\nu_k \mathrm{dA}$$
$$\text{when } x_3^R > x_{3,max}^{(0)}\,. \tag{10.43}$$

Taking the limit that $s \to j\omega$, we obtain its frequency-domain counterpart. Then, the integral operator of Eq. (10.43) is known as the inverse extrapolator (WAPENAAR and BERKHOUT, 1989, chapter VII). However the latter authors have derived the result under the assumption that the contribution of the non-propagating plane waves of the spectral decomposition of the surface sources at $\partial\mathbb{D}_1$ can be neglected. The present derivation is rigorous.

For completeness, we present the frequency-domain counterpart of the Green's states:

$$\hat{G}^q(\boldsymbol{x}^R|\boldsymbol{x}, -j\omega) = -j\omega\rho\hat{G}(\boldsymbol{x}^R - \boldsymbol{x}, -j\omega) \tag{10.44}$$

and

$$\hat{\Gamma}_k^q(\boldsymbol{x}^R|\boldsymbol{x}, -j\omega) = -\partial_k^R\hat{G}(\boldsymbol{x}^R - \boldsymbol{x}, -j\omega)\,, \tag{10.45}$$

where the anti-causal Green's function is given by

$$\hat{G}(\boldsymbol{x}^R - \boldsymbol{x}, -j\omega) = \frac{\exp(j\frac{\omega}{c}|\boldsymbol{x}^R - \boldsymbol{x}|)}{4\pi|\boldsymbol{x}^R - \boldsymbol{x}|}\,. \tag{10.46}$$

In order to apply an inverse transform of Eq. (10.43) from the $s$-domain to the time domain, we obtain a correlation result. Therefore it is necessary to take imaginary values of the Laplace transform parameter $s$. Then, by applying the standard rules for inversion from the $s$-domain to the time domain ($s \to j\omega$), the time-domain representation of the upgoing wavefield is directly obtained from Eq. (10.43) as

$$p^{up}(\boldsymbol{x}^R,t)=\int_{\boldsymbol{x}\in\partial\mathbb{D}_0}\left[\rho\frac{\partial_t v_k(\boldsymbol{x},t+\frac{|\boldsymbol{x}^R-\boldsymbol{x}|}{c})}{4\pi|\boldsymbol{x}^R-\boldsymbol{x}|}-\partial_k^R\frac{p(\boldsymbol{x},t+\frac{|\boldsymbol{x}^R-\boldsymbol{x}|}{c})}{4\pi|\boldsymbol{x}^R-\boldsymbol{x}|}\right]\nu_k \mathrm{dA}$$

$$\text{when } x_3^R > x_{3,max}^{(0)} . \tag{10.47}$$

Observe that in the time-domain result the upgoing wavefield is obtained from the time-advanced contribution of the monopole and dipole sources distributed over the interface $\partial\mathbb{D}_0$. Comparing Eqs. (10.42) and (10.47), we see that the pertaining sign in the temporal dependence of the surface sources determines the direction of extrapolation (see also SCHNEIDER, 1978).

With this result we conclude this chapter. The decomposition in up- and downgoing waves is of extreme importance in the further analysis of seismic-processing techniques. The deghosting procedure of Chapter 11 and the removal of surface related wave phenomena of Chapter 12 are based on the present decomposition results.

# Chapter 11

# Deghosting

In this chapter we discuss the procedure of deghosting. We assume that in a homogeneous subdomain $\mathbb{D}$ below a plane at $x_3 = 0$ and above an interface $\partial\mathbb{D}_g$, the acoustic wavefield is measured by a number of receivers at a horizontal level at $x_3 = x_3^R$ (see Fig. 11.1). We further assume that the total acoustic pressure vanishes at $x_3 = 0$. This is the situation as it occurs in marine seismics. The acoustic wavefield is generated by a monopole source of the volume injection type located at $\boldsymbol{x}^S$. The constitutive parameters in $\mathbb{D}$ are the constants $\rho$ and $\kappa$. The first objective is to reconstruct the upgoing wave constituent at the receiver level. This is the receiver-deghosting procedure. In fact, we want to determine the upgoing and downgoing wavefield constituents of the wavefield. This problem is very similar to the wavefield-decomposition problem of the previous chapter. The only complication is the presence of a source in the homogeneous domain $\mathbb{D}$. The total wavefield in a point $\boldsymbol{x} \in \mathbb{D}$, generated by the source at $\boldsymbol{x}^S$, is denoted as $\{\hat{p}, \hat{v}_k\}(\boldsymbol{x}|\boldsymbol{x}^S)$ and at the plane surface $x_3 = 0$ we have the boundary condition

$$\lim_{x_3 \downarrow 0} \hat{p}(\boldsymbol{x}|\boldsymbol{x}^S, s) = 0 \,. \tag{11.1}$$

We first determine the incident wavefield in the halfspace $0 < x_3 < \infty$. This is the wavefield that would be present in this halfspace, if the domain $\mathbb{D}_g$ showed no contrast with the domain $\mathbb{D}$. This wavefield is determined in Section 9.4 (cf. Eqs. (9.55)). In a point $\boldsymbol{x}^R$ we may write

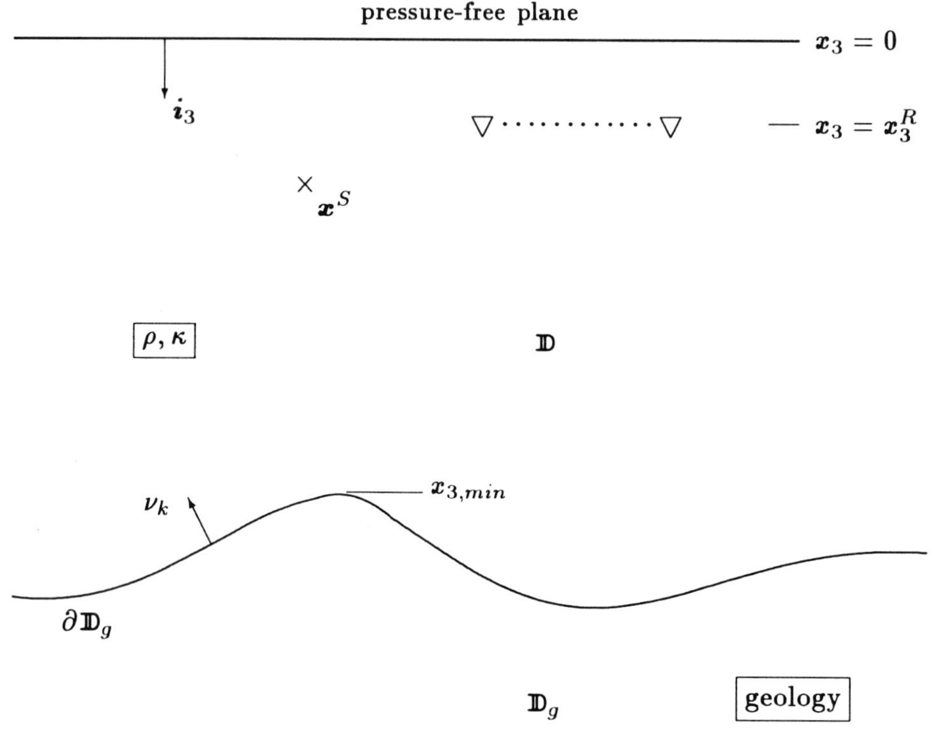


Figure 11.1. A homogeneous subdomain $\mathbb{D}$ bounded by the plane $x_3 = 0$ and the interface $\partial\mathbb{D}_g$.

$$\hat{p}^{inc,H}(\boldsymbol{x}|\boldsymbol{x}^S, s) = s\rho\hat{q}^S\hat{G}^H(\boldsymbol{x}|\boldsymbol{x}^S, s)\,, \tag{11.2}$$

where

$$\hat{G}^H(\boldsymbol{x}|\boldsymbol{x}^S, s) = \frac{\exp(-\hat{\gamma}|\boldsymbol{x}-\boldsymbol{x}^S|)}{4\pi|\boldsymbol{x}-\boldsymbol{x}^S|} - \frac{\exp(-\hat{\gamma}|\boldsymbol{x}-\boldsymbol{x}^{S^I}|)}{4\pi|\boldsymbol{x}-\boldsymbol{x}^{S^I}|}\,, \tag{11.3}$$

in which

$$\boldsymbol{x}^{S^I} = (x_1^S, x_2^S, -x_3^S) \tag{11.4}$$

denotes the image point of $\boldsymbol{x}^S$ with respect to the reflecting surface at $x_3 = 0$. Note that this wavefield $\hat{p}^{inc,H}$ vanishes at $x_3 = 0$,

$$\lim_{x_3 \downarrow 0} \hat{p}^{inc,H}(\boldsymbol{x}|\boldsymbol{x}^S, s) = 0\,. \tag{11.5}$$

We subsequently split the wavefield into an incident-wavefield constituent and a scattered-wavefield constituent,

$$\hat{p}(\boldsymbol{x}|\boldsymbol{x}^S, s) = \hat{p}^{inc,H}(\boldsymbol{x}|\boldsymbol{x}^S, s) + \hat{p}^{sct}(\boldsymbol{x}|\boldsymbol{x}^S, s)\,. \tag{11.6}$$

In view of Eqs. (11.1) and (11.5) the scattered wavefield $\hat{p}^{sct}$ vanishes at $x_3 = 0$ as well,

$$\lim_{x_3 \downarrow 0} \hat{p}^{sct}(\boldsymbol{x}|\boldsymbol{x}^S, s) = 0\,. \tag{11.7}$$

## 11.1. Decomposition based on field reciprocity

We apply the reciprocity theorem of Section 5.1 to the domain $\mathbb{D}$ bounded by $x_3 = 0$ and the interface $\partial\mathbb{D}_g$ (see Fig. 11.1). The normal $\nu_k$ to the latter interface is directed towards the domain $\mathbb{D}$. State $A$ is taken to be the scattered wavefield. The wavefield of State $B$ is generated by a point source of volume injection located at $\boldsymbol{x} = \boldsymbol{x}^R$ (cf. Table 11.1). Using the Green's states of Eq. (7.62) we arrive at

Table 11.1. States in the field reciprocity theorem

| | State $A$ (scattered field) | State $B$ (volume-injection Green's state) |
|---|---|---|
| Field state | $\{\hat{p}^{sct}, \hat{v}_k^{sct}\}(\boldsymbol{x}\|\boldsymbol{x}^S, s)$ | $\{\hat{p}^q, \hat{v}_k^q\}(\boldsymbol{x}\|\boldsymbol{x}^R, s)$ |
| Material state | $\{\rho, \kappa\}$ | $\{\rho, \kappa\}$ |
| Source state | $\{0, 0\}$ | $\{\hat{q}^B(s)\delta(\boldsymbol{x}-\boldsymbol{x}^R), 0\}$ |
| Domain $\mathbb{D}$ (see Fig. 11.1) | | |

$$\hat{p}^{sct}(\boldsymbol{x}^R|\boldsymbol{x}^S,s) = \int_{(x_1,x_2)\in\mathbb{R}^2} \hat{G}^q(\boldsymbol{x}^R|x_1,x_2,0,s)\hat{v}_3^{sct}(x_1,x_2,0|\boldsymbol{x}^S,s)\mathrm{dA}$$

$$+\int_{\boldsymbol{x}\in\partial\mathbb{D}_g} [\hat{G}^q(\boldsymbol{x}^R|\boldsymbol{x},s)\hat{v}_k^{sct}(\boldsymbol{x}|\boldsymbol{x}^S,s) + \hat{\Gamma}_k^q(\boldsymbol{x}^R|\boldsymbol{x},s)\hat{p}^{sct}(\boldsymbol{x}|\boldsymbol{x}^S,s)]\nu_k\mathrm{dA}$$

$$\text{when } \boldsymbol{x}^R \in \mathbb{D}\,, \qquad (11.8)$$

where we have used boundary equation (11.7). Further, in the derivation of Eq. (11.8) we have taken into account that contributions of the bounding surfaces at $(x_1^2 + x_2^2) \to \infty$ vanish, since the integrand of Eq. (11.8) is of Order $((x_1^2 + x_2^2)^{-1})$ as $(x_1^2 + x_2^2) \to \infty$ (cf. Section 10.1). Following the analysis of Section 10.1, Eq. (11.8) may be written as

$$\hat{p}^{sct}(\boldsymbol{x}^R|\boldsymbol{x}^S,s) = \hat{p}^{down}(\boldsymbol{x}^R|\boldsymbol{x}^S,s) + \hat{p}^{up}(\boldsymbol{x}^R|\boldsymbol{x}^S,s)\,, \qquad (11.9)$$

where

$$\hat{p}^{down}(\boldsymbol{x}^R|\boldsymbol{x}^S,s) = \int_{(x_1,x_2)\in\mathbb{R}^2} \hat{G}^q(\boldsymbol{x}^R|x_1,x_2,0,s)\hat{v}_3^{sct}(x_1,x_2,0|\boldsymbol{x}^S,s)\mathrm{dA}\,, \qquad (11.10)$$

when $0 < x_3^R < x_{3,min}$, $x_{3,min}$ is the minimum value of $x_3$ on $\partial\mathbb{D}_g$, see Fig. 11.1. The expression for the downgoing wavefield may be written as

$$F\left\{\hat{p}^{down}(x_1,x_2,x_3|\boldsymbol{x}^S,s)\right\} = \frac{\exp(-s\Gamma x_3)}{2\Gamma} F\left\{\rho\hat{v}_3^{sct}(x_1,x_2,0|\boldsymbol{x}^S,s)\right\}\,. \qquad (11.11)$$

Here, $F$ denotes the two-dimensional Fourier transformation defined by Eq. (9.27). Once the vertical particle velocity, $\hat{v}_3^{sct}(x_1,x_2,0|\boldsymbol{x}^S,s)$, at the pressure-free plane is determined, the downgoing wavefield follows from Eq. (11.11). The upgoing wavefield consists of contributions of $\mathbb{D}_g$. Since in most seismic applications the geology is not known, we are not interested in this expression for the upgoing wavefield.

## 11.2. Decomposition based on power reciprocity

We apply the reciprocity theorem of Section 5.3 to the domain $\mathbb{D}$ bounded by $x_3 = 0$ and the interface $\partial\mathbb{D}_g$ (see Fig. 11.1). The normal $\nu_k$ to the latter interface is directed towards the domain $\mathbb{D}$. State $A$ is taken to be the scattered wavefield. The wavefield of State $B$ is the anti-causal wavefield generated by a point source of volume injection located at $\boldsymbol{x} = \boldsymbol{x}^R$ (cf. Table 11.2). Using the Green's states of Eq. (7.62), in which we replace $s$ by $-s$, we arrive at

$$\hat{p}^{sct}(\boldsymbol{x}^R|\boldsymbol{x}^S, s) = -\int_{(x_1,x_2)\in\mathbb{R}^2} \hat{G}^q(\boldsymbol{x}^R|x_1, x_2, 0, -s)\hat{v}_3^{sct}(x_1, x_2, 0|\boldsymbol{x}^S, s)\mathrm{dA}$$

$$+\int_{\boldsymbol{x}\in\partial\mathbb{D}_g} [-\hat{G}^q(\boldsymbol{x}^R|\boldsymbol{x}, -s)\hat{v}_k^{sct}(\boldsymbol{x}|\boldsymbol{x}^S, s) + \hat{\Gamma}_k^q(\boldsymbol{x}^R|\boldsymbol{x}, -s)\hat{p}^{sct}(\boldsymbol{x}|\boldsymbol{x}^S, s)]\nu_k\mathrm{dA}$$

$$\text{when } \boldsymbol{x}^R \in \mathbb{D}, \qquad (11.12)$$

where we have used boundary equation (11.7). Further, in the derivation of Eq. (11.12) we have taken into account that contributions of the bounding surfaces at $(x_1^2 + x_2^2) \to \infty$ vanish, since the integrand of Eq. (11.12) is of Order $((x_1^2 + x_2^2)^{-1})$ as $(x_1^2 + x_2^2) \to \infty$ (cf. Section 10.2). Following the analysis of Section 10.2, Eq. (11.12) may be written as

$$\hat{p}^{sct}(\boldsymbol{x}^R|\boldsymbol{x}^S, s) = \hat{p}^{up}(\boldsymbol{x}^R|\boldsymbol{x}^S, s) + \hat{p}^{down}(\boldsymbol{x}^R|\boldsymbol{x}^S, s), \qquad (11.13)$$

Table 11.2. States in the power reciprocity theorem

| | State $A$ (scattered field) | State $B$ (volume-injection Green's state) |
|---|---|---|
| Field state | $\{\hat{p}^{sct}, \hat{v}_k^{sct}\}(\boldsymbol{x}|\boldsymbol{x}^S, s)$ | $\{\hat{p}^q, \hat{v}_k^q\}(\boldsymbol{x}|\boldsymbol{x}^R, -s)$ |
| Material state | $\{\rho, \kappa\}$ | $\{\rho, \kappa\}$ |
| Source state | $\{0, 0\}$ | $\{\hat{q}^B(-s)\delta(\boldsymbol{x}-\boldsymbol{x}^R), 0\}$ |
| Domain $\mathbb{D}$ (see Fig. 11.1) | | |

where

$$\hat{p}^{up}(\boldsymbol{x}^R|\boldsymbol{x}^S,s) = -\int_{(x_1,x_2)\in\mathbb{R}^2} \hat{G}^q(\boldsymbol{x}^R|\boldsymbol{x}_1,\boldsymbol{x}_2,0,-s)\hat{v}_3^{sct}(\boldsymbol{x}_1,\boldsymbol{x}_2,0|\boldsymbol{x}^S,s)\mathrm{dA}\,, \tag{11.14}$$

when $0 < \boldsymbol{x}_3^R < \boldsymbol{x}_{3,min}$, $\boldsymbol{x}_{3,min}$ is the minimum value of $\boldsymbol{x}_3$ on $\partial\mathbb{D}_g$, see Fig. 11.1. The expression for the upgoing wavefield may be written as

$$F\Big\{\hat{p}^{up}(\boldsymbol{x}_1,\boldsymbol{x}_2,\boldsymbol{x}_3|\boldsymbol{x}^S,s)\Big\} = \frac{\exp(s\Gamma\boldsymbol{x}_3)}{-2\Gamma}F\Big\{\rho\hat{v}_3^{sct}(\boldsymbol{x}_1,\boldsymbol{x}_2,0|\boldsymbol{x}^S,s)\Big\}\,. \tag{11.15}$$

Here, $F$ denotes the two-dimensional Fourier transformation defined by Eq. (9.27). Once the vertical particle velocity, $\hat{v}_3^{sct}(\boldsymbol{x}_1,\boldsymbol{x}_2,0|\boldsymbol{x}^S,s)$, at the pressure-free plane is determined, the upgoing wave constituent follows from Eq. (11.15). The downgoing wavefield consists of contributions of $\mathbb{D}_g$. Since in most seismic applications the geology is not known, we are not interested in this expression for the downgoing wavefield.

## 11.3. The surface-related vertical particle velocity

In the previous sections we have seen that the downgoing and upgoing constituents of the scattered wavefield can be obtained from the surface-related vertical particle velocity $\hat{v}_3^{sct}(\boldsymbol{x}_1,\boldsymbol{x}_2,0|\boldsymbol{x}^S,s)$. In order to determine this particle velocity, we combine Eqs. (11.11) and (11.15), while using either Eq. (11.9) or (11.13); this leads to

$$F\Big\{\hat{p}^{sct}(\boldsymbol{x}_1,\boldsymbol{x}_2,\boldsymbol{x}_3|\boldsymbol{x}^S,s)\Big\} = \frac{\sinh(s\Gamma\boldsymbol{x}_3)}{-\Gamma}F\Big\{\rho\hat{v}_3^{sct}(\boldsymbol{x}_1,\boldsymbol{x}_2,0|\boldsymbol{x}^S,s)\Big\}\,. \tag{11.16}$$

From this relation it directly follows that

$$\hat{v}_3^{sct}(\boldsymbol{x}_1,\boldsymbol{x}_2,0|\boldsymbol{x}^S,s) = \frac{-1}{\rho}F^{-1}\left\{\frac{\Gamma}{\sinh(s\Gamma\boldsymbol{x}_3)}F\Big\{\hat{p}^{sct}(\boldsymbol{x}|\boldsymbol{x}^S,s)\Big\}\right\}\,. \tag{11.17}$$

Here, $F^{-1}$ denotes the inverse Fourier transformation defined by Eq. (9.28). If the incident wavefield $\hat{p}^{inc,H}$ is known, $\hat{p}^{sct}$ follows from

$$\hat{p}^{sct}(\boldsymbol{x}|\boldsymbol{x}^S,s) = \hat{p}(\boldsymbol{x}|\boldsymbol{x}^S,s) - \hat{p}^{inc,H}(\boldsymbol{x}|\boldsymbol{x}^S,s)\,. \tag{11.18}$$

Note that $\hat{p}(\boldsymbol{x}^R|\boldsymbol{x}^S, s)$ is the measured total wavefield at the receiver positions distributed over the plane $x_3 = x_3^R$.

We note that Eqs. (11.16) and (11.17) also follow directly from an appropriate application of the field reciprocity theorem. To show this, we apply the reciprocity theorem of Section 5.1 to the homogeneous domain between $x_3 = 0$ and $x_3 = x_3^R$. As State $A$ we take the actual scattered wavefield values $\{\hat{p}^A, \hat{v}_k^A\} = \{\hat{p}^{sct}, \hat{v}_k^{sct}\}(\boldsymbol{x}|\boldsymbol{x}^S)$ and as State $B$ we take an auxiliary source-free plane wave with vanishing pressure at the level $x_3 = x_3^R$, viz.,

$$\hat{p}^B(\boldsymbol{x}, s) = \exp(js\alpha_1 x_1 + js\alpha_2 x_2)\sinh[s\Gamma(x_3^R - x_3)] . \tag{11.19}$$

Application of the reciprocity theorem of Section 5.1 to the domain $\{\boldsymbol{x} \in \mathbb{R}^3; -\infty < x_1, x_2 < \infty, 0 < x_3 < x_3^R\}$ with the states mentioned (see Table 11.3) results into

$$\int_{(x_1,x_2)\in\mathbb{R}^2} \exp(js\alpha_1 x_1 + js\alpha_2 x_2)\sinh(s\Gamma x_3^R)\hat{v}_3^{sct}(x_1, x_2, 0|\boldsymbol{x}^S, s)\mathrm{dA}$$

$$+ \int_{(x_1^R,x_2^R)\in\mathbb{R}^2} \hat{p}^{sct}(\boldsymbol{x}^R|\boldsymbol{x}^S, s)\frac{\Gamma}{\rho}\exp(js\alpha_1 x_1^R + js\alpha_2 x_2^R)\mathrm{dA} = 0 , \tag{11.20}$$

Table 11.3. States in the field reciprocity theorem

| | State $A$ (scattered field) | State $B$ (auxiliary state) |
|---|---|---|
| Field state | $\{\hat{p}^{sct}, \hat{v}_k^{sct}\}(\boldsymbol{x}|\boldsymbol{x}^S, s)$ | $\{1, \frac{-1}{s\rho}\partial_k\}\hat{p}^B(\boldsymbol{x}, s)$ |
| Material state | $\{\rho, \kappa\}$ | $\{\rho, \kappa\}$ |
| Source state | $\{0, 0\}$ | $\{0, 0\}$ |
| Domain $\{\boldsymbol{x} \in \mathbb{R}^3; -\infty < x_1, x_2 < \infty, 0 < x_3 < x_3^R\}$ (see Fig. 11.1) | | |

where we have used that $\hat{p}^{sct}(x_1, x_2, 0|\boldsymbol{x}^S, s) = 0$ at the pressure-free plane $x_3 = 0$ and $\hat{p}^B(\boldsymbol{x}^R, s) = 0$ at the receiver level. In the derivation of Eq. (11.20) we have taken into account that the integral of the bounding surface at $(x_1^2 + x_2^2) \to \infty$ vanishes. This follows immediately from the far-field approximation of Eq. (4.36) and the Order$((x_1^2 + x_2^2)^{-\frac{1}{2}})$ of the cylindrical integration of $\hat{p}^B$ as $(x_1^2+x_2^2) \to \infty$, that follows from the Riemann-Lebesgue lemma (WHITTAKER and WATSON, 1927, p. 172.)

Using the definition of the Fourier transformation of Eq. (9.27), we may write Eq. (11.20) as

$$\frac{\sinh(s\Gamma x_3^R)}{\Gamma} F\Big\{\rho\hat{v}_3^{sct}(x_1, x_2, 0|\boldsymbol{x}^S, s)\Big\} + F\Big\{\hat{p}_3^{sct}(x_1, x_2, x_3^R|\boldsymbol{x}^S, s)\Big\} = 0\,. \tag{11.21}$$

From this equation either $\hat{v}_3^{sct}(x_1, x_2, 0|\boldsymbol{x}^S, s)$ can be expressed in terms of $\hat{p}^{sct}(\boldsymbol{x}^R|\boldsymbol{x}^S, s)$ (cf. Eq. (11.17)) or $\hat{p}^{sct}(\boldsymbol{x}^R|\boldsymbol{x}^S, s)$ can be expressed in terms of $\hat{v}_3^{sct}(x_1, x_2, 0|\boldsymbol{x}^S, s)$ (cf. Eq. (11.16)).

## 11.4. Receiver deghosting

The upgoing wavefield at an arbitrary level $x_3$ is obtained from the scattered field at $x_3^R$ by combining the operations as defined by Eqs. (11.15) and (11.17) as

$$\hat{p}^{up}(x_1, x_2, x_3|\boldsymbol{x}^S, s) = F^{-1}\left\{\frac{\exp(s\Gamma x_3)}{2\sinh(s\Gamma x_3^R)} F\Big\{\hat{p}^{sct}(x_1, x_2, x_3^R|\boldsymbol{x}^S, s)\Big\}\right\}, \tag{11.22}$$

where $\hat{p}^{sct}(\boldsymbol{x}^R|\boldsymbol{x}^S, s)$ follows from the measured acoustic pressure $\hat{p}(\boldsymbol{x}^R|\boldsymbol{x}^S, s)$ at the receiver locations as

$$\hat{p}^{sct}(\boldsymbol{x}^R|\boldsymbol{x}^S, s) = \hat{p}(\boldsymbol{x}^R|\boldsymbol{x}^S, s) - \hat{p}^{inc,H}(\boldsymbol{x}^R|\boldsymbol{x}^S, s)\,. \tag{11.23}$$

The expression for the upgoing wavefield on the right-hand side of Eq. (11.22) shows that, although the scattered wavefield satisfies the general reciprocity relation of Eq. (6.9), viz. $\hat{p}^{sct}(\boldsymbol{x}^S|\boldsymbol{x}^R, s) = \hat{p}^{sct}(\boldsymbol{x}^R|\boldsymbol{x}^S, s)$, this relation does not hold for the upgoing wavefield. The incident wavefield $\hat{p}^{inc,H}(\boldsymbol{x}^R|\boldsymbol{x}^S, s)$

in the homogeneous halfspace follows from Eqs. (11.2) and (11.3). Using the plane-wave representation (cf. Eq. (4.60)) expressed as a spectral decomposition with respect to the receiver coordinates, we write

$$\hat{p}^{inc,H}(x_1, x_2, x_3^R|\boldsymbol{x}^S, s) =$$

$$\hat{W}(s)F^{-1}\left\{\frac{\exp(js\alpha_1 x_1^S + js\alpha_2 x_2^S)}{2s\Gamma}\left\{\exp[-s\Gamma|x_3^R - x_3^S|] - \exp[-s\Gamma(x_3^R + x_3^S)]\right\}\right\}. \tag{11.24}$$

The source wavelet $\hat{W}(s)$ is known as the Laplace transform of the source wavelet, which is related to the $s$-domain time rate of volume injection $\hat{q}^S$ through

$$\hat{W}(s) = s\rho\hat{q}^S(s). \tag{11.25}$$

In Eq. (11.22) we need the Fourier transformation with respect to the horizontal coordinates, for a fixed source point $\boldsymbol{x}^S$. In fact, this so-called receiver deghosting is carried out using common-source data (also called common-shot data, see Fig. 11.2). In view of further seismic processing techniques we rather want to consider the upgoing wavefield at the level $x_3 = 0$. This wavefield at $x_3 = 0$ is follows directly from Eq. (11.22) as

$$\hat{p}^{up}(x_1, x_2, 0|\boldsymbol{x}^S, s) = F^{-1}\left\{\frac{1}{2\sinh(s\Gamma x_3^R)}F\left\{\hat{p}^{sct}(x_1, x_2, x_3^R|\boldsymbol{x}^S, s)\right\}\right\}, \tag{11.26}$$

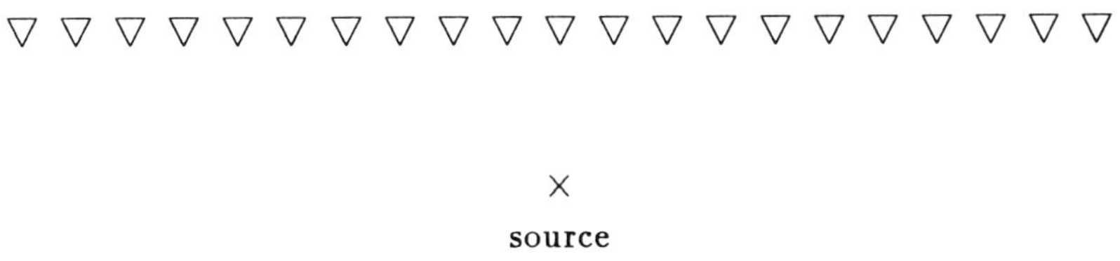

▽ ▽ ▽ ▽ ▽ ▽ ▽ ▽ ▽ ▽ ▽ ▽ ▽ ▽ ▽ ▽ ▽ ▽ ▽ ▽

×

source

Figure 11.2. Acquisition configuration to obtain common-source data.

Note that the spatial Fourier transformations are carried out with respect to the horizontal receiver coordinates of the scattered wavefield. This finalizes the discussion on the receiver deghosting. The upgoing wavefield is an intermediate wavefield and will be used as input for the source deghosting discussed in the next section.

## 11.5. Source deghosting

In this case we employ the following plane-wave representation for the incident wavefield in the homogeneous halfspace

$$\hat{p}^{inc,H}(\boldsymbol{x}^R|x_1,x_2,x_3^S,s) =$$

$$\hat{W}(s)F^{-1}\left\{\frac{\exp(jsa_1x_1^R+jsa_2x_2^R)}{2s\Gamma}\left\{\exp[-s\Gamma|x_3^R-x_3^S|]-\exp[-s\Gamma(x_3^R+x_3^S)]\right\}\right\}. \tag{11.27}$$

Obviously, in Eq. (11.24) we have an inverse spatial Fourier transform with respect to the horizontal receiver coordinates, while in Eq. (11.27) we have an inverse spatial Fourier transform with respect to the horizontal source coordinates. In the case that $x_3^R = x_{3,min}$ we obtain

$$\begin{aligned}
&\hat{p}^{inc,H}(x_1^R,x_2^R,x_{3,min}|x_1,x_2,x_3^S,s) \\
&= \hat{W}(s)F^{-1}\left\{\frac{\exp(jsa_1x_1^R+jsa_2x_2^R-s\Gamma x_{3,min})}{2s\Gamma}2\sinh(s\Gamma x_3^S)\right\} \\
&= F^{-1}\left\{2\sinh(s\Gamma x_3^S)F\left\{\hat{p}^{inc}(x_1^R,x_2^R,x_{3,min}|x_1,x_2,0)\right\}\right\},
\end{aligned} \tag{11.28}$$

in which $\hat{p}^{inc}$ is the incident wavefield in a homogeneous medium originating from the source position at $x_3^S = 0$, and incident upon the geology below $x_3 = x_{3,min}$, viz.,

$$\hat{p}^{inc}(\boldsymbol{x}^R|\boldsymbol{x}^S,s) = \hat{W}(s)G(\boldsymbol{x}^R|\boldsymbol{x}^S,s), \tag{11.29}$$

in which

$$\hat{G}(\boldsymbol{x}^R|\boldsymbol{x}^S,s) = \frac{\exp(-\hat{\gamma}|\boldsymbol{x}^R-\boldsymbol{x}^S|)}{4\pi|\boldsymbol{x}^R-\boldsymbol{x}^S|}. \tag{11.30}$$

The true incident wavefield $\hat{p}^{inc,H}$ is related to the incident wavefield $\hat{p}^{inc}$ in the spatial Fourier domain through a multiplicative factor $2\sinh(s\Gamma x_3^S)$.

By inverting the operations of Eq. (11.28), we obtain the following representation for the incident wavefield in the homogeneous medium expressed in terms of the incident wavefield in the homogeneous halfspace

$$\hat{p}^{inc}(x_1^R, x_2^R, x_{3,min}|x_1, x_2, 0, s) =$$

$$F^{-1}\left\{\frac{1}{2\sinh(s\Gamma x_3^S)} F\left\{\hat{p}^{inc,H}(x_1^R, x_2^R, x_{3,min}|x_1, x_2, x_3^S, s)\right\}\right\} . \tag{11.31}$$

The wavefield scattered by the geology is linear dependent on the incident-wavefield strength. Hence, using the linearity of the wavefield problem, we deduce that, replacing the true incident wavefield of the halfspace by the incident wavefield of the homogeneous medium, the upgoing wavefield has to be replaced by the deghosted wavefield

$$\hat{p}^{dgh}(x_1^R, x_2^R, 0|x_1, x_2, 0, s) =$$

$$F^{-1}\left\{\frac{1}{2\sinh(s\Gamma x_3^S)} F\left\{\hat{p}^{up}(x_1^R, x_2^R, 0|x_1, x_2, x_3^S, s)\right\}\right\} . \tag{11.32}$$

This so-called source deghosting is carried out using common-receiver data (see Fig. 11.3). Note that the spatial Fourier transformations are carried out with respect to the horizontal source coordinates of the upgoing wavefield of Eq. (11.24).

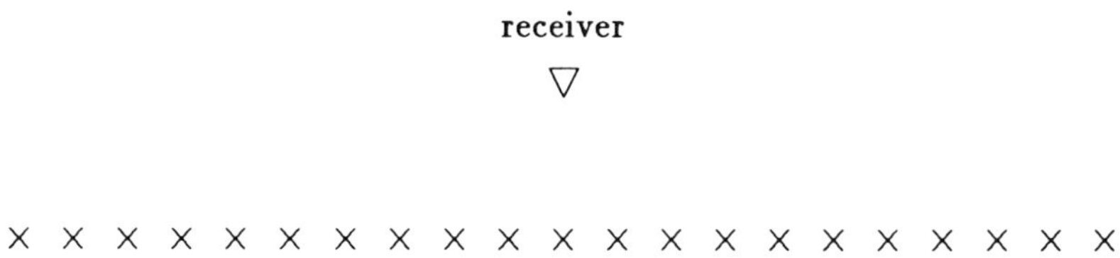

receiver

▽

× × × × × × × × × × × × × × × × × × × × ×

Figure 11.3. Acquisition configuration to obtain common-receiver data.

By cascading the receiver and source deghosting, through combining Eqs. (11.26) and (11.32), we conclude that $\hat{p}^{dgh}$ satisfies the reciprocity relation

$$\hat{p}^{dgh}(x_1^S, x_2^S, 0 | x_1^R, x_2^R, 0, s) = \hat{p}^{dgh}(x_1^R, x_2^R, 0 | x_1^S, x_2^S, 0, s) . \tag{11.33}$$

## 11.6. Deghosting in the strip configuration

To study the deghosting procedure, we check its capabilities on the two-dimensional dataset of the strip configuration embedded in the homogeneous halfspace, as it was calculated in Section 9.5.2. This dataset consists of 255 source positions with 255 receiver coordinates per source, arranged in a split-spread fashion. The two-dimensionality of the scattering problem is enforced by assuming that both the excitation and the configuration are independent of $x_2$. As a consequence we can adapt the three-dimensional analysis of the previous sections to the two-dimensional one by simply replacing the two-dimensional Fourier transforms with respect to the horizontal coordinates by the one-dimensional Fourier transform with respect to $x_1$. In particular, the incident wavefield in the homogeneous halfspace is given by

$$\hat{p}^{inc,H}(x_1, x_3^R | x_1^S, x_3^S, s) =$$

$$\hat{W}(s) F^{-1}\left\{ \frac{\exp(j s \alpha_1 x_1^S)}{2 s \Upsilon} \left\{ \exp[-s\Upsilon |x_3^R - x_3^S|] - \exp[-s\Upsilon(x_3^R + x_3^S)] \right\} \right\} , \tag{11.34}$$

where $\Upsilon = (c^{-2} + \alpha_1^2)^{\frac{1}{2}}$ and the definition of the inverse Fourier transformation $F^{-1}$ follows from Eq. (9.102). More details on the two-dimensional simplifications can be found in Chapter 9.

The actual deghosting is carried out in three stages. The first stage is the removal of the incident wavefield to create the scattered wavefield

$$\hat{p}^{sct}(x_1^R, x_3^R | x_1^S, x_3^S, s) = \hat{p}(x_1^R, x_3^R | x_1^S, x_3^S, s) - \hat{p}^{inc,H}(x_1^R, x_3^R | x_1^S, x_3^S, s) . \tag{11.35}$$

The second stage is the receiver deghosting to construct the truly upgoing wavefield at $x_3 = 0$ according to

$$\hat{p}^{up}(x_1, 0|x_1^S, x_3^S, s) = F^{-1}\left\{\frac{1}{2\sinh(s\Upsilon x_3^R)} F\left\{\hat{p}^{sct}(x_1, x_3^R|x_1^S, x_3^S, s)\right\}\right\}, \tag{11.36}$$

which is carried out in the common-source domain. The last stage is carried out in the common-receiver domain and constitutes the source deghosting and finalizes our deghosting procedure:

$$\hat{p}^{dgh}(x_1^R, 0|x_1, 0, s) = F^{-1}\left\{\frac{1}{2\sinh(s\Upsilon x_3^S)} F\left\{\hat{p}^{up}(x_1^R, 0|x_1, x_3^S, s)\right\}\right\}. \tag{11.37}$$

*The numerical procedure*

In our numerical computations we take (see Section 9.5.2)

$$s = 4 + j\omega\,. \tag{11.38}$$

Then, we may write for the vertical slowness

$$\Upsilon = \left(\frac{1}{c^2} + \frac{(s\alpha_1)^2}{(4+j\omega)^2}\right)^{\frac{1}{2}}, \quad \mathrm{Re}(\Upsilon) > 0\,. \tag{11.39}$$

Here, $s\alpha_1$ is the (real) Fourier-transform parameter of the Fourier transformation defined by Eqs. (9.101) - (9.102).

Fig. 11.4 shows the results for the compliant strip in the case that the source position is above the center of the strip. The top figure represents the scattered wavefield in the homogeneous halfspace at the level $x_3^R = 0$; for comparison we have delayed the arrival times of the scattered wavefield by inserting an extra factor $\exp(-s\Gamma x_3^S)$ in the Fourier transform of the data with respect to the source coordinates. The bottom figure shows the results after receiver and source deghosting, computed from the input data set using Eqs. (11.36) and (11.37). The resulting wavefield represents the deghosted wavefield at the receiver level $x_3^R = 0$ and source level $x_3^S = 0$. Fig. 11.5 shows the results for the compliant strip with the source position above the left edge of the strip. Finally, Figs. 11.6 and 11.7 show the pertaining results for the rigid strip.

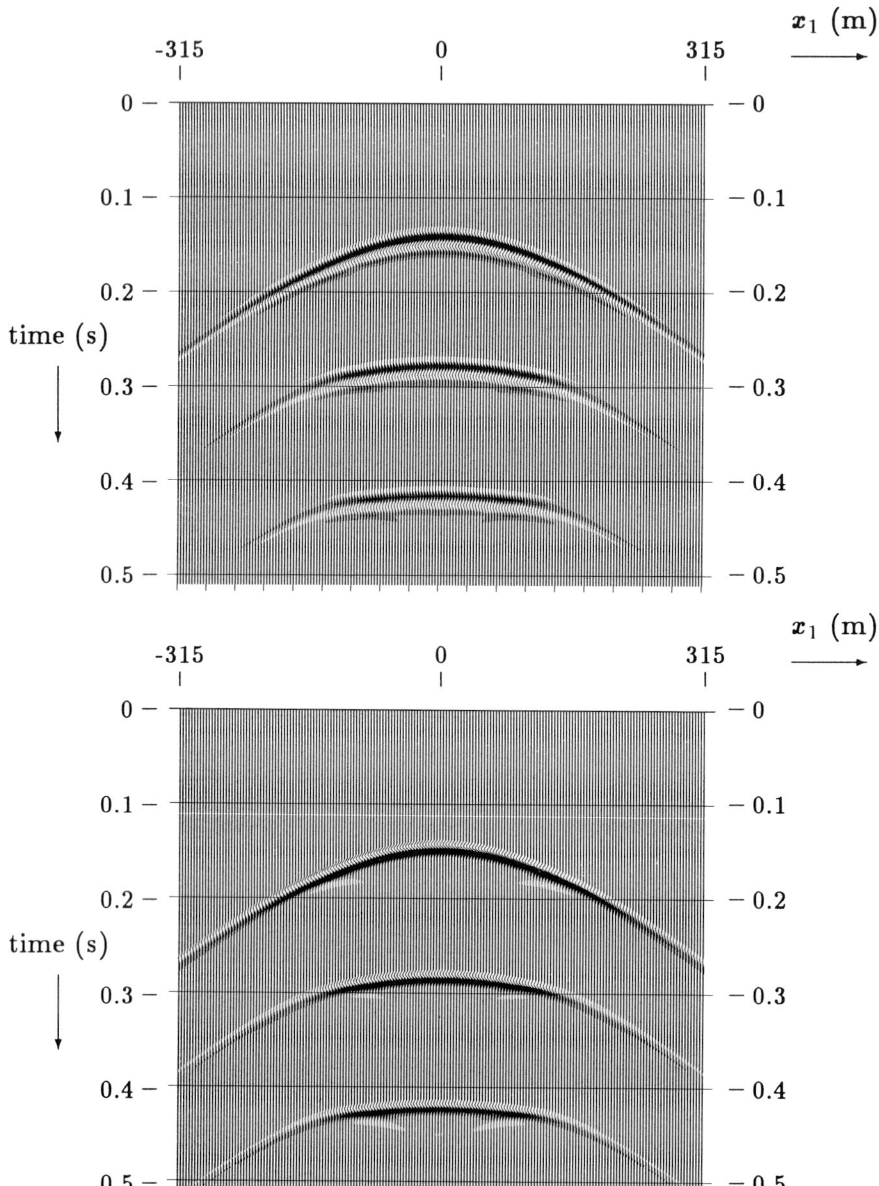


Figure 11.4. The scattered wavefield of a compliant strip in a homogeneous halfspace (top) and its deghosted wavefield (bottom). The source position is above the center of the strip ($x_1^S = 0$).

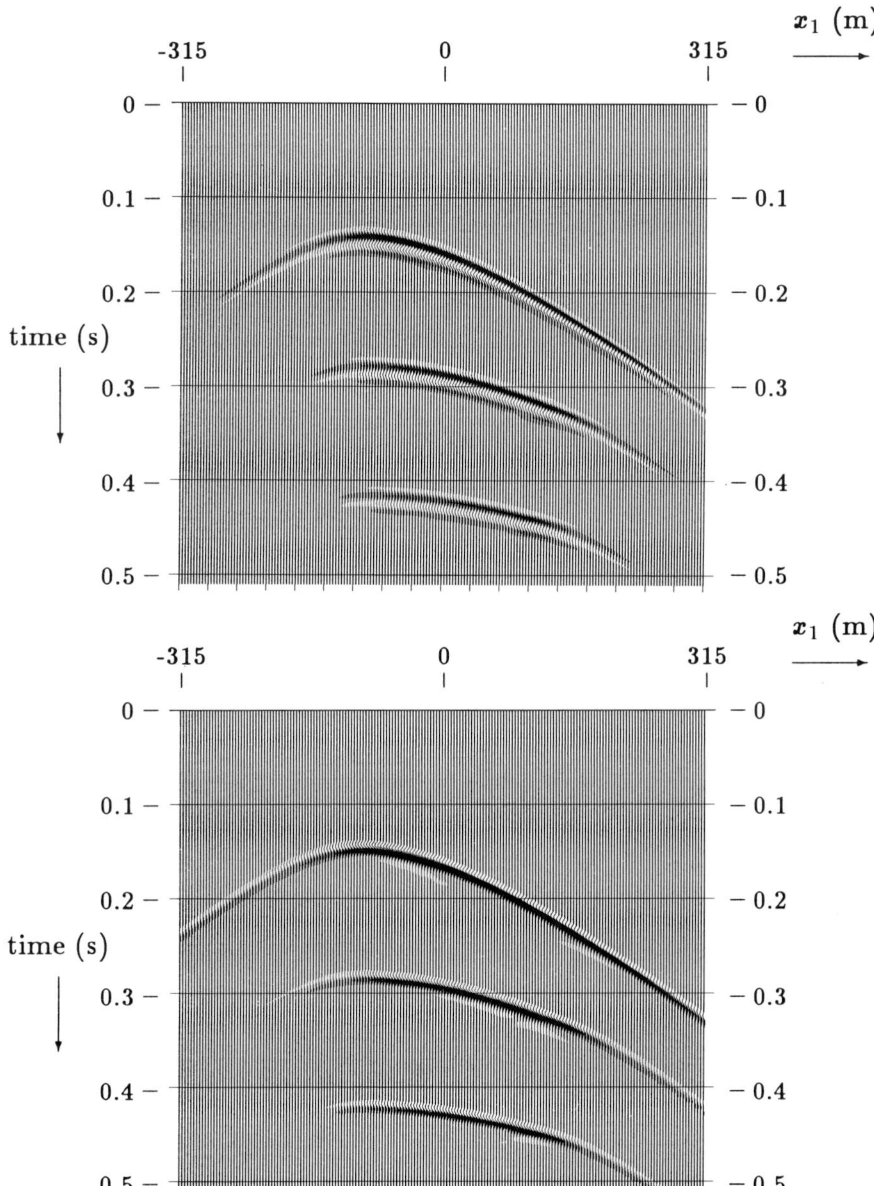


Figure 11.5. The scattered wavefield of a compliant strip in a homogeneous halfspace (top) and its deghosted wavefield (bottom). The source position is above the left edge of the strip ($x_1^S = -105$ m).

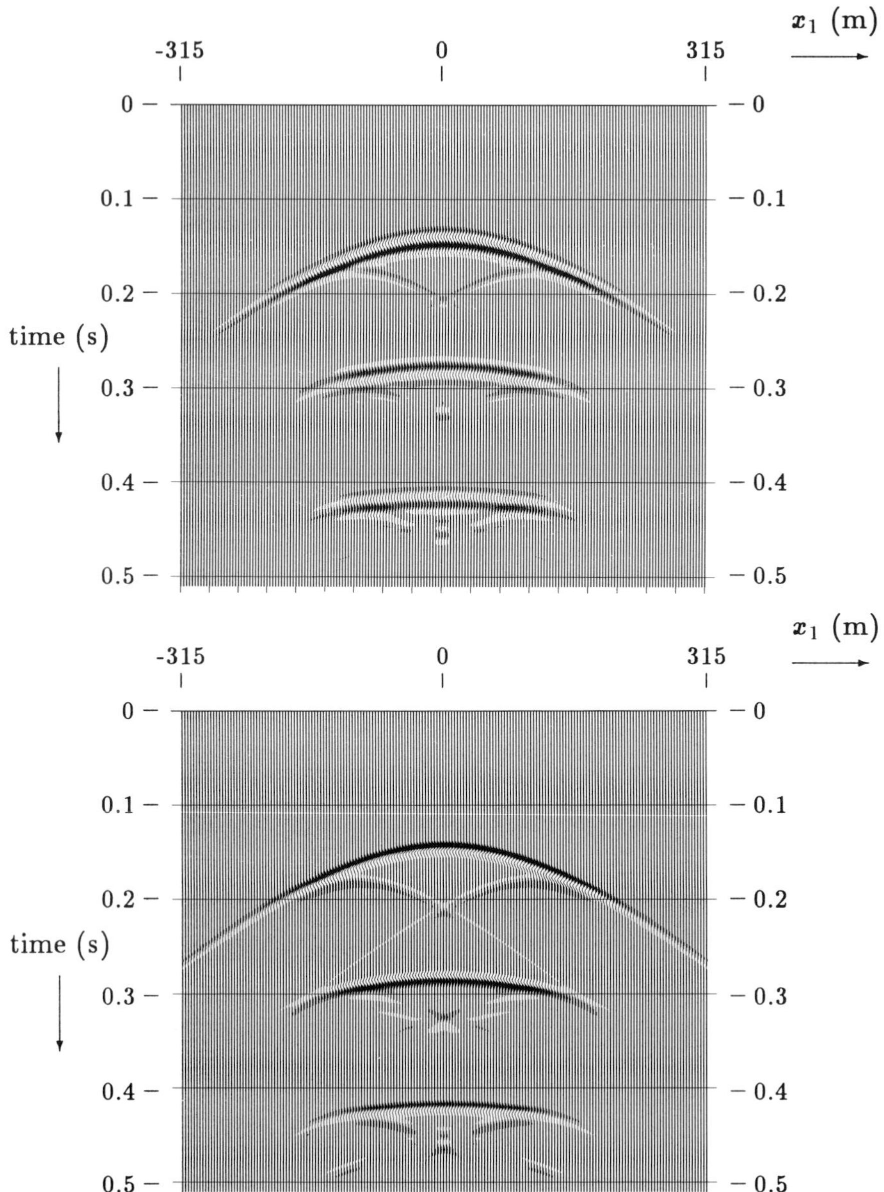


Figure 11.6. The scattered wavefield of a rigid strip in a homogeneous halfspace (top) and its deghosted wavefield (bottom). The source position is above the center of the strip ($x_1^S = 0$).

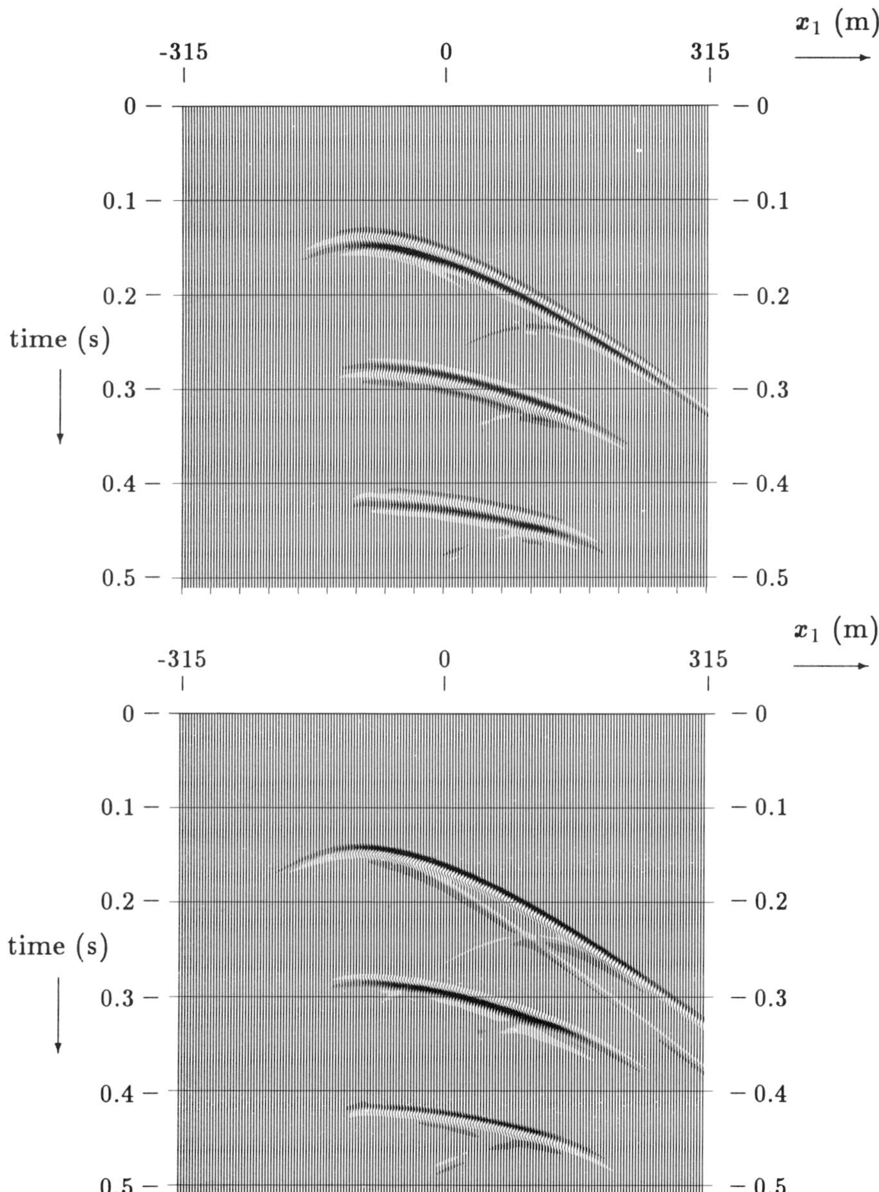


Figure 11.7. The scattered wavefield of a rigid strip in a homogeneous halfspace (top) and its deghosted wavefield (bottom). The source position is above the left edge of the strip ($x_1^S = -105$ m).

# Chapter 12

# Removal of Surface Related Wave Phenomena

The presence of surface-related wave phenomena in geophysical data as water-surface multiples in the marine case leads to problems in further analysis of the data in inversion and migration. This multiple removal has to be effected without changing any relevant subsurface information present in the recorded data. Rayleigh's reciprocity theorem furnishes the tool for this removal. In this theorem the interaction of two non-identical states is considered. One state is identified with the actual situation, while the other is the desired one: the same geology but without the water surface. This procedure does not require any information about the subsurface geology, neither structural nor material.

## 12.1. Reciprocity between the actual and desired state

In the marine case we have the situation as shown in Fig. 12.1. The domain of interest is the halfspace $\mathbb{D} = \{\boldsymbol{x} \in \mathbb{R}^3;\ -\infty < x_1, x_2 < \infty,\ 0 < x_3 < \infty\}$. This halfspace consists of the homogeneous water layer $\mathbb{D}$ and the earth geology $\mathbb{D}_g$ with boundary $\partial\mathbb{D}_g$. The acoustic pressure at the

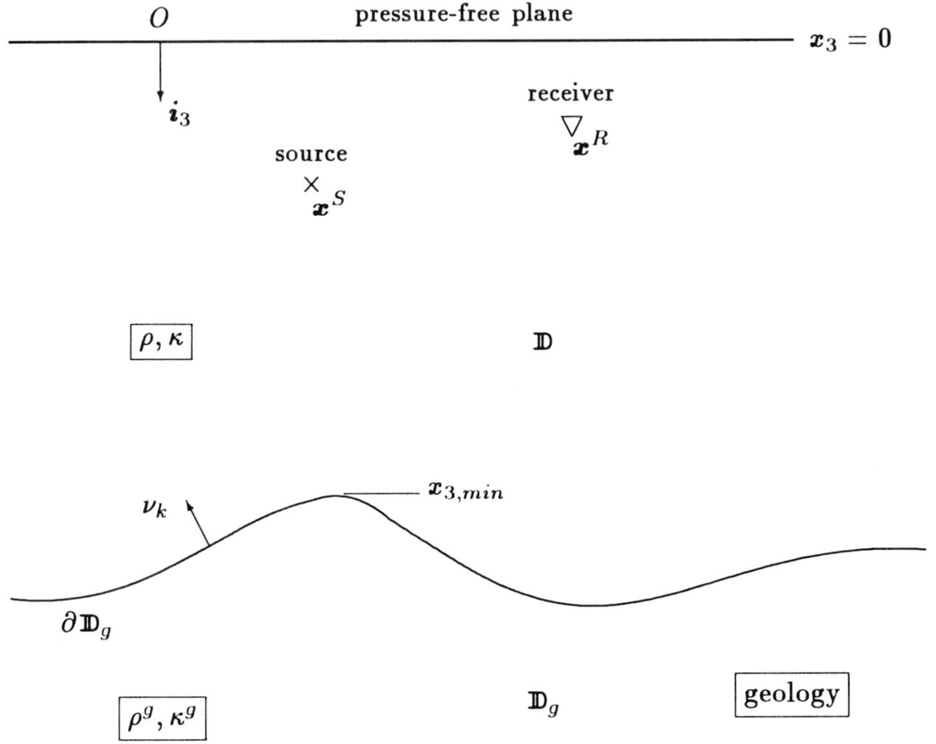


Figure 12.1. State $A$ in the actual configuration.

water surface $x_3 = 0$ is zero. The material constants in $\mathbb{D}$ are $\rho$ and $\kappa$, and the material constants in $\mathbb{D}_g$ are $\rho^g$ and $\kappa^g$. A monopole source of the volume injection type is used and is located at $\boldsymbol{x}^S$. The seismic response, the acoustic pressure, is measured at $\boldsymbol{x}^R$ below the water surface. In order to remove the effect of the water surface, we apply the reciprocity theorem of Section 5.1 to the domain $\mathbb{D} \cup \mathbb{D}_g$. State $A$ is the actual state resulting from the real seismic experiment (see Fig. 12.1). Let this wavefield be denoted as $\{\hat{p}^A, \hat{v}_k^A\} = \{\hat{p}, \hat{v}_k\}(\boldsymbol{x}|\boldsymbol{x}^S, s)$ with sources $\{\hat{q}^A, \hat{f}_k^A\} = \{\hat{q}^S(s)\delta(\boldsymbol{x}-\boldsymbol{x}^S), 0\}$, where $\hat{q}^S(s)$ is the spectrum of the volume injection source. In State $B$, the desired state, the water layer extends to $x_3 \to -\infty$ (see Fig. 12.2). In this

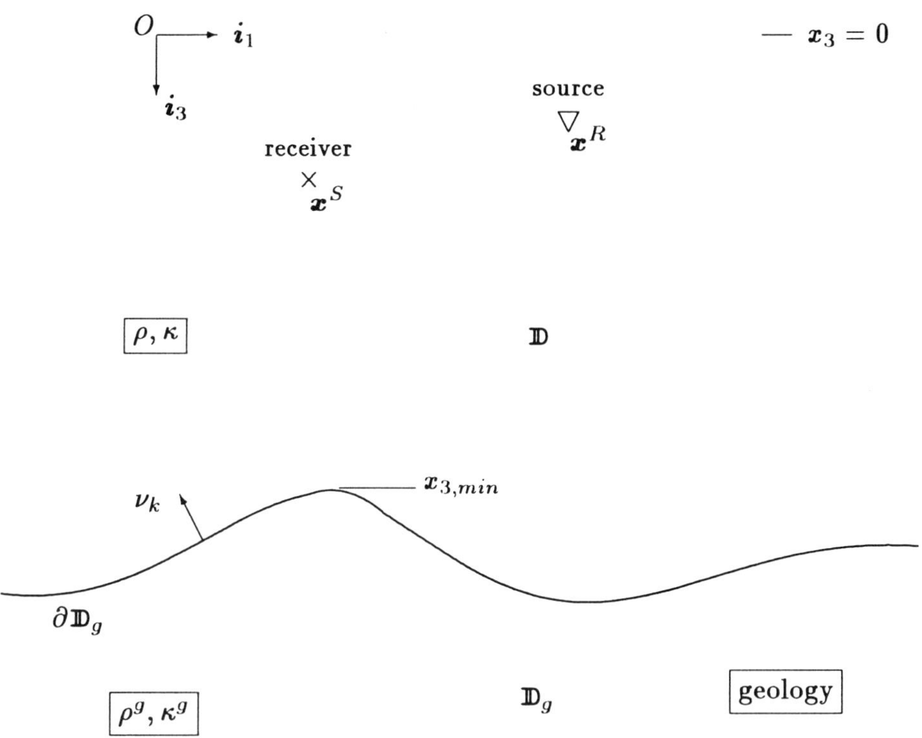


Figure 12.2. State $B$ in the desired configuration.

case $x_3 = 0$ is just an artificial boundary Further, in State $B$ we choose a point source with the spectrum $\hat{q}^S(s)$ identical to the one of the real situation, however the source is now located at the receiver location $\boldsymbol{x}^R \in \mathbb{D}$. Let this wavefield be denoted as $\{\hat{p}^B, \hat{v}_k^B\} = \{\hat{p}^d, \hat{v}_k^d\}(\boldsymbol{x}|\boldsymbol{x}^R, s)$ with sources $\{\hat{q}^B, \hat{f}_k^B\} = \{\hat{q}^S(s)\delta(\boldsymbol{x}-\boldsymbol{x}^R), 0\}$ (see Table 12.1).

With the above mentioned states, we apply the reciprocity theorem to the domain $\mathbb{D} \cup \mathbb{D}_g$ enclosed by a semi-infinite sphere $S_\Delta$ of radius $\Delta$ and center $O$ of the chosen coordinate system (see Fig. 12.3). Taking the limit $\Delta \to \infty$ we arrive at

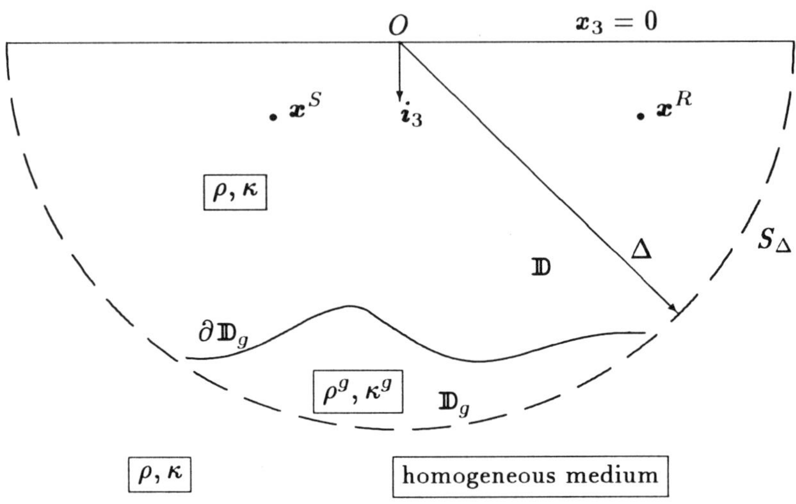


Figure 12.3. Domain of application of the reciprocity theorem.

Table 12.1. States in the field reciprocity theorem

| | State $A$ (actual wavefield) | State $B$ (desired wavefield) |
|---|---|---|
| Field state | $\{\hat{p}, \hat{v}_k\}(\boldsymbol{x}\vert\boldsymbol{x}^S, s)$ | $\{\hat{p}^d, \hat{v}_k^d\}(\boldsymbol{x}\vert\boldsymbol{x}^R, s)$ |
| Material state | $\{\rho, \kappa\}$ in $\mathbb{D}$<br>$\{\rho^g, \kappa^g\}$ in $\mathbb{D}_g$ | $\{\rho, \kappa\}$ in $\mathbb{D}$<br>$\{\rho^g, \kappa^g\}$ in $\mathbb{D}_g$ |
| Source state | $\{\hat{q}^S(s)\delta(\boldsymbol{x}-\boldsymbol{x}^S), 0\}$ | $\{\hat{q}^S(s)\delta(\boldsymbol{x}-\boldsymbol{x}^R), 0\}$ |
| Domain $\mathbb{D} \cup \mathbb{D}_g$ (see Fig. 12.3) | | |

$$\int_{(x_1,x_2)\in\mathbb{R}^2} \hat{p}^d(x_1,x_2,0|\boldsymbol{x}^R,s)\hat{v}_3(x_1,x_2,0|\boldsymbol{x}^S,s)\mathrm{dA}$$
$$= \hat{q}^S(s)\hat{p}(\boldsymbol{x}^R|\boldsymbol{x}^S,s) - \hat{q}^S(s)\hat{p}^d(\boldsymbol{x}^S|\boldsymbol{x}^R,s)\,, \tag{12.1}$$

where we have taken into account that the pressure wavefield of the actual situation vanishes at $x_3 = 0$, viz.,

$$\lim_{x_3\downarrow 0} \hat{p}(\boldsymbol{x}|\boldsymbol{x}^S,s) = 0\,. \tag{12.2}$$

Further we have assumed that outside some semi-sphere of bounded radius and center at the origin $O$ of the chosen coordinate system, the fluid is homogeneous with material constants $\rho$ and $\kappa$ (see Fig. 12.3). As a consequence the contribution of the surface integration over $S_\Delta$ vanishes when $\Delta \to \infty$ (according to the causality condition, cf. Eq. (5.9)).

## 12.2. Auxiliary reciprocity relations

In order to arrive at a reciprocity relation of the type of Eq. (12.1), in which only the scattered-wavefield quantities of the actual state

$$\{\hat{p}^{sct},\hat{v}_k^{sct}\} = \{\hat{p},\hat{v}_k\} - \{\hat{p}^{inc,H},\hat{v}_k^{inc,H}\} \tag{12.3}$$

and the reflected-wavefield quantities of the desired state

$$\{\hat{p}^r,\hat{v}_k^r\} = \{\hat{p}^d,\hat{v}_k^d\} - \{\hat{p}^{inc},\hat{v}_k^{inc}\} \tag{12.4}$$

occur, we derive some auxiliary reciprocity relations. The incident wavefield $\hat{p}^{inc,H}$ of the actual state is given by

$$\hat{p}^{inc,H}(\boldsymbol{x}|\boldsymbol{x}^S,s) = s\rho\hat{q}^S(s)\left[\frac{\exp(-\hat{\gamma}|\boldsymbol{x}-\boldsymbol{x}^S|)}{4\pi|\boldsymbol{x}-\boldsymbol{x}^S|} - \frac{\exp(-\hat{\gamma}|\boldsymbol{x}-\boldsymbol{x}^{S'}|)}{4\pi|\boldsymbol{x}-\boldsymbol{x}^{S'}|}\right]\,, \tag{12.5}$$

where $\hat{\gamma} = s(\kappa\rho)^{\frac{1}{2}} = s/c$ and $\boldsymbol{x}^{S'} = (x_1^S, x_2^S, -x_3^S)$ denotes the image point of $\boldsymbol{x}^S$ with respect to the reflecting surface at $x_3 = 0$. Note that this wavefield $\hat{p}^{inc,H}$ vanishes at $x_3 = 0$. The incident wavefield $\hat{p}^{inc}$ of the desired state is given by

$$\hat{p}^{inc}(\boldsymbol{x}|\boldsymbol{x}^R,s) = s\rho\hat{q}^S(s)\frac{\exp(-\hat{\gamma}|\boldsymbol{x}-\boldsymbol{x}^R|)}{4\pi|\boldsymbol{x}-\boldsymbol{x}^R|}\,. \tag{12.6}$$

*Reciprocity between the actual incident wavefield and the desired wavefield*

We apply the reciprocity theorem of Section 5.1 to the domain $\mathbb{D}'$ ($\boldsymbol{x} \in \mathbb{R}^3$, $-\infty < x_1, x_2 < \infty$, $-\infty < x_3 < 0$) with material constants $\rho$ and $\kappa$ (see Fig. 12.4). State $A$ is the incident wavefield $\{\hat{p}^{inc,H}, \hat{v}_k^{inc,H}\}(\boldsymbol{x}|\boldsymbol{x}^S, s)$ and State $B$ is the desired state $\{\hat{p}^d, \hat{v}_k^d\}(\boldsymbol{x}|\boldsymbol{x}^R, s)$ (see Table 12.2). In $\mathbb{D}'$, State $A$ has a source distribution $\{-\hat{q}(s)\delta(\boldsymbol{x}-\boldsymbol{x}^{S^I}), 0\}$, while State $B$ is sourcefree. With the above mentioned states, we apply the reciprocity theorem to the domain $\mathbb{D}'$ enclosed by a semi-infinite sphere $S_\Delta$ of radius $\Delta$ and center $O$ of the chosen coordinate system. Taking the limit $\Delta \to \infty$ we arrive at

$$-\int_{(x_1,x_2)\in\mathbb{R}^2} \hat{p}^d(x_1, x_2, 0|\boldsymbol{x}^R, s)\hat{v}_3^{inc,H}(x_1, x_2, 0|\boldsymbol{x}^S, s)\mathrm{dA}$$

$$= \hat{q}^S(s)\hat{p}^d(\boldsymbol{x}^{S^I}|\boldsymbol{x}^R, s)\,, \tag{12.7}$$

where we have taken into account that $\hat{p}^{inc,H}$ vanishes at $x_3 = 0$ and that the contribution of the surface integration over $S_\Delta$ vanishes when $\Delta \to \infty$ (cf. Eq. (5.9)).

Addition of the relations of Eqs. (12.1) and (12.7) yields

$$\int_{(x_1,x_2)\in\mathbb{R}^2} \hat{p}^d(x_1, x_2, 0|\boldsymbol{x}^R, s)\hat{v}_3^{sct}(x_1, x_2, 0|\boldsymbol{x}^S, s)\mathrm{dA}$$

$$= \hat{q}^S(s)\hat{p}^{sct}(\boldsymbol{x}^R|\boldsymbol{x}^S, s) - \hat{q}^S(s)[\hat{p}^r(\boldsymbol{x}^S|\boldsymbol{x}^R, s) - \hat{p}^r(\boldsymbol{x}^{S^I}|\boldsymbol{x}^R, s)]\,, \tag{12.8}$$

where we have used

$$\hat{p}^d(\boldsymbol{x}^S|\boldsymbol{x}^R, s) - \hat{p}^d(\boldsymbol{x}^{S^I}|\boldsymbol{x}^R, s)$$

$$= \hat{p}^r(\boldsymbol{x}^S|\boldsymbol{x}^R, s) - \hat{p}^r(\boldsymbol{x}^{S^I}|\boldsymbol{x}^R, s) + \hat{p}^{inc}(\boldsymbol{x}^S|\boldsymbol{x}^R, s) - \hat{p}^{inc}(\boldsymbol{x}^{S^I}|\boldsymbol{x}^R, s)$$

$$= \hat{p}^r(\boldsymbol{x}^S|\boldsymbol{x}^R, s) - \hat{p}^r(\boldsymbol{x}^{S^I}|\boldsymbol{x}^R, s) + \hat{p}^{inc,H}(\boldsymbol{x}^S|\boldsymbol{x}^R, s)\,, \tag{12.9}$$

and Eqs. (12.3) - (12.6).

*Reciprocity between the actual wavefield and the desired wavefield*

We apply the reciprocity theorem of Section 5.1 to the domain $\mathbb{D} \cup \mathbb{D}_g$ (see Fig. 12.5). State $A$ is the actual state $\{\hat{p}, \hat{v}_k\}(\boldsymbol{x}|\boldsymbol{x}^S, s)$ and State $B$ is

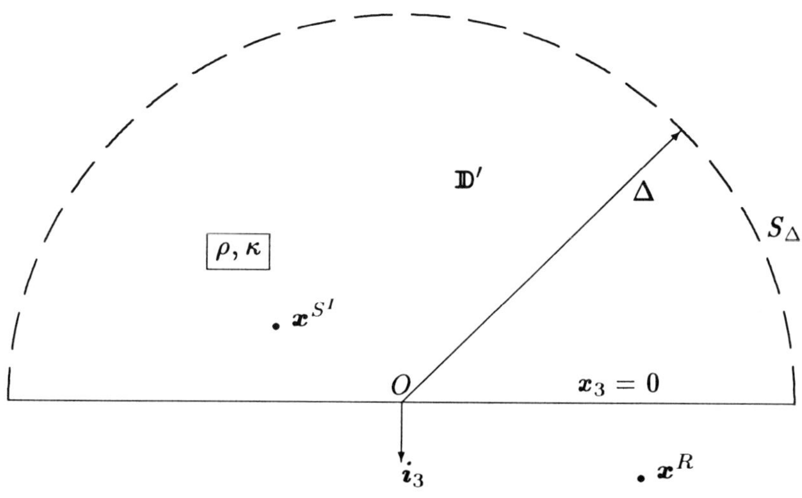


Figure 12.4. Domain of application of the reciprocity theorem.

Table 12.2. States in the field reciprocity theorem

| | State $A$ (actual incident wavefield) | State $B$ (desired wavefield) |
|---|---|---|
| Field state | $\{\hat{p}^{inc,H}, \hat{v}_k^{inc,H}\}(\boldsymbol{x}\vert\boldsymbol{x}^S, s)$ | $\{\hat{p}^d, \hat{v}_k^d\}(\boldsymbol{x}\vert\boldsymbol{x}^R, s)$ |
| Material state | $\{\rho, \kappa\}$ | $\{\rho, \kappa\}$ |
| Source state | $\{-\hat{q}^S(s)\delta(\boldsymbol{x}-\boldsymbol{x}^{S'}), 0\}$ | $\{0, 0\}$ |
| Domain $\mathbb{D}'$ $\{\boldsymbol{x} \in \mathbb{R}^3, -\infty < x_1, x_2 < \infty, -\infty < x_3 < 0\}$ (see Fig. 12.4) | | |

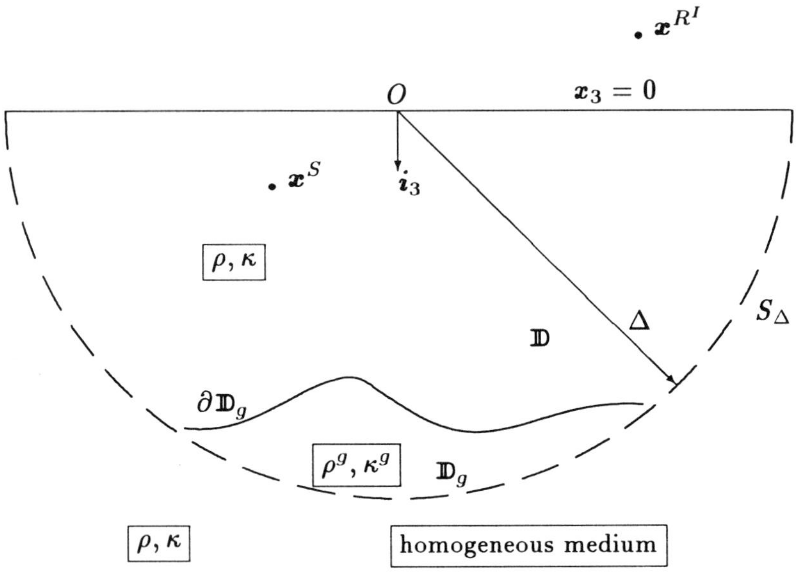


Figure 12.5. Domain of application of the reciprocity theorem.

Table 12.3. States in the field reciprocity theorem

| | State $A$ (actual wavefield) | State $B$ (desired wavefield) |
|---|---|---|
| Field state | $\{\hat{p}, \hat{v}_k\}(\boldsymbol{x}\vert\boldsymbol{x}^S, s)$ | $\{\hat{p}^d, \hat{v}_k^d\}(\boldsymbol{x}\vert\boldsymbol{x}^{R^I}, s)$ |
| Material state | $\{\rho, \kappa\}$ in $\mathbb{D}$<br>$\{\rho^g, \kappa^g\}$ in $\mathbb{D}_g$ | $\{\rho, \kappa\}$ in $\mathbb{D}$<br>$\{\rho^g, \kappa^g\}$ in $\mathbb{D}_g$ |
| Source state | $\{\hat{q}^S(s)\delta(\boldsymbol{x}-\boldsymbol{x}^S), 0\}$ | $\{0, 0\}$ |
| Domain $\mathbb{D} \cup \mathbb{D}_g$ (see Fig. 12.5) | | |

the desired wavefield with source distribution $\{\hat{q}^B, \hat{f}_k^B\} = \{\hat{q}^S(s)\delta(\boldsymbol{x}-\boldsymbol{x}^{R^I}), 0\}$. Note that the source point, $\boldsymbol{x}^{R^I} = (x_1^R, x_2^R, -x_3^R)$, is the image of the actual receiver position (see Table 12.3). With the above mentioned states, we apply the reciprocity theorem to the domain $\mathbb{D} \cup \mathbb{D}_g$ enclosed by a semi-infinite sphere $S_\Delta$ of radius $\Delta$ and center $O$ of the chosen coordinate system. Taking the limit $\Delta \to \infty$ we arrive at

$$\int_{(x_1,x_2)\in\mathbb{R}^2} \hat{p}^d(x_1, x_2, 0|\boldsymbol{x}^{R^I}, s)\hat{v}_3(x_1, x_2, 0|\boldsymbol{x}^S, s)\mathrm{dA} = -\hat{q}^S(s)\hat{p}^d(\boldsymbol{x}^S|\boldsymbol{x}^{R^I}, s)\,, \tag{12.10}$$

where we have taken into account that the pressure wavefield of the actual situation vanishes at $x_3 = 0$. Further we have assumed that outside some semi-sphere of bounded radius and center at the origin $O$ of the chosen coordinate system, the fluid is homogeneous with material constants $\rho$ and $\kappa$ (see Fig. 12.5). As a consequence the contribution of the surface integration over $S_\Delta$ vanishes when $\Delta \to \infty$ (cf. Eq. (5.9)).

*Reciprocity between the actual incident wavefield and the desired wavefield*

We apply the reciprocity theorem of Section 5.1 to the domain $\mathbb{D}'$ ($\boldsymbol{x} \in \mathbb{R}^3$, $-\infty < x_1, x_2 < \infty$, $-\infty < x_3 < 0$) with material constants $\rho$ and $\kappa$ (see Fig. 12.6). State $A$ is the incident wavefield $\{\hat{p}^{inc,H}, \hat{v}_k^{inc,H}\}(\boldsymbol{x}|\boldsymbol{x}^S, s)$ and State $B$ is the desired state $\{\hat{p}^d, \hat{v}_k^d\}(\boldsymbol{x}|\boldsymbol{x}^{R^I}, s)$ (see Table 12.4). In $\mathbb{D}'$, State $A$ has a source distribution $\{-\hat{q}(s)\delta(\boldsymbol{x}-\boldsymbol{x}^{S^I}), 0\}$, while State $B$ has a source distribution $\{\hat{q}(s)\delta(\boldsymbol{x}-\boldsymbol{x}^{R^I}), 0\}$. With the above mentioned states, we apply the reciprocity theorem to the domain $\mathbb{D}'$ enclosed by a semi-infinite sphere $S_\Delta$ of radius $\Delta$ and center $O$ of the chosen coordinate system. Taking the limit $\Delta \to \infty$ we arrive at

$$-\int_{(x_1,x_2)\in\mathbb{R}^2} \hat{p}^d(x_1, x_2, 0|\boldsymbol{x}^{R^I}, s)\hat{v}_3^{inc,H}(x_1, x_2, 0|\boldsymbol{x}^S, s)\mathrm{dA} = \hat{q}^S(s)\hat{p}^{inc,H}(\boldsymbol{x}^{R^I}|\boldsymbol{x}^S, s) + \hat{q}^S(s)\hat{p}^d(\boldsymbol{x}^{S^I}|\boldsymbol{x}^{R^I}, s)\,, \tag{12.11}$$

where we have taken into account that $\hat{p}^{inc,H}$ vanishes at $x_3 = 0$ and that the contribution of the surface integration over $S_\Delta$ vanishes when $\Delta \to \infty$ (cf. Eq. (5.9)).

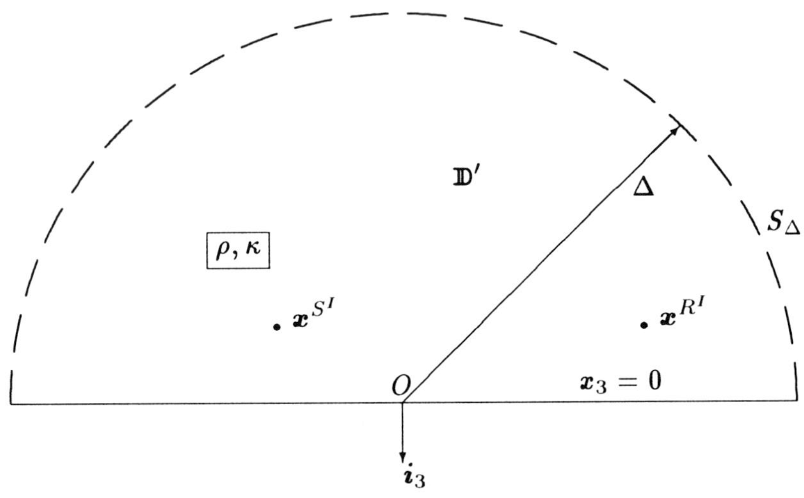


Figure 12.6. Domain of application of the reciprocity theorem.

Table 12.4. States in the field reciprocity theorem

| | State $A$ (actual incident wavefield) | State $B$ (desired wavefield) |
|---|---|---|
| Field state | $\{\hat{p}^{inc,H}, \hat{v}_k^{inc,H}\}(\boldsymbol{x}\vert\boldsymbol{x}^S, s)$ | $\{\hat{p}^d, \hat{v}_k^d\}(\boldsymbol{x}\vert\boldsymbol{x}^{R^I}, s)$ |
| Material state | $\{\rho, \kappa\}$ | $\{\rho, \kappa\}$ |
| Source state | $\{-\hat{q}^S(s)\delta(\boldsymbol{x}-\boldsymbol{x}^{S^I}), 0\}$ | $\{\hat{q}^S(s)\delta(\boldsymbol{x}-\boldsymbol{x}^{R^I}), 0\}$ |
| Domain $\mathbb{D}'$ $\{\boldsymbol{x} \in \mathbb{R}^3, -\infty < x_1, x_2 < \infty, -\infty < x_3 < 0\}$ (see Fig. 12.6) | | |

Addition of the relations of Eqs. (12.10) and (12.11) yields

$$\int_{(x_1,x_2)\in\mathbb{R}^2} \hat{p}^d(\boldsymbol{x}_1,\boldsymbol{x}_2,0|\boldsymbol{x}^{R^I},s)\hat{v}_3^{sct}(\boldsymbol{x}_1,\boldsymbol{x}_2,0|\boldsymbol{x}^S,s)\mathrm{dA}$$
$$= -\hat{q}^S(s)[\hat{p}^r(\boldsymbol{x}^S|\boldsymbol{x}^{R^I},s) - \hat{p}^r(\boldsymbol{x}^{S^I}|\boldsymbol{x}^{R^I},s)]\,, \tag{12.12}$$

where we have used

$$\hat{p}^d(\boldsymbol{x}^S|\boldsymbol{x}^{R^I},s) - \hat{p}^d(\boldsymbol{x}^{S^I}|\boldsymbol{x}^{R^I},s)$$
$$= \hat{p}^r(\boldsymbol{x}^S|\boldsymbol{x}^{R^I},s) - \hat{p}^r(\boldsymbol{x}^{S^I}|\boldsymbol{x}^{R^I},s) + \hat{p}^{inc}(\boldsymbol{x}^S|\boldsymbol{x}^{R^I},s) - \hat{p}^{inc}(\boldsymbol{x}^{S^I}|\boldsymbol{x}^{R^I},s)$$
$$= \hat{p}^r(\boldsymbol{x}^S|\boldsymbol{x}^{R^I},s) - \hat{p}^r(\boldsymbol{x}^{S^I}|\boldsymbol{x}^{R^I},s) + \hat{p}^{inc,H}(\boldsymbol{x}^{R^I}|\boldsymbol{x}^S,s)\,, \tag{12.13}$$

and Eqs. (12.3) - (12.6).

In the reciprocity relations of Eqs. (12.8) and (12.12) still the total wavefield quantities of the desired wavefield occur. We therefore subtract Eq. (12.12) from Eq. (12.8), and since $\hat{p}^{inc,H}(x_1,x_2,0|\boldsymbol{x}^R,s) = 0$, we use

$$\hat{p}^d(\boldsymbol{x}_1,\boldsymbol{x}_2,0|\boldsymbol{x}^R,s) - \hat{p}^d(\boldsymbol{x}_1,\boldsymbol{x}_2,0|\boldsymbol{x}^{R^I},s)$$
$$= \hat{p}^r(\boldsymbol{x}_1,\boldsymbol{x}_2,0|\boldsymbol{x}^R,s) - \hat{p}^r(\boldsymbol{x}_1,\boldsymbol{x}_2,0|\boldsymbol{x}^{R^I},s) \tag{12.14}$$

to arrive at

$$\int_{(x_1',x_2')\in\mathbb{R}^2} [\hat{p}^r(\boldsymbol{x}_1',\boldsymbol{x}_2',0|\boldsymbol{x}^R,s) - \hat{p}^r(\boldsymbol{x}_1',\boldsymbol{x}_2',0|\boldsymbol{x}^{R^I},s)]\hat{v}_3^{sct}(\boldsymbol{x}_1',\boldsymbol{x}_2',0|\boldsymbol{x}^S,s)\mathrm{dA}$$
$$= \hat{q}^S(s)\hat{p}^{sct}(\boldsymbol{x}^R|\boldsymbol{x}^S,s) - \hat{q}^S(s)[\hat{p}^r(\boldsymbol{x}^S|\boldsymbol{x}^R,s) - \hat{p}^r(\boldsymbol{x}^S|\boldsymbol{x}^{R^I},s)]$$
$$+\hat{q}^S(s)[\hat{p}^r(\boldsymbol{x}^{S^I}|\boldsymbol{x}^R,s) - \hat{p}^r(\boldsymbol{x}^{S^I}|\boldsymbol{x}^{R^I},s)]\,. \tag{12.15}$$

Equation (12.15) is the required relation where only the scattered-wavefield quantities of the actual wavefield and the reflected-wavefield quantities of the desired wavefield occur. In the left-hand side of Eq. (12.15) we observe that the reflected wavefield depends only on the receiver positions $(\boldsymbol{x}_1',\boldsymbol{x}_2',0)$ and the source positions $\boldsymbol{x}^R$ and $\boldsymbol{x}^{R^I}$, while in its right-hand side the reflected wavefield depends on the receiver positions $\boldsymbol{x}^S$ and $\boldsymbol{x}^{S^I}$ and the source positions $\boldsymbol{x}^R$ and $\boldsymbol{x}^{R^I}$. In the next section we align the different vertical positions.

## 12.3. Alignment of vertical positions

On account of the general reciprocity relation of Eq. (6.9) between a transmitting and a receiving transducer of the volume injection type we immediately deduce

$$\hat{q}^S(s)\hat{p}^d(\boldsymbol{x}^S|\boldsymbol{x}^R,s) = \hat{q}^S(s)\hat{p}^d(\boldsymbol{x}^R|\boldsymbol{x}^S,s)\,. \tag{12.16}$$

Since the incident-wavefield constituent of the desired wavefield also fulfills a similar reciprocity relation, we observe that the reflected wavefield satisfies the reciprocity relation

$$\hat{q}^S(s)\hat{p}^r(\boldsymbol{x}^S|\boldsymbol{x}^R,s) = \hat{q}^S(s)\hat{p}^r(\boldsymbol{x}^R|\boldsymbol{x}^S,s)\,. \tag{12.17}$$

Using the plane-wave-decomposition properties in a homogeneous section (see Eq. (10.9)), we observe that

$$\hat{p}^r(x_1,x_2,x_3^R|\boldsymbol{x}^S,s) = F^{-1}\Big\{\exp(s\Gamma x_3^R)F\Big\{\hat{p}^r(x_1,x_2,0|\boldsymbol{x}^S,s)\Big\}\Big\} \tag{12.18}$$

and using the reciprocity relation of Eq. (12.17), we have

$$\hat{p}^r(\boldsymbol{x}^S|x_1,x_2,x_3^R,s) = F^{-1}\Big\{\exp(s\Gamma x_3^R)F\Big\{\hat{p}^r(x_1,x_2,0|\boldsymbol{x}^S,s)\Big\}\Big\}\,. \tag{12.19}$$

Here, $F$ denotes the two-dimensional Fourier transformation defined by Eq. (9.27) and $F^{-1}$ denotes its inverse defined by Eq. (9.28). Similarly, we have

$$\hat{p}^r(\boldsymbol{x}^S|x_1,x_2,-x_3^R,s) = F^{-1}\Big\{\exp(-s\Gamma x_3^R)F\Big\{\hat{p}^r(x_1,x_2,0|\boldsymbol{x}^S,s)\Big\}\Big\}\,, \tag{12.20}$$

hence, subtraction of these two results leads to

$$\begin{aligned}&\hat{p}^r(\boldsymbol{x}^S|x_1,x_2,x_3^R,s) - \hat{p}^r(\boldsymbol{x}^S|x_1,x_2,-x_3^R,s)\\ &\quad = F^{-1}\Big\{2\sinh(s\Gamma x_3^R)F\Big\{\hat{p}^r(x_1,x_2,0|\boldsymbol{x}^S,s)\Big\}\Big\}\end{aligned} \tag{12.21}$$

and thus

$$\begin{aligned}&\hat{p}^r(x_1',x_2',0|x_1,x_2,x_3^R,s) - \hat{p}^r(x_1',x_2',0|x_1,x_2,-x_3^R,s)\\ &\quad = F^{-1}\Big\{2\sinh(s\Gamma x_3^R)F\{\hat{p}^r(x_1,x_2,0|x_1',x_2',0,s)\}\Big\}\,.\end{aligned} \tag{12.22}$$

Using these expressions in Eq. (12.15) and applying to both sides a Fourier transformation with respect to $x_1^R, x_2^R$, dividing the result by $2\sinh(s\Gamma x_3^R)$ and taking an inverse transformation, we obtain

$$\int_{(x_1',x_2')\in\mathbb{R}^2} \hat{p}^r(x_1^R, x_2^R, 0|x_1', x_2', 0, s)\hat{v}_3^{sct}(x_1', x_2', 0|\boldsymbol{x}^S, s)\mathrm{dA}$$

$$= \hat{q}^S(s)\hat{p}^{up}(x_1^R, x_2^R, 0|\boldsymbol{x}^S, s) \tag{12.23}$$

$$-\hat{q}^S(s)[\hat{p}^r(x_1^R, x_2^R, 0|\boldsymbol{x}^S, s) - \hat{p}^r(x_1^R, x_2^R, 0|\boldsymbol{x}^{S'}, s)]\,,$$

where $\hat{p}^{up}$ is given by Eq. (11.26). Similar to Eq. (12.21) we have

$$\hat{p}^r(x_1^R, x_2^R, 0|x_1, x_2, x_3^S, s) - \hat{p}^r(x_1^R, x_2^R, 0|x_1, x_2, -x_3^S, s)$$

$$= F^{-1}\Big\{2\sinh(s\Gamma x_3^S)F\Big\{\hat{p}^r(x_1^R, x_2^R, 0|x_1, x_2, 0, s)\Big\}\Big\}\,. \tag{12.24}$$

Using this expression in the right-hand side of Eq. (12.23), applying to both sides a Fourier transformation with respect to $x_1^S, x_2^S$, dividing the result by $2\sinh(s\Gamma x_3^S)$ and taking an inverse transformation, we obtain

$$\int_{(x_1',x_2')\in\mathbb{R}^2} \hat{p}^r(x_1^R, x_2^R, 0|x_1', x_2', 0, s)\hat{V}(x_1', x_2'|x_1^S, x_2^S, s)\mathrm{dA}$$

$$= \hat{q}^S(s)\hat{p}^{dgh}(x_1^R, x_2^R, 0|x_1^S, x_2^S, 0, s) - \hat{q}^S(s)\hat{p}^r(x_1^R, x_2^R, 0|x_1^S, x_2^S, 0, s)\,, \tag{12.25}$$

where $\hat{p}^{dgh}$ is given by Eq. (11.32) and

$$\hat{V}(x_1', x_2'|x_1, x_2, s) = F^{-1}\left\{\frac{1}{2\sinh(s\Gamma x_3^S)}F\Big\{\hat{v}_3^{sct}(x_1', x_2', 0|x_1, x_2, x_3^S, s)\Big\}\right\}, \tag{12.26}$$

where (cf. Eq. (11.15))

$$\hat{v}_3^{sct}(x_1, x_2, 0|\boldsymbol{x}^S, s) = F^{-1}\left\{\frac{-2\Gamma}{\rho}F\Big\{\hat{p}^{up}(x_1, x_2, 0|\boldsymbol{x}^S, s)\Big\}\right\}\,. \tag{12.27}$$

Interchanging the order of application of the forward and inverse Fourier transformations of Eq. (12.26) with the ones of Eq. (12.27) and using the definition of $\hat{p}^{dgh}$ of Eq. (11.32), we then obtain the alternative expression

$$\hat{V}(x_1, x_2|x_1^S, x_2^S, s) = F^{-1}\left\{\frac{-2\Gamma}{\rho}F\Big\{\hat{p}^{dgh}(x_1, x_2, 0|x_1^S, x_2^S, 0, s)\Big\}\right\}\,. \tag{12.28}$$

Equation (12.25) is a linear integral equation of the second kind for the unknown $\hat{p}^r(x_1^R, x_2^R, 0|x_1^S, x_2^S, 0, s)$ provided that the source spectrum $\hat{q}^S$ is known. For a discussion on the relevance of the source wavelet we refer to ZIOLKOWSKI (1991). VERSCHUUR *et al.* (1992) has suggested an adaptive multiple-removal procedure, in which the multiple removal and the wavelet estimation are carried out simultaneously using some minimum-energy criterion. Further $\hat{V}(x_1^R, x_2^R|x_1^S, x_2^S, s)$ represents the kernel of the integral equation, while $\hat{q}^S(s)\hat{p}^{dgh}(x_1^R, x_2^R, 0|x_1^S, x_2^S, 0, s)$ represents the known term of the integral equation. Note that the integral variables in the integral equation (12.25) are the horizontal coordinates of the various source locations, while the receiver position of the desired wavefield is fixed. Hence, the solution is obtained in the common-receiver domain. The kernel and the known term of the integral equation are both expressed in terms of $\hat{p}^{dgh}$.

*Integral equation for the particle velocity*

Equation (12.25) constitutes an integral equation of the second kind for the acoustic pressure $\hat{p}^r$. A simpler integral equation may be obtained by considering the particle velocity (cf. Eq. (7.70))

$$\hat{v}_3^r(x_1, x_2, 0|x_1^S, x_2^S, 0, s) = \lim_{x_3 \downarrow 0} \frac{-1}{s\rho}\partial_3 \hat{p}^r(\boldsymbol{x}, |x_1^S, x_2^S, 0, s)\,. \tag{12.29}$$

Using the Fourier transformation with respect to the horizontal receiver coordinates and Eq. (12.18), the expression for the particle velocity $\hat{v}_3^r$ may be written as

$$\hat{v}_3^r(x_1, x_2, 0|x_1^S, x_2^S, 0, s) = F^{-1}\left\{\frac{-\Gamma}{\rho}F\left\{\hat{p}^r(x_1, x_2, 0|x_1^S, x_2^S, 0, s)\right\}\right\}\,. \tag{12.30}$$

Application of the Fourier transformation with respect to $x_1^R, x_2^R$ to both sides of Eq. (12.25), multiplying the result with $-\Gamma/\rho$ and taking an inverse Fourier transformation, we arrive at

$$\begin{aligned}\int_{(x_1', x_2') \in \mathbb{R}^2} & \hat{v}_3^r(x_1^R, x_2^R, 0|x_1', x_2', 0, s)\hat{V}(x_1', x_2'|x_1^S, x_2^S, s)\mathrm{dA} \\ &= \tfrac{1}{2}\hat{q}^S(s)\hat{V}(x_1^R, x_2^R|x_1^S, x_2^S, s) - \hat{q}^S(s)\hat{v}_3^r(x_1^R, x_2^R, 0|x_1^S, x_2^S, 0, s)\,,\end{aligned} \tag{12.31}$$

in which we have used the expression for $\hat{V}$, Eq. (12.28). This integral is simpler than the one of Eq. (12.25) in the respect that the known term, for a fixed receiver position, is a subset of the space to which the kernel functions belong. Once we have solved the integral equation (12.31), the acoustic pressure $\hat{p}^r$ follows from

$$\hat{p}^r(x_1, x_2, 0|x_1^S, x_2^S, 0, s) = F^{-1}\left\{\frac{-\rho}{\Gamma} F\left\{\hat{v}_3^r(x_1, x_2, 0|x_1^S, x_2^S, 0, s)\right\}\right\} . \quad (12.32)$$

## 12.4. Actual multiple-removal procedure

In the following analysis we concentrate on the integral equation for the acoustic pressure that follows from Eq. (12.25) as

$$\hat{p}^r(x_1^R, x_2^R, 0|x_1^S, x_2^S, 0, s) = \hat{p}^{dgh}(x_1^R, x_2^R, 0|x_1^S, x_2^S, 0, s)$$
$$- \int_{(x_1', x_2') \in \mathbb{R}^2} \hat{p}^r(x_1^R, x_2^R, 0|x_1', x_2', 0, s)\hat{V}^{dcv}(x_1', x_2'|x_1^S, x_2^S, s)\mathrm{dA} , \quad (12.33)$$

in which $\hat{V}^{dcv}$ is the deconvolved counterpart of $\hat{V}$, viz.,

$$\hat{V}^{dcv}(x_1', x_2'|x_1^S, x_2^S, s) = \frac{1}{\hat{q}^S(s)}\hat{V}(x_1', x_2'|x_1^S, x_2^S, s) . \quad (12.34)$$

The next step is to transform Eq. (12.33) back to the time-domain,

$$p^r(x_1^R, x_2^R, 0|x_1^S, x_2^S, 0, t) = \chi_{T_1}(t)p^{dgh}(x_1^R, x_2^R, 0|x_1^S, x_2^S, 0, t)$$
$$- \int_{(x_1', x_2') \in \mathbb{R}^2} C_t\{p^r, V^{dcv}\}\mathrm{dA} , \quad (12.35)$$

where

$$C_t\{p^r, V^{dcv}\} = \int_{t'=t_1}^{t} p^r(x_1^R, x_2^R, 0|x_1', x_2', 0, t-t')V^{dcv}(x_1', x_2'|x_1^S, x_2^S, t')\mathrm{d}t' \quad (12.36)$$

and $\chi_{T_1}(t)$ is the characteristic function of the set $T_1$,

$$\chi_{T_1}(t) = \{1, \tfrac{1}{2}, 0\} \quad \text{when } t \in \{T_1, \partial T_1, T_1'\} , \quad (12.37)$$

with $\partial T_1 = t_1$ and $T_1'$ is the complement of $T_1$ in $\mathbb{R}$. The time interval $T_1$ needs some explanation. We have assumed that the seismic experiment starts to act at $t = 0$. In view of causality we observe that $p^{dgh}$ is a wavefield that has travelled at least from the plane $x_3 = 0$ to and from the plane $x_3 = x_{3,min}$ (see Fig. 12.2), hence

$$T_1 = \left\{ t \in \mathbb{R},\ t > t_1 = \frac{2\, x_{3,min}}{c} \right\} . \tag{12.38}$$

In view of the time-domain equivalent of Eq. (12.28) the same observation holds for both $V$ and $V^{dcv}$. Therefore, the lower bound of integration of the convolution in Eq. (12.36) is set at $t' = t_1$. The upper bound of integration follows directly from the causality of $p^r$. In order to solve the integral equation (12.35), we apply a Neumann expansion of the type (see Section 2.5)

$$p^r(x_1^R, x_2^R, 0|x_1^S, x_2^S, 0, t) = \sum_{n=0}^{\infty} p_n^r(x_1^R, x_2^R, 0|x_1^S, x_2^S, 0, t) , \tag{12.39}$$

where the first term of this series is given by

$$p_0^r(x_1^R, x_2^R, 0|x_1^S, x_2^S, 0, t) = \chi_{T_1}(t) p^{dgh}(x_1^R, x_2^R, 0|x_1^S, x_2^S, 0, t) , \tag{12.40}$$

and the $n^{\text{th}}$ term follows from the $(n-1)^{\text{th}}$ one as

$$p_n^r(x_1^R, x_2^R, 0|x_1^S, x_2^S, 0, t) = - \int_{(x_1', x_2') \in \mathbb{R}^2} \mathrm{dA}$$

$$\int_{t'=t_1}^{t} p_{n-1}^r(x_1^R, x_2^R, 0|x_1', x_2', 0, t-t') V^{dcv}(x_1', x_2'|x_1^S, x_2^S, t') \mathrm{d}t' . \tag{12.41}$$

As a next step we proof by induction that $p_n^r$ is zero for $t < (n+1)t_1$. It is certainly true for $n = 0$ according to Eq. (12.40). Let us assume that $p_{n-1}^r$ is zero for $t < nt_1$, $n > 0$; then, $p_{n-1}^r(x_1^R, x_2^R, 0|x_1', x_2', 0, t-t')$ vanishes for $t - t' < nt_1$. Since the integration variable $t' > t_1$, we conclude that $p_{n-1}^r(x_1^R, x_2^R, 0|x_1', x_2', 0, t-t')$ vanishes for $t < (n+1)t_1$, and as a consequence of Eq. (12.41), that $p_n^r(x_1^R, x_2^R, 0|x_1', x_2', 0, t)$ vanishes for $t < (n+1)t_1$. This allows us to rewrite Eq. (12.39) as

$$p^r(x_1^R, x_2^R, 0|x_1^S, x_2^S, 0, t) = \sum_{n=0}^{\infty} \chi_{T_{(n+1)}}(t) p_n^r(x_1^R, x_2^R, 0|x_1^S, x_2^S, 0, t) , \tag{12.42}$$

where

$$T_{n+1} = \left\{ t \in \mathbb{R},\ t > t_{n+1} = (n+1)t_1 = \frac{2(n+1)\, x_{3,min}}{c} \right\} . \tag{12.43}$$

For a finite time interval of observation, $0 < t < t_{N+1}$, the summation of Eq. (12.42) is confined to $0 \leq n \leq N-1$, and the Neumann expansion is convergent. Therefore, we replace the infinite summation in Eq. (12.42) by a finite one,

$$p^r(x_1^R, x_2^R, 0|x_1^S, x_2^S, 0, t) = \sum_{n=0}^{N-1} p_n^r(x_1^R, x_2^R, 0|x_1^S, x_2^S, 0, t) ,$$
$$\text{for } 0 < t < t_{N+1} = \frac{2(N+1)\, x_{3,min}}{c} . \tag{12.44}$$

The evaluation of the temporal convolution is a simple algebraic multiplication in the Laplace-transform domain. Therefore we transform Eqs. (12.44) and (12.41) to this domain, resulting into

$$\hat{p}^r(x_1^R, x_2^R, 0|x_1^S, x_2^S, 0, s) = \sum_{n=0}^{N-1} \hat{p}_n^r(x_1^R, x_2^R, 0|x_1^S, x_2^S, 0, s) , \tag{12.45}$$

where the first term of this series is given by

$$\hat{p}_0^r(x_1^R, x_2^R, 0|x_1^S, x_2^S, 0, s) = \hat{p}^{dgh}(x_1^R, x_2^R, 0|x_1^S, x_2^S, 0, s) , \tag{12.46}$$

and the $n^{\text{th}}$ follows from the $(n-1)^{\text{th}}$ one as

$$\hat{p}_n^r(x_1^R, x_2^R, 0|x_1^S, x_2^S, 0, s) =$$
$$- \int_{(x_1', x_2') \in \mathbb{R}^2} \hat{p}_{n-1}^r(x_1^R, x_2^R, 0|x_1', x_2', 0, s) \frac{1}{\hat{q}^S(s)} \hat{V}(x_1', x_2'|x_1^S, x_2^S, s) \mathrm{dA} , \tag{12.47}$$

where we have used Eq. (12.34). Note that we may not conclude that the expression of Eq. (12.45), for some value of $s$, is convergent when $N$ tends to infinity. Only, the result of the time-domain expression in a finite time interval (see Eq. (12.42)) is convergent when $N$ tends to infinity. After the completion of the summation in the right-hand side of Eq. (12.45), the result can be transformed back to the time domain yielding the pressure wavefield $p^r(x_1^R, x_2^R, 0|x_1^S, x_2^S, 0, t)$.

## 12.5. Multiple removal in the strip configuration

As a test case for the multiple-removal procedure, we choose the two-dimensional dataset of the strip configuration embedded in the homogeneous halfspace as it was calculated in Section 9.5.2 for the compliant and rigid strip, respectively This dataset consists of 255 source positions with 255 receivers per source arranged in a split-spread fashion. Since we are dealing with a two-dimensional wavefield problem, all the relevant operators and wavefield representations are replaced by their two-dimensional equivalents. In particular the desired reflected wavefield in the Laplace-transform domain follows from the transformed Neumann expansion

$$\hat{p}^r(x_1^R, 0|x_1^S, 0, s) = \sum_{n=0}^{N-1} \hat{p}_n^r(x_1^R, 0|x_1^S, 0, s)\,, \tag{12.48}$$

where the expansion terms follows from the recursion formula

$$\hat{p}_0^r(x_1^R, 0|x_1^S, 0) = \hat{p}^{dgh}(x_1^R, 0|x_1^S, 0, s)\,, \tag{12.49}$$

$$\hat{p}_n^r(x_1^R, 0|x_1^S, 0, s) = \\ -\int_{x_1' \in \mathbb{R}} \hat{p}_{n-1}^r(x_1^R, 0|x_1', 0, s)\frac{1}{\hat{W}(s)} s\rho\hat{V}(x_1', x_2'|x_1^S, x_2^S, s)\mathrm{dA}\,, \\ n = 1, \cdots, N-1, \tag{12.50}$$

and the source wavelet

$$\hat{W}(s) = s\rho\hat{q}^S(s)\,. \tag{12.51}$$

The deghosted wavefield in the expression for the first term of the Neumann expansion, Eq. (12.49), follows from the relevant calculations in Section 11.6. For each source position, the kernel $s\rho\hat{V}$ follows from the deghosted wavefield as (cf. Eq. (12.28))

$$s\rho\hat{V}(x_1|x_1^S, s) = F^{-1}\Big\{-2s\Upsilon\, F\Big\{\hat{p}^{dgh}(x_1, 0|x_1^S, 0, s)\Big\}\Big\}\,, \tag{12.52}$$

where $\Upsilon = (c^{-2} + \alpha_1^2)^{\frac{1}{2}}$ and the one-dimensional Fourier transformation $F$ and its inverse $F^{-1}$ are defined by Eqs. (9.101) and (9.102), respectively. Note that this kernel is calculated in the common-source domain (also called common-shot domain).

Inspection of the structure of the recursion formula of Eq. (12.50) learns that, in order to keep the process operative, in each iteration we have to calculate the wavefield quantities for all relevant source positions (for a chosen receiver position). After we have carried out this procedure in the common-receiver domain, the reflected part of the desired wavefield $\hat{p}^r(x_1^R, 0|x_1^S, 0, s)$ is obtained for one particular horizontal receiver position $x_1^R$ at the level $x_3^R = 0$ and all relevant horizontal source positions $x_1^S$ at the level $x_3^S = 0$. The results in the common-source domain are obtained by using the reciprocity relation of Eq. (12.17) and noting that the vertical levels and the volume injection functions in the two cases are the same; hence

$$\hat{p}^r(x_1^S, 0|x_1^R, 0, s) = \hat{p}^r(x_1^R, 0|x_1^S, 0, s)\,. \tag{12.53}$$

This implies that our computed results in the common-receiver domain are equal to the ones in the common-source domain just by interchanging the role of source and receiver (see Fig. 12.7). Finally the common-source domain data are transferred back to the time domain.

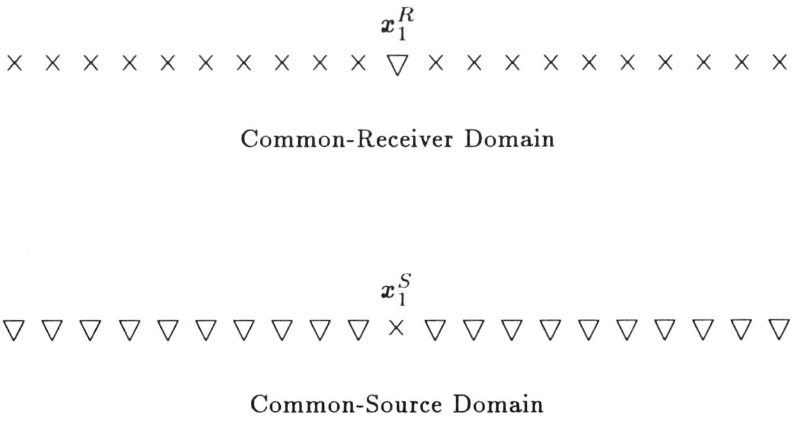

$x_1^R$

× × × × × × × × × × ▽ × × × × × × × × × ×

Common-Receiver Domain

$x_1^S$

▽ ▽ ▽ ▽ ▽ ▽ ▽ ▽ ▽ ▽ × ▽ ▽ ▽ ▽ ▽ ▽ ▽ ▽ ▽ ▽

Common-Source Domain

Figure 12.7. Acquisition configuration in the common-receiver domain and in the common-source domain, respectively.

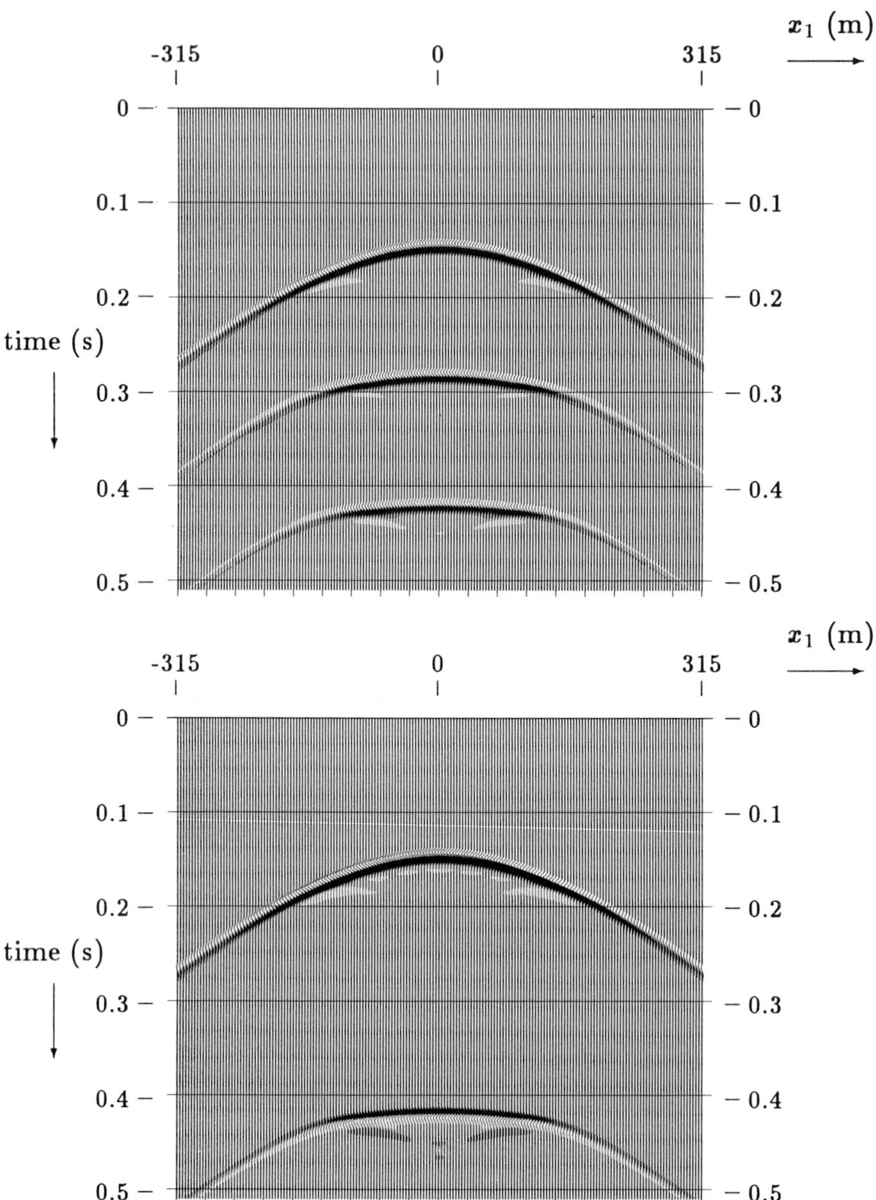


Figure 12.8. The deghosted wavefield of a compliant strip in a homogeneous halfspace (top) and the result after the first-order multiple elimination (bottom). The source position is above the center of the strip ($x_1^S = 0$).

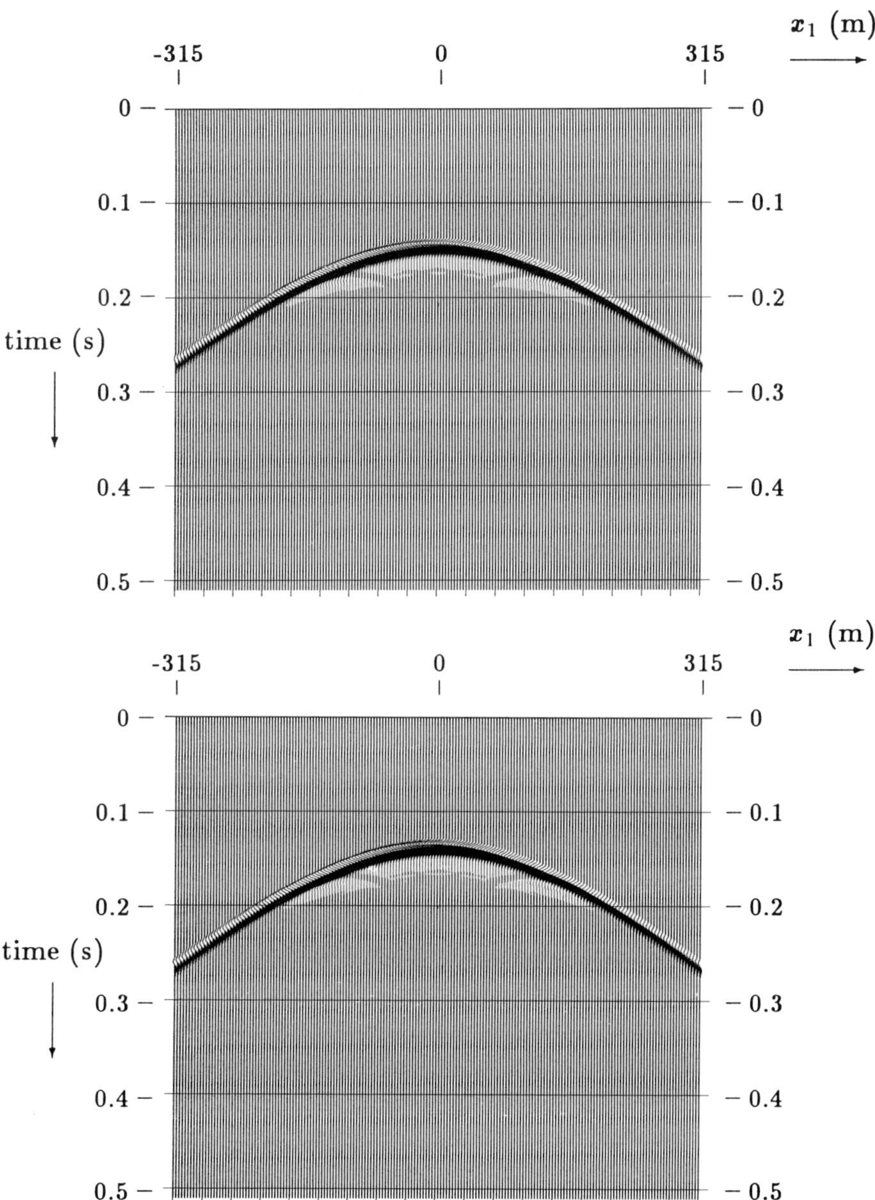


Figure 12.9. The wavefield of Fig. 12.8 after the second-order multiple elimination (top) and the wavefield, at $x_3^R = x_3^S = 0$, of a compliant strip in a homogeneous embedding (bottom).

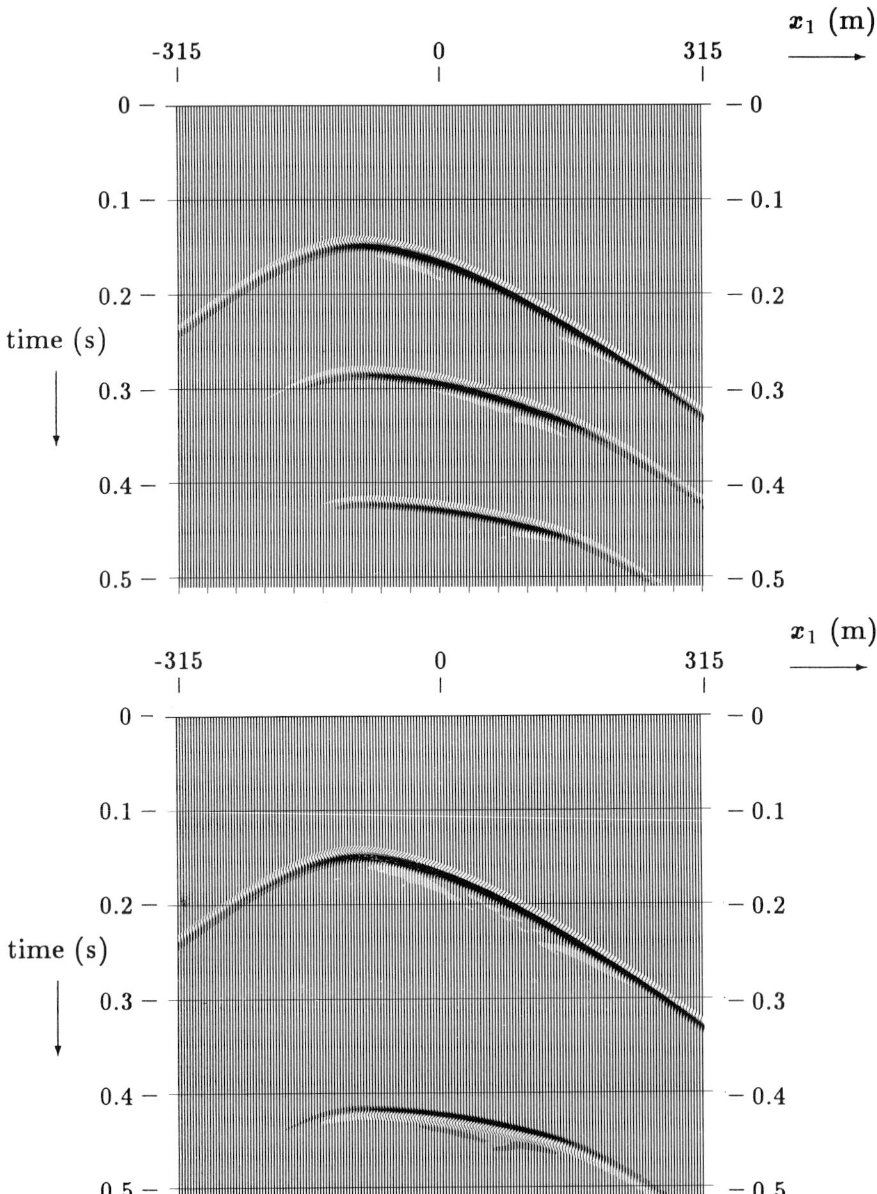


Figure 12.10. The deghosted wavefield of a compliant strip in a homogeneous half-space (top) and the result after the first-order multiple elimination (bottom). The source position is above the left edge of the strip ($x_1^S = -105$ m).

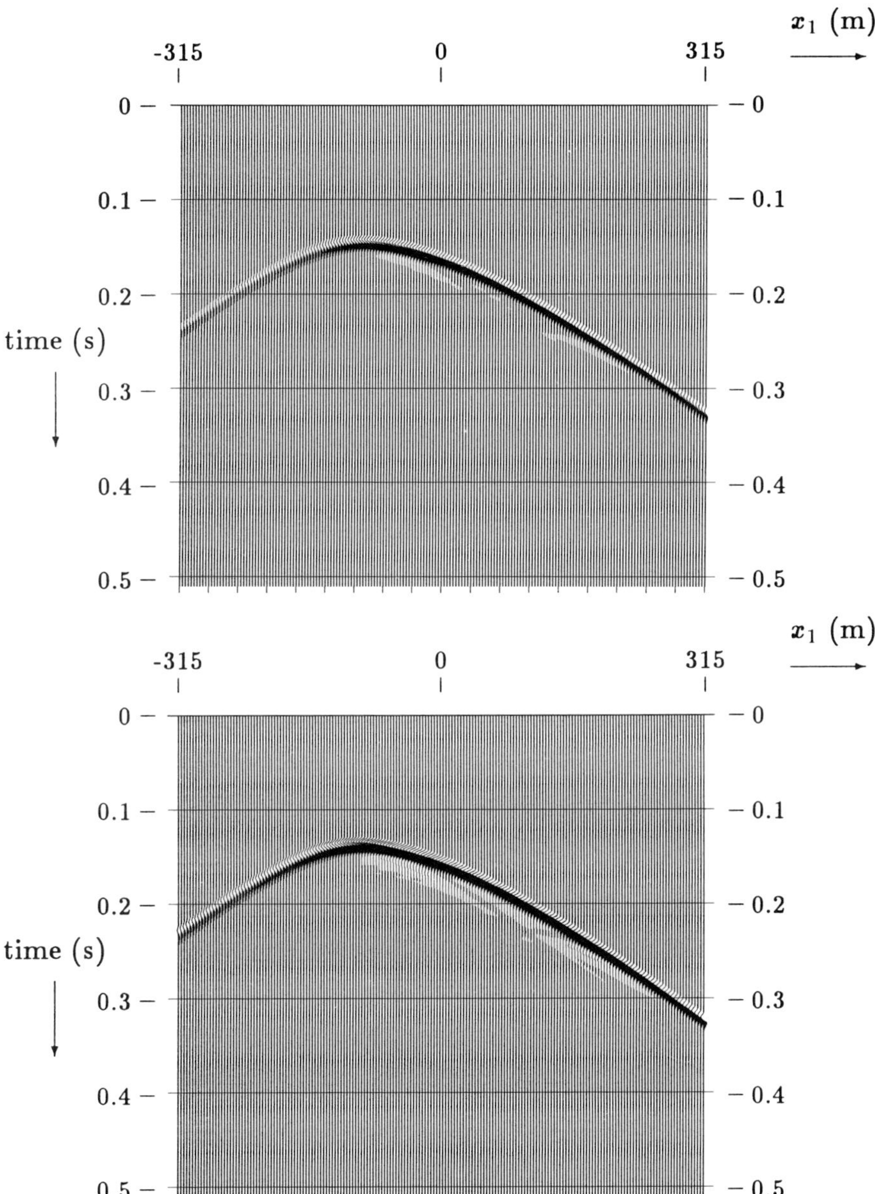


Figure 12.11. The wavefield of Fig. 12.10 after the second-order multiple elimination (top) and the wavefield, at $x_3^R = x_3^S = 0$, of a compliant strip in a homogeneous embedding (bottom).

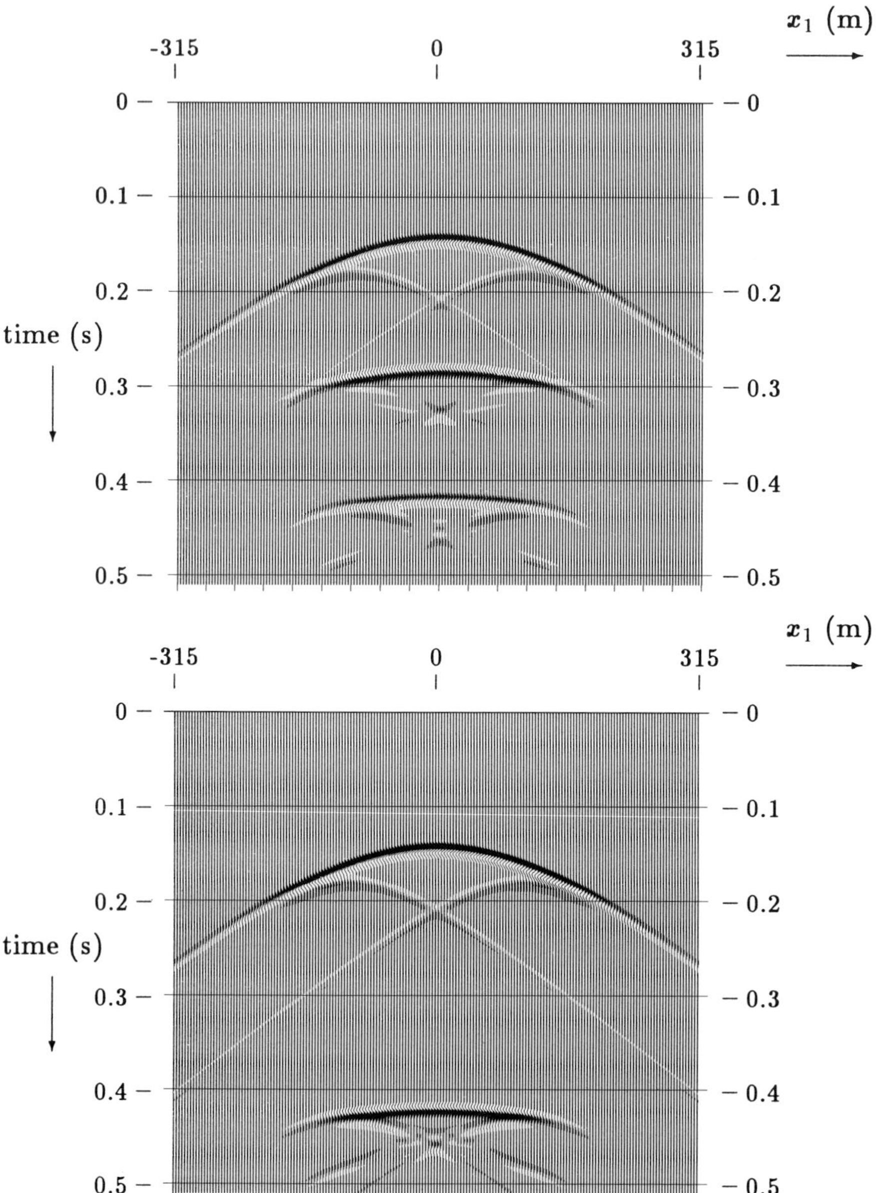


Figure 12.12. The deghosted wavefield of a rigid strip in a homogeneous halfspace (top) and the result after the first-order multiple elimination (bottom). The source position is above the center of the strip ($x_1^S = 0$).

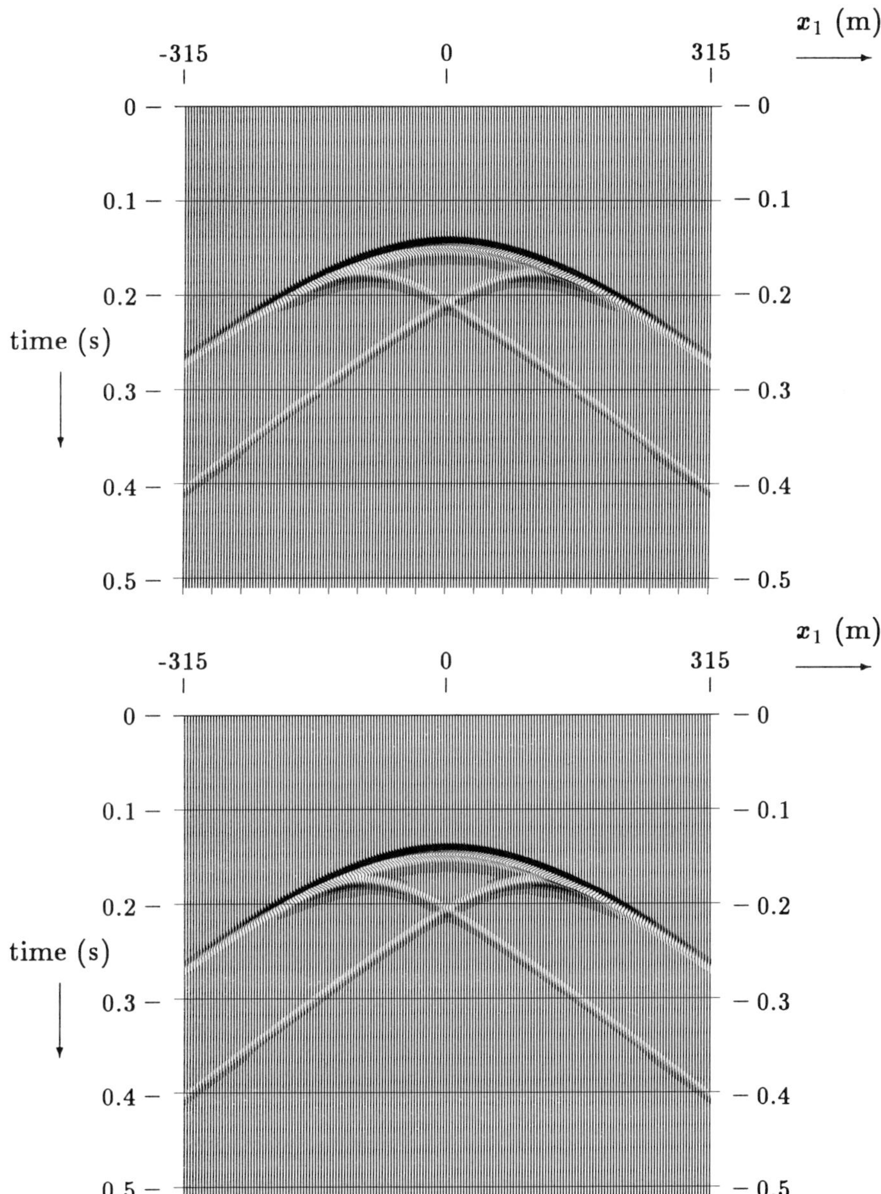


Figure 12.13. The wavefield of Fig. 12.12 after the second-order multiple elimination (top) and the wavefield, at $x_3^R = x_3^S = 0$, of a rigid strip in a homogeneous embedding (bottom).

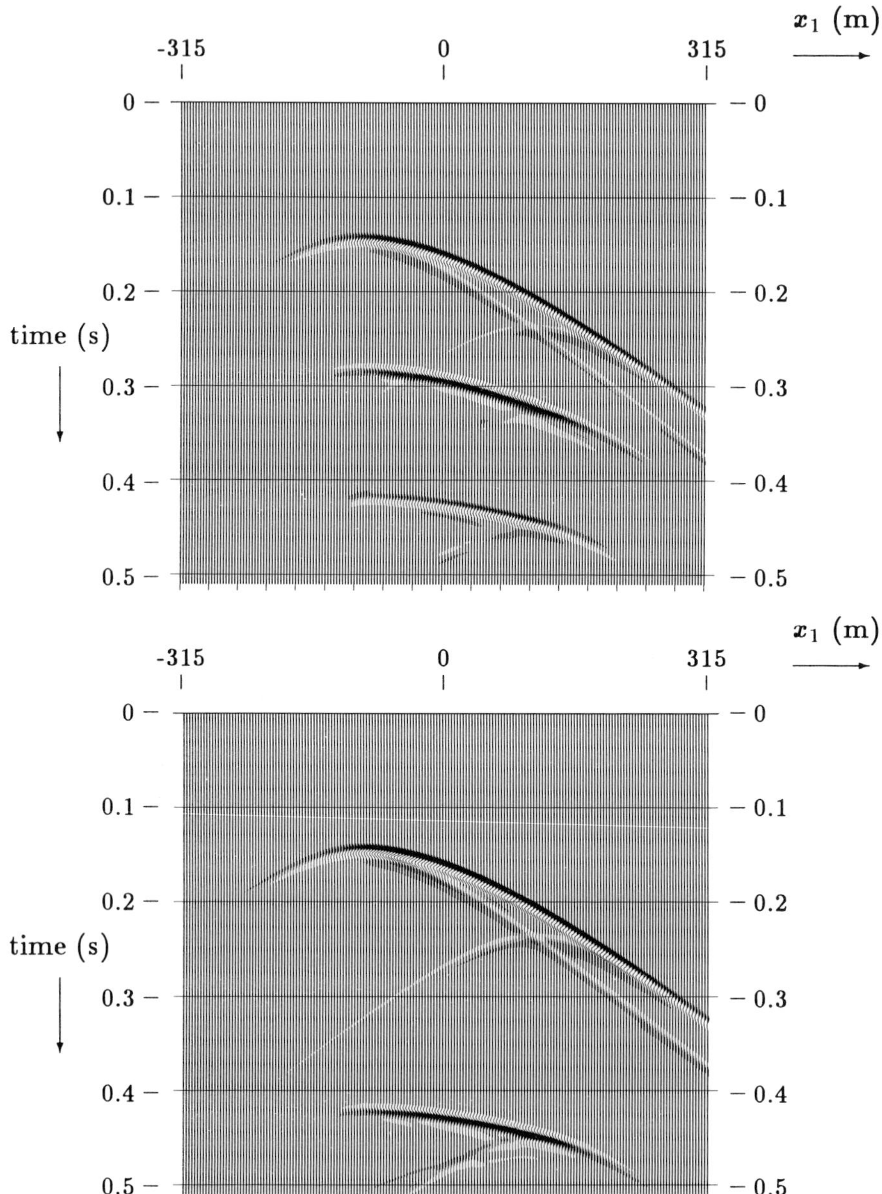


Figure 12.14. The deghosted wavefield of a rigid strip in a homogeneous halfspace (top) and the result after the first-order multiple elimination (bottom). The source position is above the left edge of the strip ($x_1^S = -105$ m).

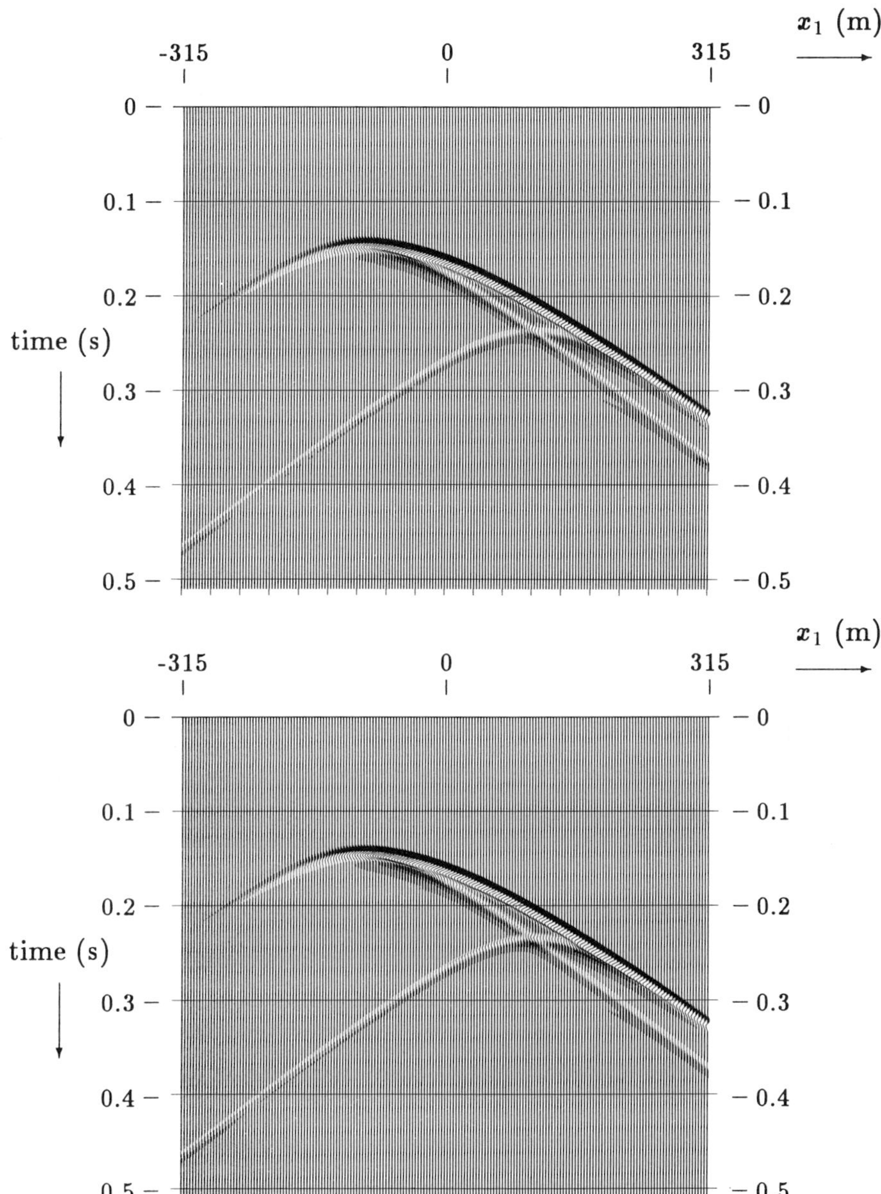


Figure 12.15. The wavefield of Fig. 12.14 after the second-order multiple elimination (top) and the wavefield, at $x_3^R = x_3^S = 0$, of a rigid strip in a homogeneous embedding (bottom).

*The numerical procedure*

In our numerical computations we take (see Section 9.5.2)

$$s = 4 + j\omega \, . \tag{12.54}$$

Then, we may write for the vertical slowness

$$\Upsilon = \left( \frac{1}{c^2} + \frac{(s\alpha_1)^2}{(4+j\omega)^2} \right)^{\frac{1}{2}} , \quad \mathrm{Re}(\Upsilon) > 0 \, . \tag{12.55}$$

Here, $s\alpha_1$ is the (real) Fourier transform parameter of the Fourier transform defined by Eqs. (9.101) and (9.102). The integral in the recursion formula of Eq. (12.50) is implemented numerically as a trapezoidal rule.

Fig. 12.8 shows the results for the compliant strip with source position above the center of the strip. The top figure shows the first term of the Neumann expression, the actual deghosted wavefield as it was calculated in Section 11.6. The bottom figure shows the results after one iteration of the Neumann expansion (with $N = 2$). Clearly, the first-order multiple has been removed. Also observe the sign reversal of the second-order multiple due to the negative sign in the right-hand side of Eq. (12.50). Finally, the top figure of Fig. 12.9 shows the result of the Neumann expansion with $N = 3$. All multiples have been removed; this is due to the fact that in the time window of 0.5 s only two multiples were present. The bottom figure of Fig. 12.9 shows the result of the calculation of Section 9.5.2 for the compliant strip in the homogeneous embedding. This is the actual required result. Comparison of the top and bottom figures of Fig. 12.9 illustrates the excellent performance of the multiple-removal operation. Figs. 12.10 and 12.11 show the results of the multiple removal for the compliant strip, but now the source position is chosen above the left edge of the strip. Also in this case we observe good convergence of the Neumann expansion and an excellent agreement with the required result. In Figs. 12.12 - 12.15, similar results are presented for the rigid strip.

In our numerical treatment of the multiple removal scheme we have chosen the Neumann expansion to solve the integral equation (12.33) in the frequency domain. We have opted for this solution because the convergence of the Neumann expansion is guaranteed by causality arguments in the time domain. In this way, the algorithm performance of the successive elimination of the multiples is monitored. However, an alternative procedure would have been to compute the desired reflected wavefield directly from the discrete version of the integral equation (12.33) by means of matrix inversion (VAN BORSELEN *et al.*, 1991).

The deghosting procedure of Chapter 11 and the removal procedure of surface-related wave phenomena of the present chapter encompass the necessary preprocessing steps. The numerical results obtained in this chapter are collected in a dataset and will be used as input for the imaging procedure of the next chapter.

# Chapter 13

# Boundary Imaging

In this chapter we discuss a method to image a boundary discontinuity of the earth parameters. In our analysis we distinguish between imaging and inversion. We consider the imaging of a boundary discontinuity as a part of inversion in the sense that it tries to delineate accurately the scattering horizons, if they are constructively present in the seismogram. This is possible because the velocity contrast is present in the arrival times and in the amplitude. Imaging of a boundary discontinuity only concentrates on the time delay. After this imaging process one hopes that the subsurface geometry of the geology as far as it is sufficiently described by the velocity structure, is known. Then, seismic inversion as a following-up process can use the results as a priori information in the reconstruction procedure of the constitutive parameters.

As starting point we take the boundary-integral representation. In the latter representation we employ a locally-plane-reflector approximation both for the acoustic pressure and the particle velocity. This high-frequency approximation does not change the imaging. In the wavefield, we separate the horizontal phase contribution from the vertical one by employing a Radon transformation with respect to the horizontal source and receiver coordinates only. As next step we switch to the frequency domain (imaginary axis in the complex Laplace-transform domain) and the vertical slowness is taken as independent variable. In that case stationary-phase arguments show that the results can be classified as a spatial Fourier transform in the frequency domain. In the space-time domain we obtain a function with a laterally variant vertical travel time that represents a space-time image of the profile of the interface as the envelope of the first arrivals.

## 13.1. The boundary-integral representation

We consider the wavefield to be reflected by a non-planar interface between a homogeneous medium $\mathbb{D}$ and an inhomogeneous geology $\mathbb{D}_g$. The non-planar interface $\partial\mathbb{D}_g$ is considered as a perturbation of a plane surface and is described as

$$x_3 = h(x_1, x_2) , \tag{13.1}$$

where $h$ is an arbitrary positive (smooth) function of $x_1$ and $x_2$, and is denoted as the profile function to be imaged (see Fig. 13.1). The material constants in $\mathbb{D}$ are $\rho$ and $\kappa$, and the material parameters in $\mathbb{D}_g$ are $\rho^g = \rho^g(\boldsymbol{x})$ and $\kappa^g = \kappa^g(\boldsymbol{x})$. The impulsive wave motion generated by a point source at $\boldsymbol{x}^S = \{x_1^S, x_2^S, 0\}$ starts to act at $t = 0$. The receiver is located at $\boldsymbol{x}^R = \{x_1^R, x_2^R, 0\}$.

In the $s$-domain, the source wavefield at the observation point $\boldsymbol{x}$ originating from a monopole source at $\boldsymbol{x}^S$ is then denoted as $\{\hat{p}^{inc}, \hat{v}_k^{inc}\}(\boldsymbol{x}|\boldsymbol{x}^S, s)$ and is given by

$$\hat{p}^{inc}(\boldsymbol{x}|\boldsymbol{x}^S, s) = s\rho\hat{q}^S(s)\hat{G}(\boldsymbol{x}-\boldsymbol{x}^S, s) , \tag{13.2}$$

$$\hat{v}_k^{inc}(\boldsymbol{x}|\boldsymbol{x}^S, s) = -\hat{q}^S(s)\partial_k\hat{G}(\boldsymbol{x}-\boldsymbol{x}^S, s) , \tag{13.3}$$

where the Green's function

$$\hat{G}(\boldsymbol{x}, s) = \frac{\exp(-\frac{s}{c}|\boldsymbol{x}|)}{4\pi|\boldsymbol{x}|}, \quad c = (\rho\kappa)^{-\frac{1}{2}} . \tag{13.4}$$

The wavefield reflected by the interface $\partial\mathbb{D}_g$ is given by $\{\hat{p}^r, \hat{v}_k^r\}(\boldsymbol{x}|\boldsymbol{x}^S, s)$. The reflected wavefield is considered to be the wavefield generated by secondary surface sources (monopole and dipole sources) at the interface $\partial\mathbb{D}_g$ and it can be expressed in terms of these surface sources as (cf. Eq. (8.40))

$$\hat{p}^r(\boldsymbol{x}^R|\boldsymbol{x}^S, s) = \int_{\boldsymbol{x}\in\partial\mathbb{D}_g} [\hat{G}^q\,\hat{v}_k^r(\boldsymbol{x}|\boldsymbol{x}^S, s) + \hat{\Gamma}_k^q\,\hat{p}^r(\boldsymbol{x}|\boldsymbol{x}^S, s)]\nu_k \mathrm{dA} , \tag{13.5}$$

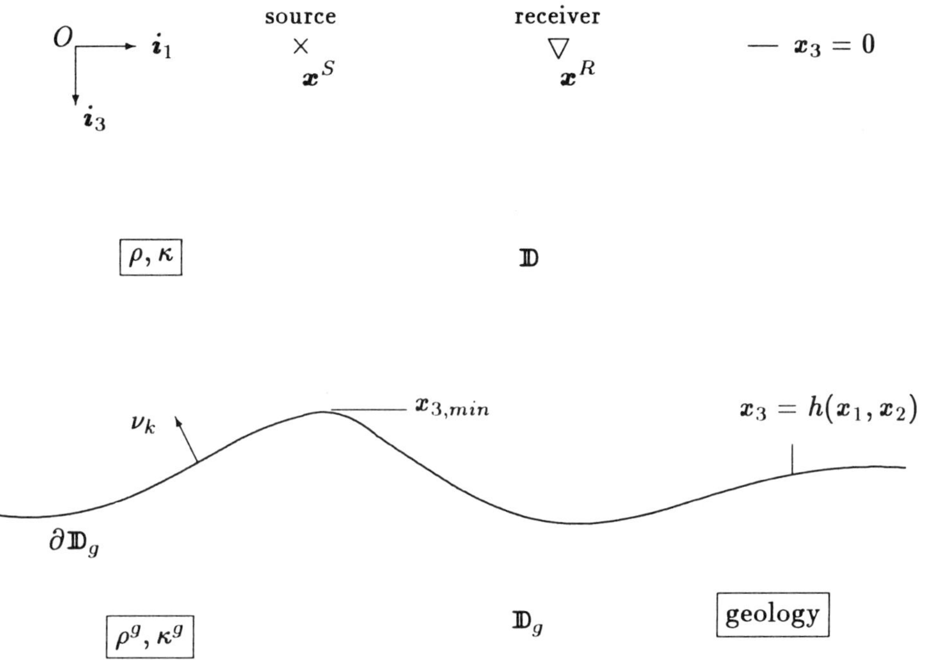


Figure 13.1. The reflector configuration.

in which the Green's states $\hat{G}^q$ and $\hat{\Gamma}^q_k$ are given by

$$\hat{G}^q(\boldsymbol{x}^R|\boldsymbol{x}, s) = s\rho\hat{G}(\boldsymbol{x}-\boldsymbol{x}^R, s) \tag{13.6}$$

and

$$\hat{\Gamma}^q_k(\boldsymbol{x}^R|\boldsymbol{x}, s) = \partial_k\hat{G}(\boldsymbol{x}-\boldsymbol{x}^R, s)\,. \tag{13.7}$$

Further, $\nu_k$ denotes the normal at $\partial\mathbb{D}_g$ (see Fig. 13.1).

For large values of $|s|$ we can employ a locally-plane-reflector approximation. This approximation is based on the assumption that the interface $\partial\mathbb{D}_g$ is a smooth surface with a continuous normal $\nu_k$. The approximation assumes that the scale on which the wavefield varies in space is so small that compared with this the radius of curvature of the interface is infinitely large. Then, around a point, $\boldsymbol{x} \in \partial\mathbb{D}_g$, the interface can be considered as a locally plane interface between two locally homogeneous media. We further assume that the local wavefield takes on the values that would result from the reflection and transmission of the incident wavefield at this plane interface. Then, the reflected wavefield on $\partial\mathbb{D}_g$ can be expressed in terms of the incident wavefield as

$$\hat{p}^r(\boldsymbol{x}|\boldsymbol{x}^S, s) = \eta\,\hat{p}^{inc}(\boldsymbol{x}|\boldsymbol{x}^S, s), \quad \boldsymbol{x} \in \partial\mathbb{D}_g\,, \tag{13.8}$$

in which the reflection factor $\eta$ is given by (ČERVENÝ, 1989)

$$\eta = \frac{\rho^g(\boldsymbol{x})c^g(\boldsymbol{x})\cos(\theta) - \rho c\cos(\theta_g)}{\rho^g(\boldsymbol{x})c^g(\boldsymbol{x})\cos(\theta) + \rho c\cos(\theta_g)}\,, \tag{13.9}$$

with $c^g(\boldsymbol{x}) = [\rho^g(\boldsymbol{x})\kappa^g(\boldsymbol{x})]^{\frac{1}{2}}$, and where

$$\cos(\theta) = \frac{-\nu_k(x_k - x_k^S)}{|\boldsymbol{x} - \boldsymbol{x}^S|} \tag{13.10}$$

and where $\theta^g = \theta^g(\boldsymbol{x})$ follows from Snell's law as

$$\frac{\sin(\theta^g)}{c^g(\boldsymbol{x})} = \frac{\sin(\theta)}{c}\,. \tag{13.11}$$

We observe that $\eta = \eta(\boldsymbol{x}|\boldsymbol{x} - \boldsymbol{x}^S)$ is a real, $s$-independent, proportionality function of $\boldsymbol{x}$ and $\boldsymbol{x} - \boldsymbol{x}^S$. Equivalently, we can also express the normal component of the particle velocity of the reflected wavefield as

$$\nu_k\hat{v}_k^r(\boldsymbol{x}|\boldsymbol{x}^S, s) = -\zeta\,\nu_k\hat{v}_k^{inc}(\boldsymbol{x}|\boldsymbol{x}^S, s)\,, \quad \boldsymbol{x} \in \partial\mathbb{D}_g\,. \tag{13.12}$$

The function $\zeta = \zeta(\boldsymbol{x}|\boldsymbol{x}–\boldsymbol{x}^S)$ is a real, $s$-independent proportionality function of $\boldsymbol{x}$ and $\boldsymbol{x} - \boldsymbol{x}^S$. Its relation with respect to $\eta$ will be discussed below.

Using these approximations and Eqs. (13.2) - (13.3) in the integral representation of Eq. (13.5) we obtain

$$\hat{p}^r(\boldsymbol{x}^R|\boldsymbol{x}^S, s) = \hat{W}(s)\int_{\boldsymbol{x}\in\partial\mathbb{D}_g}[\zeta(\boldsymbol{x}|\boldsymbol{x} - \boldsymbol{x}^S)\hat{G}(\boldsymbol{x} - \boldsymbol{x}^R, s)\partial_k\hat{G}(\boldsymbol{x} - \boldsymbol{x}^S, s)$$
$$+\eta(\boldsymbol{x}|\boldsymbol{x} - \boldsymbol{x}^S)\hat{G}(\boldsymbol{x} - \boldsymbol{x}^S, s)\partial_k\hat{G}(\boldsymbol{x} - \boldsymbol{x}^R, s)]\nu_k\mathrm{dA}, \tag{13.13}$$

where $\hat{W}(s) = s\rho\hat{q}^S(s)$ is the source wavelet. On account of the reciprocity relation

$$\hat{p}^r(\boldsymbol{x}^R|\boldsymbol{x}^S, s) = \hat{p}^r(\boldsymbol{x}^S|\boldsymbol{x}^R, s)\,, \tag{13.14}$$

we conclude from Eq. (13.13) that

$$\zeta(\boldsymbol{x}|\boldsymbol{x}-\boldsymbol{x}^R) = \eta(\boldsymbol{x}|\boldsymbol{x}-\boldsymbol{x}^S), \quad \zeta(\boldsymbol{x}|\boldsymbol{x}-\boldsymbol{x}^S) = \eta(\boldsymbol{x}|\boldsymbol{x}-\boldsymbol{x}^R). \tag{13.15}$$

The partial derivative operating on the Green's function $\hat{G}(\boldsymbol{x}, s)$ is obtained as

$$\partial_k \hat{G}(\boldsymbol{x}, s) = -\frac{s}{c}\hat{G}(\boldsymbol{x}, s)\left(1 + \frac{c}{s|\boldsymbol{x}|}\right)\partial_k|\boldsymbol{x}|, \tag{13.16}$$

so that Eq. (13.13) simplifies to

$$\hat{p}^r(\boldsymbol{x}^R|\boldsymbol{x}^S, s) = \frac{s}{c}\hat{W}(s)\int_{\boldsymbol{x}\in\partial\mathbb{D}_g} \hat{A}\,\hat{G}(\boldsymbol{x}-\boldsymbol{x}^R, s)\hat{G}(\boldsymbol{x}-\boldsymbol{x}^S, s)\mathrm{dA}, \tag{13.17}$$

where the function $\hat{A} = \hat{A}(\boldsymbol{x}, \boldsymbol{x}-\boldsymbol{x}^R, \boldsymbol{x}-\boldsymbol{x}^S, s)$ is given by

$$\begin{aligned}\hat{A} = &-\zeta(\boldsymbol{x}|\boldsymbol{x}-\boldsymbol{x}^S)\left(1 + \frac{c}{s|\boldsymbol{x}-\boldsymbol{x}^S|}\right)\nu_k\partial_k|\boldsymbol{x}-\boldsymbol{x}^S| \\ &-\eta(\boldsymbol{x}|\boldsymbol{x}-\boldsymbol{x}^S)\left(1 + \frac{c}{s|\boldsymbol{x}-\boldsymbol{x}^R|}\right)\nu_k\partial_k|\boldsymbol{x}-\boldsymbol{x}^R|.\end{aligned} \tag{13.18}$$

The function $\hat{A}$ becomes independent of $s$ as $|s| \to \infty$, viz.,

$$A = -\zeta(\boldsymbol{x}|\boldsymbol{x}-\boldsymbol{x}^S)\,\nu_k\partial_k|\boldsymbol{x}-\boldsymbol{x}^S| - \eta(\boldsymbol{x}|\boldsymbol{x}-\boldsymbol{x}^S)\,\nu_k\partial_k|\boldsymbol{x}-\boldsymbol{x}^R|. \tag{13.19}$$

Using the reciprocity relation of Eq. (13.15), the function $A$ may be written in an alternative form

$$A = -\zeta(\boldsymbol{x}|\boldsymbol{x}-\boldsymbol{x}^S)\,\nu_k\partial_k|\boldsymbol{x}-\boldsymbol{x}^S| - \zeta(\boldsymbol{x}|\boldsymbol{x}-\boldsymbol{x}^R)\,\nu_k\partial_k|\boldsymbol{x}-\boldsymbol{x}^R|. \tag{13.20}$$

We consider a seismic experiment as the collection of measurements of the seismic signal at different receiver positions. This procedure is repeated for different source positions. This means that the values of $\hat{p}^r(\boldsymbol{x}^R|\boldsymbol{x}^S, s)$ are the (to the $s$-domain transformed) recorded signals for all $\boldsymbol{x}^R$ and $\boldsymbol{x}^S$. Our objective is to use Eq. (13.17) to delineate the interface $\partial\mathbb{D}_g$. To meet our objective we apply a spatial Fourier transformation with respect to the horizontal source coordinates followed by a spatial Fourier transformation with respect to the horizontal receiver coordinates.

## 13.2. Fourier transform with respect to source coordinates

We first define a Fourier transform with respect to the horizontal source coordinates, viz.

$$\bar{p}^r(\boldsymbol{x}^R|js\boldsymbol{\alpha}^S,s) = \int_{(x_1^S,x_2^S)\in\mathbb{R}^2} \exp(js\alpha_1^S x_1^S + js\alpha_2^S x_2^S)\hat{p}^r(\boldsymbol{x}^R|\boldsymbol{x}^S,s)\mathrm{dA}\,, \tag{13.21}$$

where $\boldsymbol{\alpha}^S = \{\alpha_1^S, \alpha_2^S\}$ is a two-dimensional vector, while $s\alpha_1^S$ and $s\alpha_2^S$ are real transform parameters. In our further analysis, we take both $s$ and $\boldsymbol{\alpha}^S$ to be real. Application of the Fourier transform to Eq. (13.17) yields

$$\bar{p}^r(\boldsymbol{x}^R|js\boldsymbol{\alpha}^S,s) = \frac{s}{c}\hat{W}(s)\int_{\boldsymbol{x}\in\partial\mathbb{D}_g} \overline{B}\,\hat{G}(\boldsymbol{x}-\boldsymbol{x}^R,s)\exp(js\alpha_1^S x_1 + js\alpha_2^S x_2)\mathrm{dA}\,, \tag{13.22}$$

with

$$\overline{B} = \int_{(x_1^S,x_2^S)\in\mathbb{R}^2} \frac{A\exp\left[-s\left(j\alpha_1^S(x_1-x_1^S)+j\alpha_2^S(x_2-x_2^S)+|\boldsymbol{x}-\boldsymbol{x}^S|/c\right)\right]}{4\pi|\boldsymbol{x}-\boldsymbol{x}^S|}\mathrm{dA}\,. \tag{13.23}$$

Our objective is to evaluate this expression of $\overline{B}$ for large values of $s$. The most convenient way to perform the asymptotic evaluation is the steepest-descent method. In order to find the steepest-descent path of this multiple integral, we deform the integration surface such that the expression in the argument of the exponential function in the right-hand side of Eq. (13.23) becomes real. Therefore, we introduce the polar coordinates in the $(\alpha_1^S, \alpha_2^S)$-plane through

$$\alpha_1^S = \kappa^S\cos(\phi^S), \quad \alpha_2^S = \kappa^S\sin(\phi^S), \quad \kappa^S = \left[(\alpha_1^S)^2+(\alpha_2^S)^2\right]^{\frac{1}{2}}\,, \tag{13.24}$$

where $0 \le \kappa^S < \infty$, $0 \le \phi^S < 2\pi$. Next we replace the variables of integrations $x_1^S$ and $x_2^S$ in Eq. (13.23) by $y^S$ and $z^S$ through

$$\begin{aligned} x_1 - x_1^S &= -jy^S\cos(\phi^S) - z^S\sin(\phi^S)\,, \\ x_2 - x_2^S &= -jy^S\sin(\phi^S) + z^S\cos(\phi^S)\,, \end{aligned} \tag{13.25}$$

which entails $(x_1 - x_1^S)^2 + (x_2 - x_2^S)^2 = (z^S)^2 - (y^S)^2$, $\mathrm{d}x_1^S \mathrm{d}x_2^S = -j\mathrm{d}y^S \mathrm{d}z^S$ and $j\alpha_1^S(x_1 - x_1^S) + j\alpha_2^S(x_2 - x_2^S) = \kappa^S y^S$. Hence, Eq. (13.23) is replaced by

$$\overline{B} = \frac{1}{4\pi j} \int_{-\infty}^{\infty} \mathrm{d}z^S \int_{-j\infty}^{j\infty} \frac{A \exp\left[-s\left(\kappa^S y^S + R^S/c\right)\right]}{R^S} \mathrm{d}y^S , \tag{13.26}$$

where

$$R^S = \left[x_3^2 + (z^S)^2 - (y^S)^2\right]^{\frac{1}{2}} . \tag{13.27}$$

Now we use a procedure similar to the one in the Cagniard-de Hoop method. We extend the integrand of Eq. (13.26) in the complex $y^S$-plane. Since we want to keep the integrand single valued, we introduce a branch cut for the square-root expression $R^S$ along $\{[x_3^2 + (z^S)^2]^{\frac{1}{2}} < |\mathrm{Re}(y^S)| < \infty$, $\mathrm{Im}(y^S) = 0\}$ and keep $\mathrm{Re}(R^S) \geq 0$ in the entire cut $y^S$-plane (see Fig. 13.2). The imaginary $y^S$-axis is now deformed into a steepest-descent contour

$$\kappa^S y^S + \frac{R^S}{c} = \tau^S , \tag{13.28}$$

where $\tau^S$ is real and positive. This contour is located in the right-half of the $y^S$-plane and is a branch of a hyperbola. The parametric representation

$$y^S = \frac{\kappa^S \tau^S + j\frac{1}{c}\left\{(\tau^S)^2 - [T^S(z^S)]^2\right\}^{\frac{1}{2}}}{\frac{1}{c^2} + (\kappa^S)^2} , \tag{13.29}$$

with

$$T^S(z^S) = \left[\frac{1}{c^2} + (\kappa^S)^2\right]^{\frac{1}{2}} \left[x_3^2 + (z^S)^2\right]^{\frac{1}{2}} , \tag{13.30}$$

denotes the part of the hyperbola in the upper half of the $y^S$-plane, while $y^S = y^{S\star}$ (the star denotes complex conjugation) represents the lower half. The Jacobian of this transformation from $y^S$ to $\tau^S$ is found as

$$\frac{\partial y^S}{\partial \tau^S} = \frac{\kappa^S + j\frac{\tau^S}{c}\left\{(\tau^S)^2 - [T^S(z^S)]^2\right\}^{-\frac{1}{2}}}{\frac{1}{c^2} + (\kappa^S)^2} = j\frac{R^S}{\left\{(\tau^S)^2 - [T^S(z^S)]^2\right\}^{\frac{1}{2}}} . \tag{13.31}$$

After this deformation, Eq. (13.26) is rewritten as

$$\overline{B} = \frac{1}{4\pi} \int_{-\infty}^{\infty} \mathrm{d}z^S \int_{T^S(z^S)}^{\infty} \exp(-s\tau^S) \frac{A(z^S, y^S) + A(z^S, y^{S\star})}{\left\{(\tau^S)^2 - [T^S(z^S)]^2\right\}^{\frac{1}{2}}} \mathrm{d}\tau^S , \tag{13.32}$$

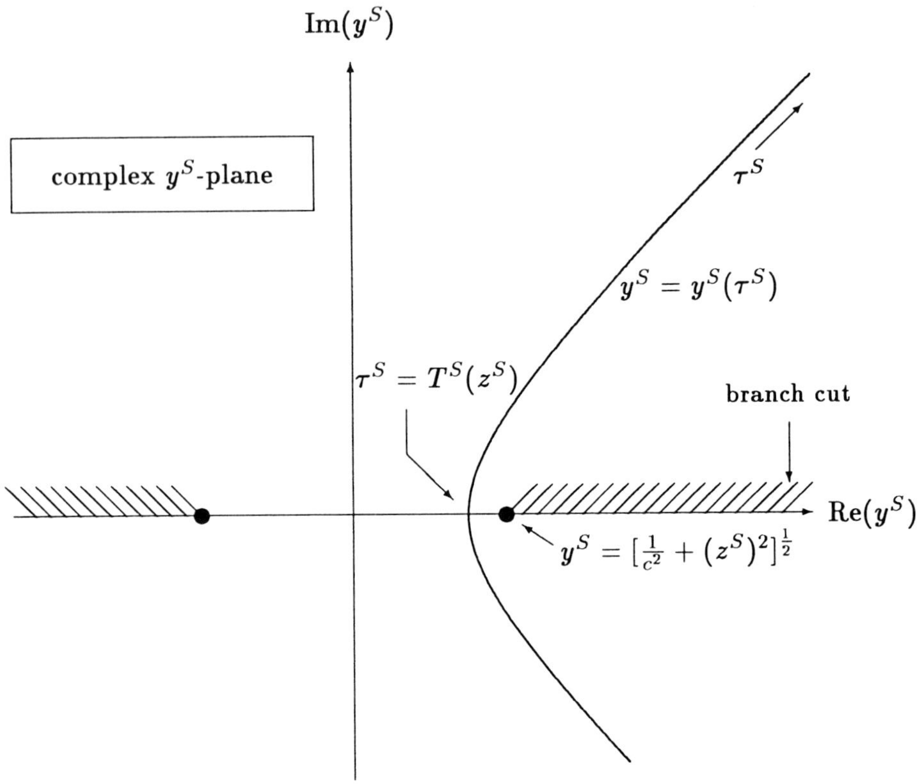


Figure 13.2. Complex $y^S$-plane with Cagniard-de Hoop path $y^S(\tau^S)$.

where $y^S$ is a function of $\tau^S$ via Eq. (13.29). Next we interchange the order of integrations leading to

$$\overline{B} = \frac{1}{4\pi\left[\frac{1}{c^2}+(\kappa^S)^2\right]^{\frac{1}{2}}} \int_{T_0^S}^{\infty} \exp(-s\tau^S)\mathrm{d}\tau^S \int_{-Z^S(\tau^S)}^{Z^S(\tau^S)} \frac{A(z^S,y^S)+A(z^S,y^{S\star})}{\left\{[Z^S(\tau^S)]^2-(z^S)^2\right\}^{\frac{1}{2}}}\mathrm{d}z^S, \tag{13.33}$$

where

$$T_0^S = T^S(0) = \left[\frac{1}{c^2} + (\kappa^S)^2\right]^{\frac{1}{2}} x_3 \tag{13.34}$$

and

$$Z^S(\tau^S) = \left[ \frac{(\tau^S)^2}{\frac{1}{c^2} + (\kappa^S)^2} - x_3^2 \right]^{\frac{1}{2}} . \tag{13.35}$$

Introduction of the new variable of integration, $z^S = Z^S(\tau^S)\sin(\psi^S)$, leads to

$$\overline{B} = \frac{1}{4\pi \left[ \frac{1}{c^2} + (\kappa^S)^2 \right]^{\frac{1}{2}}} \int_{T_0^S}^{\infty} \exp(-s\tau^S) \mathrm{d}\tau^S \int_{-\frac{1}{2}\pi}^{\frac{1}{2}\pi} [A(z^S, y^S) + A(z^S, y^{S\star})] \mathrm{d}\psi^S . \tag{13.36}$$

This expression can be seen as a Laplace transform of the type

$$\overline{B}(s) = \int_{T_0^S}^{\infty} \exp(-s\tau^S) \bar{b}(\tau^S) \mathrm{d}\tau^S , \tag{13.37}$$

where

$$\bar{b}(\tau^S) = \frac{1}{4\pi \left[ \frac{1}{c^2} + (\kappa^S)^2 \right]^{\frac{1}{2}}} \int_{-\frac{1}{2}\pi}^{\frac{1}{2}\pi} [A(z^S, y^S) + A(z^S, y^{S\star})] \mathrm{d}\psi^S . \tag{13.38}$$

From the initial-value theorem of Eq. (1.18) we conclude that

$$\lim_{s \to \infty} \exp(sT_0^S)\, s\overline{B}(s) = \lim_{\tau^S \downarrow T_0^S} \bar{b}(\tau^S) . \tag{13.39}$$

When $\tau^S \downarrow T_0^S$, from Eqs. (13.30) and (13.34), we observe that $z^S \downarrow 0$ and hence

$$\begin{aligned} \lim_{\tau^S \downarrow T_0^S} \bar{b}(\tau^S) &= \frac{1}{4\pi \left[ \frac{1}{c^2} + (\kappa^S)^2 \right]^{\frac{1}{2}}} \int_{-\frac{1}{2}\pi}^{\frac{1}{2}\pi} [A(0, y^S) + A(0, y^{S\star})] \mathrm{d}\psi^S \\ &= \frac{1}{2\Gamma^S} A|_{z^S=0,\, \tau^S = T_0^S} , \end{aligned} \tag{13.40}$$

where we have used that the integrand is independent of $\psi^S$ and where

$$\Gamma^S = \left[ \frac{1}{c^2} + (\alpha_1^S)^2 + (\alpha_2^S)^2 \right]^{\frac{1}{2}} . \tag{13.41}$$

Taking $s \to \infty$, Eqs. (13.39) and (13.40) yield

$$\overline{B} = \frac{\exp(-sT_0^S)}{2s\Gamma^S} A|_{z^S=0,\,\tau^S=T_0^S}, \tag{13.42}$$

where

$$T_0^S = \Gamma^S x_3 = \Gamma^S h. \tag{13.43}$$

In order to evaluate the expression for $A$, we first calculate the expression for $\nu_k\partial_k|\boldsymbol{x}-\boldsymbol{x}^S|$. Using $z^S=0$ and $\tau^S = T_0^S$ we observe that

$$x_1 - x_1^S = -\frac{j\alpha_1^S h}{\Gamma^S}, \quad x_2 - x_2^S = -\frac{j\alpha_2^S h}{\Gamma^S}, \quad x_3 = h, \tag{13.44}$$

$$|\boldsymbol{x}-\boldsymbol{x}^S| = R^S = \frac{h}{c\Gamma^S}, \tag{13.45}$$

and

$$\nu_k\partial_k|\boldsymbol{x}-\boldsymbol{x}^S| = \frac{(x_1-x_1^S)\partial_1 h + (x_2-x_2^S)\partial_2 h - x_3}{|\boldsymbol{x}-\boldsymbol{x}^S|\,[(\partial_1 h)^2+(\partial_2 h)^2+1]^{\frac{1}{2}}}, \tag{13.46}$$

where $\partial_1 h$ and $\partial_2 h$ are the spatial derivatives of $h$ with respect to $x_1$ and $x_2$, respectively. We finally arrive at

$$\begin{aligned} A|_{z^S=0,\,\tau^S=T_0^S} = {} & c\zeta(\boldsymbol{x}|\frac{-j\alpha_1^S h}{\Gamma^S}, \frac{-j\alpha_2^S h}{\Gamma^S}, h)\,\frac{j\alpha_1^S\partial_1 h + j\alpha_2^S\partial_2 h + \Gamma^S}{[(\partial_1 h)^2+(\partial_2 h)^2+1]^{\frac{1}{2}}} \\ & - \zeta(\boldsymbol{x}|\boldsymbol{x}-\boldsymbol{x}^R)\nu_k\partial_k|\boldsymbol{x}-\boldsymbol{x}^R| \quad \text{for } s \to \infty. \end{aligned} \tag{13.47}$$

## 13.3. Fourier transform with respect to receiver coordinates

We subsequently define a Fourier transform with respect to the horizontal receiver coordinates, viz.

$$\overline{\overline{p}}^r(js\boldsymbol{\alpha}^R|js\boldsymbol{\alpha}^S, s) = \int_{(x_1^R,x_2^R)\in\mathbb{R}^2} \exp(js\alpha_1^R x_1^R + js\alpha_2^R x_2^R)\overline{p}^r(\boldsymbol{x}^R|js\boldsymbol{\alpha}^S, s)\mathrm{dA}, \tag{13.48}$$

where $\boldsymbol{\alpha}^R = \{\alpha_1^R, \alpha_2^R\}$ is a two-dimensional vector, while $s\alpha_1^R$ and $s\alpha_2^R$ are real transform parameters. We again take both $s$ and $\boldsymbol{\alpha}^R$ to be real. Application of the Fourier transform to Eq. (13.22) yields

$$\bar{\bar{p}}^r(js\boldsymbol{\alpha}^R|js\boldsymbol{\alpha}^S,s) = \frac{s}{c}\hat{W}(s)\int_{\boldsymbol{x}\in\partial\mathbb{D}_g}\bar{\bar{C}}\exp[js(\alpha_1^R+\alpha_1^S)x_1+js(\alpha_2^R+\alpha_2^S)x_2]\mathrm{dA}, \tag{13.49}$$

with

$$\bar{\bar{C}} = \int_{(x_1^R,x_2^R)\in\mathbb{R}^2}\frac{\overline{B}\exp\left[-s\left(j\alpha_1^R(x_1-x_1^R)+j\alpha_2^R(x_2-x_2^R)+|\boldsymbol{x}-\boldsymbol{x}^R|/c\right)\right]}{4\pi|\boldsymbol{x}-\boldsymbol{x}^R|}\mathrm{dA}. \tag{13.50}$$

Our objective is to evaluate this expression of $\bar{\bar{C}}$ for large values of $s$. The most convenient way to perform the asymptotic evaluation is the steepest-descent method. In order to find the steepest-descent path of this multiple integral, we deform the integration surface such that the expression in the argument of the exponential function in the right-hand side of Eq. (13.50) becomes real. Therefore, we introduce the polar coordinates in the $(\alpha_1^R, \alpha_2^R)$-plane through

$$\alpha_1^R = \kappa^R\cos(\phi^R), \quad \alpha_2^R = \kappa^R\sin(\phi^R), \quad \kappa^R = \left[(\alpha_1^R)^2+(\alpha_2^R)^2\right]^{\frac{1}{2}}, \tag{13.51}$$

where $0 \leq \kappa^R < \infty$, $0 \leq \phi^R < 2\pi$. Next we replace the variables of integrations $x_1^R$ and $x_2^R$ in Eq. (13.50) by $y^R$ and $z^R$ through

$$\begin{aligned} x_1 - x_1^R &= -jy^R\cos(\phi^R) - z^R\sin(\phi^R), \\ x_2 - x_2^R &= -jy^R\sin(\phi^R) + z^R\cos(\phi^R), \end{aligned} \tag{13.52}$$

which entails $(x_1-x_1^R)^2+(x_2-x_2^R)^2 = (z^R)^2-(y^R)^2$, $\mathrm{d}x_1^R\mathrm{d}x_2^R = -j\mathrm{d}y^R\mathrm{d}z^R$ and $j\alpha_1^R(x_1-x_1^R)+j\alpha_2^R(x_2-x_2^R) = \kappa^R y^R$. Hence, Eq. (13.50) is replaced by

$$\bar{\bar{C}} = \frac{1}{4\pi j}\int_{-\infty}^{\infty}\mathrm{d}z^R\int_{-j\infty}^{j\infty}\frac{\overline{B}\exp\left[-s\left(\kappa^R y^R+R^R/c\right)\right]}{R^R}\mathrm{d}y^R, \tag{13.53}$$

where

$$R^R = \left[x_3^2+(z^R)^2-(y^R)^2\right]^{\frac{1}{2}}. \tag{13.54}$$

Now we use a procedure similar to the one in the Cagniard-de Hoop method. We extend the integrand of Eq. (13.53) in the complex $y^R$-plane. Since we want to keep the integrand single valued, we introduce a branch cut for the square-root expression $R^R$ along $\{[x_3^2 + (z^R)^2]^{\frac{1}{2}} < |\mathrm{Re}(y^R)| < \infty$, $\mathrm{Im}(y^R) = 0\}$ and keep $\mathrm{Re}(R^R) \geq 0$ in the entire cut $y^R$-plane (see Fig. 13.3). The imaginary $y^R$-axis is now deformed into a steepest-descent contour

$$\kappa^R y^R + \frac{R^R}{c} = \tau^R , \tag{13.55}$$

where $\tau^R$ is real and positive. This contour is located in the right-half of the $y^R$-plane and is a branch of a hyperbola. The parametric representation

$$y^R = \frac{\kappa^R \tau^R + j\frac{1}{c}\left\{(\tau^R)^2 - [T^R(z^R)]^2\right\}^{\frac{1}{2}}}{\frac{1}{c^2} + (\kappa^R)^2} , \tag{13.56}$$

with

$$T^R(z^R) = \left[\frac{1}{c^2} + (\kappa^R)^2\right]^{\frac{1}{2}} \left[x_3^2 + (z^R)^2\right]^{\frac{1}{2}} , \tag{13.57}$$

denotes the part of the hyperbola in the upper half of the $y^R$-plane, while $y^R = y^{R\star}$ (the star denotes complex conjugation) represents the lower half. The Jacobian of this transformation from $y^R$ to $\tau^R$ is found as

$$\frac{\partial y^R}{\partial \tau^R} = \frac{\kappa^R + j\frac{\tau^R}{c}\left\{(\tau^R)^2 - [T^R(z^R)]^2\right\}^{-\frac{1}{2}}}{\frac{1}{c^2} + (\kappa^R)^2} = j\frac{R^R}{\left\{(\tau^R)^2 - [T^R(z^R)]^2\right\}^{\frac{1}{2}}} . \tag{13.58}$$

After this deformation Eq. (13.53) is rewritten as

$$\overline{\overline{C}} = \frac{1}{4\pi}\int_{-\infty}^{\infty} \mathrm{d}z^R \int_{T^R(z^R)}^{\infty} \exp(-s\tau^R)\frac{\overline{B}(z^R, y^R) + \overline{B}(z^R, y^{R\star})}{\left\{(\tau^R)^2 - [T^R(z^R)]^2\right\}^{\frac{1}{2}}}\mathrm{d}\tau^R , \tag{13.59}$$

where $y^R$ is a function of $\tau^R$ via Eq. (13.56). Next we interchange the order of integrations leading to

$$\overline{\overline{C}} = \frac{1}{4\pi\left[\frac{1}{c^2} + (\kappa^R)^2\right]^{\frac{1}{2}}}\int_{T_0^R}^{\infty} \exp(-s\tau^R)\mathrm{d}\tau^R \int_{-Z^R(\tau^R)}^{Z^R(\tau^R)} \frac{\overline{B}(z^R, y^R) + \overline{B}(z^R, y^{R\star})}{\left\{[Z^R(\tau^R)]^2 - (z^R)^2\right\}^{\frac{1}{2}}}\mathrm{d}z^R , \tag{13.60}$$

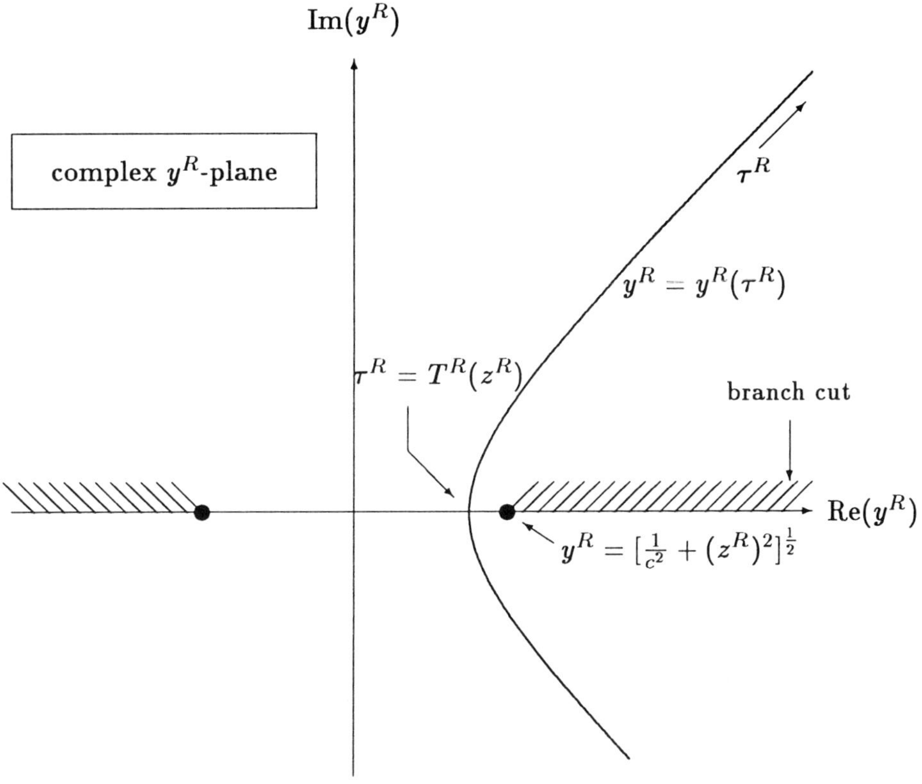


Figure 13.3. Complex $y^R$-plane with Cagniard-de Hoop path $y^R(\tau^R)$.

where

$$T_0^R = T^R(0) = \left[\frac{1}{c^2} + (\kappa^R)^2\right]^{\frac{1}{2}} x_3 \tag{13.61}$$

and

$$Z^R(\tau^R) = \left[\frac{(\tau^R)^2}{\dfrac{1}{c^2} + (\kappa^R)^2} - x_3^2\right]^{\frac{1}{2}} . \tag{13.62}$$

Introduction of the new variable of integration, $z^R = Z^R(\tau^R)\sin(\psi^R)$, leads to

$$\overline{\overline{C}} = \frac{1}{4\pi\left[\frac{1}{c^2}+(\kappa^R)^2\right]^{\frac{1}{2}}}\int_{T_0^R}^{\infty}\exp(-s\tau^R)\mathrm{d}\tau^R\int_{-\frac{1}{2}\pi}^{\frac{1}{2}\pi}[\overline{B}(z^R,y^R)+\overline{B}(z^R,y^{R\star})]\mathrm{d}\psi^R. \tag{13.63}$$

This expression can be seen as a Laplace transform of the type

$$\overline{\overline{C}}(s) = \int_{T_0^R}^{\infty}\exp(-s\tau^R)\overline{\overline{c}}(\tau^R)\mathrm{d}\tau^R\,, \tag{13.64}$$

where

$$\overline{\overline{c}}(\tau^R) = \frac{1}{4\pi\left[\frac{1}{c^2}+(\kappa^R)^2\right]^{\frac{1}{2}}}\int_{-\frac{1}{2}\pi}^{\frac{1}{2}\pi}[\overline{B}(z^R,y^R)+\overline{B}(z^R,y^{R\star})]\mathrm{d}\psi^R\,. \tag{13.65}$$

From the initial-value theorem of Eq. (1.18) we conclude that

$$\lim_{s\to\infty}\exp(sT_0^R)\,s\overline{\overline{C}}(s) = \lim_{\tau^R\downarrow T_0^R}\overline{\overline{c}}(\tau^R)\,. \tag{13.66}$$

When $\tau^R \downarrow T_0^R$, from Eqs. (13.57) and (13.61), we observe that $z^R \downarrow 0$ and hence

$$\begin{aligned}\lim_{\tau^R\downarrow T_0^R}\overline{\overline{c}}(\tau^R) &= \frac{1}{4\pi\left[\frac{1}{c^2}+(\kappa^R)^2\right]^{\frac{1}{2}}}\int_{-\frac{1}{2}\pi}^{\frac{1}{2}\pi}[\overline{B}(0,y^R)+\overline{B}(0,y^{R\star})]\mathrm{d}\psi^R \\ &= \frac{1}{2\Gamma^R}\overline{B}|_{z^R=0,\,\tau^R=T_0^R}\,,\end{aligned} \tag{13.67}$$

where we have used that the integrand is independent of $\psi^R$ and where

$$\Gamma^R = \left[\frac{1}{c^2}+(\alpha_1^R)^2+(\alpha_2^R)^2\right]^{\frac{1}{2}}\,. \tag{13.68}$$

Taking $s\to\infty$, Eqs. (13.66) and (13.67) yield

$$\overline{\overline{C}} = \frac{\exp(-sT_0^R)}{2s\Gamma^R}\overline{B}|_{z^R=0,\,\tau^R=T_0^R}\,, \tag{13.69}$$

with

$$T_0^R = \Gamma^R x_3 = \Gamma^R h\,. \tag{13.70}$$

From Eqs. (13.42) and (13.69) it follows that

$$\overline{\overline{C}} = \frac{\exp[-s(T_0^R + T_0^S)]}{4s\Gamma^R s\Gamma^S} A|_{z^S=0,\, \tau^S=T_0^S;\; z^R=0,\, \tau^R=T_0^R} \,. \tag{13.71}$$

In order to evaluate the expression for $A$, we first calculate the expression for $\nu_k \partial_k |\boldsymbol{x} - \boldsymbol{x}^R|$. Using $z^R = 0$ and $\tau^R = T_0^R$ we observe that

$$x_1 - x_1^R = -\frac{j\alpha_1^R h}{\Gamma^R}, \quad x_2 - x_2^R = -\frac{j\alpha_2^R h}{\Gamma^R}, \quad x_3 = h\,, \tag{13.72}$$

$$|\boldsymbol{x} - \boldsymbol{x}^R| = R^R = \frac{h}{c\Gamma^R}\,, \tag{13.73}$$

and

$$\nu_k \partial_k |\boldsymbol{x} - \boldsymbol{x}^R| = \frac{(x_1 - x_1^R)\partial_1 h + (x_2 - x_2^R)\partial_2 h - x_3}{|\boldsymbol{x} - \boldsymbol{x}^R|\,[(\partial_1 h)^2 + (\partial_2 h)^2 + 1]^{\frac{1}{2}}}\,, \tag{13.74}$$

where $\partial_1 h$ and $\partial_2 h$ are the spatial derivatives of $h$ with respect to $x_1$ and $x_2$, respectively. Hence,

$$\begin{aligned} A|_{z^S=0,\, \tau^S=T_0^S;\; z^R=0,\, \tau^R=T_0^R} &= c\zeta(\boldsymbol{x}|\frac{-j\alpha_1^S h}{\Gamma^S}, \frac{-j\alpha_2^S h}{\Gamma^S}, h)\, \frac{j\alpha_1^S \partial_1 h + j\alpha_2^S \partial_2 h + \Gamma^S}{[(\partial_1 h)^2 + (\partial_2 h)^2 + 1]^{\frac{1}{2}}} \\ &+ c\zeta(\boldsymbol{x}|\frac{-j\alpha_1^R h}{\Gamma^R}, \frac{-j\alpha_2^R h}{\Gamma^R}, h)\, \frac{j\alpha_1^R \partial_1 h + j\alpha_2^R \partial_2 h + \Gamma^R}{[(\partial_1 h)^2 + (\partial_2 h)^2 + 1]^{\frac{1}{2}}}. \end{aligned} \tag{13.75}$$

We now rewrite the expression for $\overline{\overline{C}}$ of Eq. (13.71) as

$$\overline{\overline{C}} = \frac{E \exp[-s(\Gamma^R + \Gamma^S)h]}{4s^2 \Gamma^R \Gamma^S [(\partial_1 h)^2 + (\partial_2 h)^2 + 1]^{\frac{1}{2}}}\,, \tag{13.76}$$

where

$$\begin{aligned} E &= c\zeta(\boldsymbol{x}|\frac{-j\alpha_1^S h}{\Gamma^S}, \frac{-j\alpha_2^S h}{\Gamma^S}, h)\, [j\alpha_1^S \partial_1 h + j\alpha_2^S \partial_2 h + \Gamma^S] \\ &+ c\zeta(\boldsymbol{x}|\frac{-j\alpha_1^R h}{\Gamma^R}, \frac{-j\alpha_2^R h}{\Gamma^R}, h)\, [j\alpha_1^R \partial_1 h + j\alpha_2^R \partial_2 h + \Gamma^R]\,. \end{aligned} \tag{13.77}$$

Note that $E$ is a function independent of $s$, while $\zeta$ is a function of $x_1$ and $x_2$, only. Hence, $E = E(x_1, x_2; j\boldsymbol{\alpha}^R, j\boldsymbol{\alpha}^S)$.

Combining the results of the transform with respect to the receiver and source coordinates and using $\mathrm{dA} = [(\partial_1 h)^2 + (\partial_2 h)^2 + 1]^{\frac{1}{2}} \mathrm{d}x_1 \mathrm{d}x_2$ for the integration along $\partial \mathbb{D}_g$, we arrive at

$$\vec{\overline{p}}(js\boldsymbol{\alpha}^R|js\boldsymbol{\alpha}^S, s) = \frac{\hat{W}(s)}{4sc\Gamma^R\Gamma^S}\int_{(x_1,x_2)\in\mathbb{R}^2}$$

$$E\exp[js(\alpha_1^R+\alpha_1^S)x_1 + js(\alpha_2^R+\alpha_2^S)x_2 - s(\Gamma^R+\Gamma^S)h]\mathrm{dA}\,,$$

$$\text{for } s \to \infty\,. \qquad (13.78)$$

With this result we finish our analysis for real and positive $s$. The remaining asymptotic analysis is more convenient in the angular-frequency domain.

## 13.4. Angular-frequency-domain analysis

Let us assume that the Laplace transforms of the causal functions $f(\boldsymbol{x}, t)$ and $g(\boldsymbol{x}, t)$ satisfy the relation

$$\hat{f}(\boldsymbol{x}, s) = \hat{g}(\boldsymbol{x}, s), \quad \text{real } s, \quad s \to \infty\,, \qquad (13.79)$$

then, in view of the initial-value theorem of the Laplace transform, we also have

$$\hat{f}(\boldsymbol{x}, j\omega) = \hat{g}(\boldsymbol{x}, j\omega), \quad \text{real } \omega, \quad \omega \to \infty\,. \qquad (13.80)$$

Hence, Eq. (13.78) holds not only for real $s \to \infty$, but also in the case that $s \to j\omega$ and $\omega \to \infty$, provided that $s\boldsymbol{\alpha}^R$ and $s\boldsymbol{\alpha}^S$ are kept real valued.

We further introduce the horizontal offset coordinates $x_1^O$ and $x_2^O$ as

$$x_1^O = x_1^R - x_1^S, \quad x_2^O = x_2^R - x_2^S\,, \qquad (13.81)$$

and in the spatial Fourier domain the coordinates $\boldsymbol{p} = \{p_1, p_2\}$ and $\boldsymbol{q} = \{q_1, q_2\}$ as

$$\boldsymbol{p} = j\boldsymbol{\alpha}^R, \quad \boldsymbol{q} = j(\boldsymbol{\alpha}^R + \boldsymbol{\alpha}^S)\,. \qquad (13.82)$$

Substituting these new coordinates in Eqs. (13.21) and (13.48) and taking $s = j\omega$, we obtain

$$\overline{\overline{p}}^r(j\omega \boldsymbol{p}|j\omega(\boldsymbol{q}-\boldsymbol{p}), j\omega) = \int_{(x_1^S, x_2^S)\in\mathbb{R}^2} \exp(j\omega q_1 x_1^S + j\omega q_2 x_2^S)\mathrm{dA}$$

$$\int_{(x_1^O, x_2^O)\in\mathbb{R}^2} \exp(j\omega p_1 x_1^O + j\omega p_2 x_2^O)\hat{p}^r(x_1^O + x_1^S, x_2^O + x_2^S, 0|\boldsymbol{x}^S, j\omega)\mathrm{dA}\,. \tag{13.83}$$

Equation (13.83) constitutes the temporal Fourier transform of the Radon type with respect to the source coordinates and the offset coordinates. In view of Eqs. (13.21) and (13.48), we may replace Eq. (13.78) by

$$\overline{\overline{p}}^r(j\omega \boldsymbol{p}|j\omega(\boldsymbol{q}-\boldsymbol{p}), j\omega) =$$

$$\frac{\hat{W}(j\omega)}{4j\omega c\Gamma^R\Gamma^S}\int_{(x_1,x_2)\in\mathbb{R}^2} E(x_1, x_2; \boldsymbol{p}, \boldsymbol{q}-\boldsymbol{p}) \exp(j\omega q_1 x_1 + j\omega q_2 x_2 - 2j\omega\varphi h)\mathrm{dA}\,. \tag{13.84}$$

Here, we have introduced the vertical slowness $\varphi = \varphi(\boldsymbol{p}, \boldsymbol{q})$ as

$$\varphi = \frac{1}{2}\left[\frac{1}{c^2} - p_1^2 - p_2^2\right]^{\frac{1}{2}} + \frac{1}{2}\left[\frac{1}{c^2} - (q_1 - p_1)^2 - (q_2 - p_2)^2\right]^{\frac{1}{2}}\,. \tag{13.85}$$

Note that in the high-frequency representation of Eq. (13.84) the structural information of the interface is present as phase information in the exponential function and as amplitude information in the functional dependence of $E$ (cf. Eq. (13.77)). In the act of imaging we are visualizing the structural information in the phase function. In order to come to an operational format, we observe that the integral in the right-hand side of Eq. (13.84) resembles a Fourier transform of the Radon type, provided that we select those values of $\boldsymbol{p}$ and $\boldsymbol{q}$ for which the vertical slowness $\varphi$ is constant. In the subsequent sections we will develop an imaging procedure along these lines. As starting point, we introduce the vertical slowness $\varphi$ as a new independent variable. For given (real) values of $q_1$, $q_2$ and $\varphi$, we determine the real solutions $\{p_1, p_2\}$ of Eq. (13.85). Since we require $\varphi$ to be real, the square roots in the latter equation have to be real too. Hence, the solutions $\{p_1, p_2\}$ are constrained by

$$p_1^2 + p_2^2 \leq \frac{1}{c^2}\,, \tag{13.86}$$

$$(q_1-p_1)^2 + (q_2-p_2)^2 \;\; \leq \;\; \frac{1}{c^2} \, . \tag{13.87}$$

Carrying out two successive squaring operations on Eq. (13.85), the solutions $\{p_1, p_2\}$ of Eq. (13.85) follow from

$$\varphi^2\left(p_1-\frac{q_1}{2}\right)^2+\varphi^2\left(p_2-\frac{q_2}{2}\right)^2+\left[\frac{q_1}{2}(p_1-\frac{q_1}{2})+\frac{q_2}{2}(p_2-\frac{q_2}{2})\right]^2$$
$$+\varphi^2\left(\varphi^2-\frac{1}{c^2}+\frac{q_1^2}{4}+\frac{q_2^2}{4}\right)=0\,. \tag{13.88}$$

Since the first three terms on the left-hand side of Eq. (13.88) are non-negative, the fourth term has to be non-positive. This limits the values of $\{q_1, q_2\}$ to the circular domain $\mathcal{Q}(\varphi)$,

$$\mathcal{Q}(\varphi) = \{\boldsymbol{q} \in \mathbb{R}^2;\; \frac{q_1^2}{4} + \frac{q_2^2}{4} \leq \frac{1}{c^2} - \varphi^2\} \, . \tag{13.89}$$

For values of $\boldsymbol{q}$ located inside this domain $\mathcal{Q} = \mathcal{Q}(\varphi)$ and for values of $\varphi \leq c^{-1}$, real solutions $\boldsymbol{p} = (p_1, p_2)$ of Eq. (13.88) exist and we define the domain $\mathcal{P} = \mathcal{P}(\boldsymbol{q}, \varphi)$ as the pertaining solution space. In this domain we perform the imaging procedure.

## 13.5. Imaging

After introducing $\varphi$ as new independent variable, we have observed that the solutions $\boldsymbol{p}$ of Eq. (13.85), for given values of $\boldsymbol{q}$ and $\varphi$, are located in the domain $\mathcal{P}(\boldsymbol{q}, \varphi)$. Of all these solutions, we take some average solution. For a particular value of $\boldsymbol{q}$ and $\varphi$, we take a weighted averaging of the quantity of Eq. (13.84) over the domain $\mathcal{P} = \mathcal{P}(\boldsymbol{q}, \varphi)$. Therefore, we multiply both sides of Eq. (13.84) with the factor $2j\omega\Gamma^R\Gamma^S/\varphi$ and integrate over the domain $\mathcal{P}$. This factor is for later convenience. The result is

$$\bar{p}^{wght}(\boldsymbol{q}, \varphi, j\omega) = \hat{W}(j\omega) \int_{(x_1,x_2)\in\mathbb{R}^2} E^{wght} \exp[j\omega(q_1 \boldsymbol{x}_1 + q_2 \boldsymbol{x}_2 - 2\varphi h)] \mathrm{dA} \, , \tag{13.90}$$

for $\boldsymbol{q} \in \mathcal{Q}(\varphi)$, where

$$\overline{p}^{wght}(\boldsymbol{q}, \varphi, j\omega) = \frac{2j\omega}{\varphi} \frac{\int_{\boldsymbol{p}\in\mathcal{P}(\boldsymbol{q},\varphi)} \Gamma^R(\boldsymbol{p})\Gamma^S(\boldsymbol{q}-\boldsymbol{p})\,\overline{\overline{p}}(j\omega\boldsymbol{p}, j\omega(\boldsymbol{q}-\boldsymbol{p}), j\omega)\mathrm{dA}}{\int_{\boldsymbol{p}\in\mathcal{P}(\boldsymbol{q},\varphi)} \mathrm{dA}}, \tag{13.91}$$

and

$$E^{wght}(\boldsymbol{x}_1, \boldsymbol{x}_2; \boldsymbol{q}, \varphi) = \frac{\int_{\boldsymbol{p}\in\mathcal{P}(\boldsymbol{q},\varphi)} E(\boldsymbol{x}_1, \boldsymbol{x}_2; \boldsymbol{p}, \boldsymbol{q}-\boldsymbol{p})\mathrm{dA}}{2c\varphi \int_{\boldsymbol{p}\in\mathcal{P}(\boldsymbol{q},\varphi)} \mathrm{dA}}. \tag{13.92}$$

For a particularly chosen value of $\varphi$, we observe from Eq. (13.92) that $E^{wght} = E^{wght}(\boldsymbol{x}_1, \boldsymbol{x}_2; \boldsymbol{q}, \varphi)$ is a function of $\boldsymbol{x}_1, \boldsymbol{x}_2$ and an algebraic function of $\boldsymbol{q}$. It does not depend on $\omega$. When $\omega$ tends to infinity the contribution of the integral on the right-hand side of Eq. (13.90) comes from the stationary point $\{\boldsymbol{x}_1, \boldsymbol{x}_2\} = \{\boldsymbol{x}_1', \boldsymbol{x}_2'\}$ following from

$$q_1 - 2\varphi\partial_1 h(\boldsymbol{x}_1, \boldsymbol{x}_2) = 0, \quad q_2 - 2\varphi\partial_2 h(\boldsymbol{x}_1, \boldsymbol{x}_2) = 0\,. \tag{13.93}$$

These relations connect the $q_1$ and $q_2$ values to the profile gradient. The stationary point is a function of $\boldsymbol{q}$. We may replace the integration variable $\{\boldsymbol{x}_1, \boldsymbol{x}_2\}$ in the expression of $E^{wght}$ by the stationary point $\{\boldsymbol{x}_1', \boldsymbol{x}_2'\}$. This does not change the result of the integral on the right-hand side of Eq. (13.90) when $\omega$ tends to infinity. Then, for some particular value of $\varphi$, $E^{wght}$ is a function of $\boldsymbol{q}$ only. But in that case we can reuse the expressions of $\boldsymbol{q}$ of Eq. (13.93) and we may define a function $E^{image} = E^{image}(\boldsymbol{x}_1, \boldsymbol{x}_2; \varphi)$ as

$$E^{image}(\boldsymbol{x}_1, \boldsymbol{x}_2; \varphi) = E^{wght}_{|q_1=2\varphi\partial_1 h,\, q_2=2\varphi\partial_2 h}\,. \tag{13.94}$$

We assume that the values of $q_1$ and $q_2$ used in Eq. (13.94) are contained in the domain $\mathcal{Q}(\varphi)$. In view of Eq. (13.93) this means that the maximum profile gradient to be imaged is restricted. Then, Eq. (13.90) is replaced by

$$\overline{p}^{wght}(\boldsymbol{q}, \varphi, j\omega)$$
$$= \hat{W}(j\omega) \int_{(x_1,x_2)\in\mathbb{R}^2} E^{image}(\boldsymbol{x}_1, \boldsymbol{x}_2; \varphi)\exp(j\omega q_1\boldsymbol{x}_1 + j\omega q_2\boldsymbol{x}_2 - 2j\omega\varphi\, h)\mathrm{dA}, \tag{13.95}$$

for $\omega \to \infty$ and for $\boldsymbol{q} \in \mathcal{Q}(\varphi)$. For points $\boldsymbol{q}$ outside $\mathcal{Q}(\varphi)$, the right-hand side of Eq. (13.95) vanishes when $\omega$ tends to infinity, because we have assumed that the values of $\boldsymbol{q}$ related to the stationary point of the integral are contained in the domain $\mathcal{Q}(\varphi)$. The right-hand side of Eq. (13.95) represents a spatial Fourier transform with respect to the spatial coordinates $x_1$ and $x_2$ with transform parameters $j\omega q_1$ and $j\omega q_2$. The inverse spatial Fourier transformation yields

$$\hat{p}^{image}(x_1, x_2, j\omega; \varphi) = E^{image}(x_1, x_2; \varphi)\, \hat{W}(j\omega) \exp(-2j\omega\varphi\, h), \quad \omega \to \infty\,, \tag{13.96}$$

with

$$\hat{p}^{image}(x_1, x_2, j\omega; \varphi) = \left(\frac{\omega}{2\pi}\right)^2 \int_{\boldsymbol{q}\in\mathcal{Q}(\varphi)} \exp(-j\omega q_1 x_1 - j\omega q_2 x_2) \bar{p}^{wght}(\boldsymbol{q}, \varphi, j\omega) \mathrm{dA}\,. \tag{13.97}$$

Equation (13.97) indicates how the recorded data have to be mapped. We subsequently transform Eq. (13.96) back to the time domain. We observe that Eq. (13.96) holds for $\omega \to \infty$, therefore we have

$$\lim_{t \downarrow T_h} p^{image}(x_1, x_2, t; \varphi) = E^{image}(x_1, x_2; \varphi)\, W(t - T_h)\,, \tag{13.98}$$

where the laterally variant vertical travel time $T_h = T_h(x_1, x_2; \varphi)$ is given by

$$T_h(x_1, x_2; \varphi) = 2\varphi\, h(x_1, x_2)\,. \tag{13.99}$$

We observe that the space-time function of Eq. (13.98) starts to act at the instant $T_h = 2\varphi\, h(x_1, x_2)$, so that we have a space-time image of the profile of the interface as the envelope of the arrivals of the reflected causal wavelet function. For a complete image of the profile of the interface it is necessary that the domain $\mathcal{Q} = \mathcal{Q}(\varphi)$ contains all the stationary points given by $q_1 = 2\varphi\partial_1 h$, $q_2 = 2\varphi\partial_2 h$. Hence, from Eqs. (13.89) and (13.93) it follows

$$\sup_{(x_1, x_2)\in\mathbb{R}^2} [(\partial_1 h)^2 + (\partial_2 h)^2] \leq \frac{1}{c^2\varphi^2} - 1\,. \tag{13.100}$$

In order to image large partial derivatives of the profile, we observe that we have to take small values of $\varphi$. This is achieved for large values of $\boldsymbol{p}$ and/or $\boldsymbol{q}$. In practical seismic measurements, the data for these values of $\boldsymbol{p}$ and $\boldsymbol{q}$ are not available and the imaging of large partial derivatives of the profile is restricted.

## 13.6. Analytic solution for a planar reflector

In this section we illustrate the imaging procedure by considering the trivial case that $\partial\mathbb{D}_g$ is a planar, perfectly reflecting, interface at $x_3 = h$ (see Fig. 13.4). We consider either the case that this plane is perfectly compliant, where the acoustic pressure vanishes,

$$\lim_{x_3 \uparrow h} \hat{p}(\boldsymbol{x}|\boldsymbol{x}^S, s) = 0 \,, \tag{13.101}$$

or the case that this plane is perfectly rigid, where the acoustic particle velocity vanishes; this is equivalent with a vanishing normal derivative of the acoustic pressure,

$$\lim_{x_3 \uparrow h} \partial_3 \hat{p}(\boldsymbol{x}|\boldsymbol{x}^S, s) = 0 \,. \tag{13.102}$$

The problem of reflection of a wavefield by a plane interface can be solved exactly. At the reflector the incident wavefield follows from Eqs. (13.2), (13.4) and (4.60) as

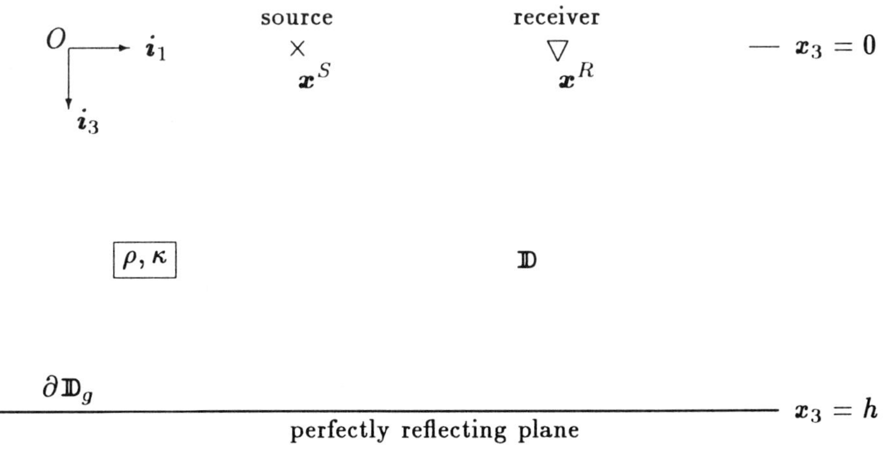


Figure 13.4. The plane-reflector configuration.

$$\hat{p}^{inc}(x_1, x_2, h|\boldsymbol{x}^S, s) = \hat{W}(s)\frac{1}{(2\pi)^2}\int_{(s\alpha_1,s\alpha_2)\in\mathbb{R}^2}$$

$$\frac{\exp[-js\alpha_1(x_1-x_1^S) - js\alpha_2(x_2-x_2^S) - s\Gamma h]}{2s\Gamma}\mathrm{dA}\,, \tag{13.103}$$

where $\Gamma = (c^{-2} + \alpha_1^2 + \alpha_2^2)^{\frac{1}{2}}$. The reflected wavefield is then given by

$$\hat{p}^r(x_1, x_2, x_3|\boldsymbol{x}^S, s) = \mp\hat{W}(s)\frac{1}{(2\pi)^2}\int_{(s\alpha_1,s\alpha_2)\in\mathbb{R}^2}$$

$$\frac{\exp[-js\alpha_1(x_1-x_1^S) - js\alpha_2(x_2-x_2^S) - s\Gamma(2h-x_3)]}{2s\Gamma}\mathrm{dA}\,. \tag{13.104}$$

It is easy to observe that the total wavefield $\hat{p} = \hat{p}^{inc} + \hat{p}^r$ satisfies the boundary conditions of Eqs. (13.101) and (13.102). The $-$ sign holds for the case of the compliant interface and the $+$ sign holds for the rigid interface.

Taking $s \to j\omega$ and introducing the horizontal offset coordinates $x_1^O = x_1 - x_1^S$, $x_2^O = x_2 - x_2^S$, and in the spatial Fourier domain the coordinates $p_1 = j\alpha_1$, $p_2 = j\alpha_2$, the reflected wavefield at $x_3 = 0$ is written as

$$\hat{p}^r(x_1^O + x_1^S, x_2^O + x_2^S, 0|\boldsymbol{x}^S, j\omega)$$

$$= \mp\frac{\hat{W}(j\omega)}{j\omega}\left(\frac{\omega}{2\pi}\right)^2\int_{(p_1,p_2)\in\mathbb{R}^2}\frac{\exp(-j\omega p_1 x_1^O - j\omega p_2 x_2^O - 2j\omega\Gamma h)}{2\Gamma}\mathrm{dA}\,, \tag{13.105}$$

where the vertical slowness $\Gamma$ is now given by

$$\Gamma = \left[\frac{1}{c^2} - p_1^2 - p_2^2\right]^{\frac{1}{2}}. \tag{13.106}$$

Next, we transform the data to the $(j\omega\boldsymbol{p}, j\omega\boldsymbol{q})$-domain using Eq. (13.83). We arrive at

$$\overline{\overline{p}}^r(j\omega\boldsymbol{p}, j\omega(\boldsymbol{q}-\boldsymbol{p}), j\omega) = \mp\frac{\hat{W}(j\omega)}{j\omega}\left(\frac{2\pi}{\omega}\right)^2\delta(\boldsymbol{q})\frac{\exp(-2j\omega\Gamma h)}{2\Gamma}. \tag{13.107}$$

In view of the behavior of the delta function that is only operative at $q_1 = q_2 = 0$, we may replace $\Gamma$ by

$$\varphi = \left[\frac{1}{c^2} - p_1^2 - p_2^2\right]^{\frac{1}{2}}, \tag{13.108}$$

that follows from Eq. (13.85) by setting $q_1 = q_2 = 0$. Hence, Eq. (13.107) immediately transfers into

$$\bar{\bar{p}}^{r}(j\omega \boldsymbol{p}, j\omega(\boldsymbol{q}-\boldsymbol{p}), j\omega) = \mp \frac{\hat{W}(j\omega)}{j\omega}\left(\frac{2\pi}{\omega}\right)^2 \delta(\boldsymbol{q})\frac{\exp(-2j\omega\varphi h)}{2\varphi}. \tag{13.109}$$

We subsequently take $\varphi$ as independent variable. We then observe that the expression in the right-hand side of Eq. (13.109) is independent of $\boldsymbol{p}$. Performing the weighted averaging (see Eq. (13.91)), of all possible solutions $\{p_1, p_2\}$ of Eq. (13.108), results in the same expression but multiplied with the factor $2j\omega\Gamma^R\Gamma^S/\varphi$. Since this factor is equal to $2j\omega\varphi$ for $q_1 = q_2 = 0$, we arrive at

$$\bar{p}^{wght}(\boldsymbol{q}, \varphi, \omega) = \mp \hat{W}(j\omega)\left(\frac{2\pi}{\omega}\right)^2 \delta(\boldsymbol{q}) \exp(-2j\omega\varphi h). \tag{13.110}$$

The resulting data $\bar{p}^{wght}(\boldsymbol{q}, \varphi, \omega)$ are stacked using Eq. (13.97). We obtain

$$\hat{p}^{image}(x_1, x_2, j\omega; \varphi) = \mp \hat{W}(j\omega) \exp(-2j\omega\varphi h). \tag{13.111}$$

As last step we apply an inverse temporal Fourier transform and arrive at

$$p^{image}(x_1, x_2, t; \varphi) = \mp W(t - T_h), \tag{13.112}$$

where

$$T_h = 2\varphi h. \tag{13.113}$$

Observe that the right-hand side of Eq. (13.112) is independent of $x_1$ and $x_2$ and starts to act at the instant $T_h$, so that we have a space-time image of the planar reflector as the arrival of the reflected causal wavelet function.

## 13.7. Imaging of a disk

As second imaging example, we consider the problem of an infinitely thin disk. The disk is either perfectly compliant or perfectly rigid (and immovable). The disk occupies the domain

$$\mathbb{D}_g = \{\boldsymbol{x} \in \mathbb{R}^3;\ (x_1, x_2) \in \Lambda_g,\ x_3 = h\}. \tag{13.114}$$

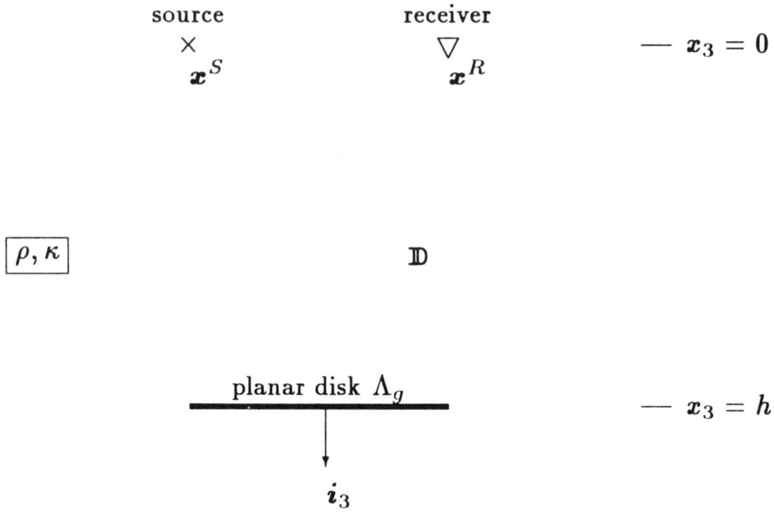

source receiver

$\times$ $\bigtriangledown$ — $x_3 = 0$

$x^S$ $x^R$

$\rho, \kappa$ $\mathbb{D}$

planar disk $\Lambda_g$ — $x_3 = h$

$i_3$

Figure 13.5. The disk configuration.

where $\Lambda_g$ is a bounded domain in the $(x_1, x_2)$-plane (see Fig. 13.5). The integral representation for the reflected wavefield is given as (cf. Eqs. (9.7) and (9.16))

$$\hat{p}^r(\boldsymbol{x}^R|\boldsymbol{x}^S, s) = s\rho \int_{\boldsymbol{x}\in\Lambda_g} \hat{G}(\boldsymbol{x}^R - \boldsymbol{x}, s)\partial\hat{q}(\boldsymbol{x}|\boldsymbol{x}^S, s)\mathrm{dA} , \tag{13.115}$$

in the case of a perfectly compliant lamina, and (cf. Eqs. (9.11) and (9.17))

$$\hat{p}^r(\boldsymbol{x}^R|\boldsymbol{x}^S, s) = -\int_{\boldsymbol{x}\in\Lambda_g} \partial_3^R\hat{G}(\boldsymbol{x}^R - \boldsymbol{x}, s)\partial\hat{f}_3(\boldsymbol{x}|\boldsymbol{x}^S, s)\mathrm{dA} , \tag{13.116}$$

in the case of an immovable rigid disk, where the equivalent surface density of injected volume time rate is defined as

$$\partial\hat{q}(\boldsymbol{x}|\boldsymbol{x}^S, s) = \lim_{\varepsilon\downarrow 0}[\hat{v}_3(\boldsymbol{x}+\varepsilon\boldsymbol{i}_3|\boldsymbol{x}^S, s) - \hat{v}_3(\boldsymbol{x}-\varepsilon\boldsymbol{i}_3|\boldsymbol{x}^S, s)]|_{\boldsymbol{x}\in\Lambda_g} , \tag{13.117}$$

and the equivalent surface-force density as

$$\partial\hat{f}_3(\boldsymbol{x}|\boldsymbol{x}^S, s) = \lim_{\varepsilon\downarrow 0}[\hat{p}(\boldsymbol{x}+\varepsilon\boldsymbol{i}_3|\boldsymbol{x}^S, s) - \hat{p}(\boldsymbol{x}-\varepsilon\boldsymbol{i}_3|\boldsymbol{x}^S, s)]|_{\boldsymbol{x}\in\Lambda_g} . \tag{13.118}$$

In Chapter 9 we have solved the problem of the reflection of a wavefield by a disk with the integral-equation method. The integral equation has been solved numerically for a strip. In Section 13.9 we shall use the numerically obtained reflection data to image the strip. In order to interpret the results of the imaging procedure, we consider an approximate solution for large values of $|s|$; the latter is furnished by the (modified) Kirchhoff approximation. The Kirchhoff approximation applies to the scattering by impenetrable objects. On the *illuminated* part of the disk, the local wavefield is assumed to take on the values that would result from the reflection of the incident wavefield against a perfectly reflecting plane. On the *dark* part of the disk, the wavefield is assumed to be negligibly small. As a consequence of this approximation the equivalent surface densities are written as

$$\partial\hat{q}(\boldsymbol{x}|\boldsymbol{x}^S, s) = -2\hat{v}_3^{inc}(\boldsymbol{x}|\boldsymbol{x}^S, s) = 2\hat{q}^S(s)\partial_3\hat{G}(\boldsymbol{x}-\boldsymbol{x}^S, s)\,, \tag{13.119}$$

and the equivalent surface-force density as

$$\partial\hat{f}_3(\boldsymbol{x}|\boldsymbol{x}^S, s) = -2\hat{p}^{inc}(\boldsymbol{x}|\boldsymbol{x}^S, s) = -2s\rho\hat{q}^S(s)\hat{G}(\boldsymbol{x}-\boldsymbol{x}^S, s)\,, \tag{13.120}$$

in which we have used the expression for the incident wavefield of Eqs. (13.2) and (13.3). Substituting these approximations into the integral representation for the reflected wavefield, we obtain

$$\hat{p}^r(\boldsymbol{x}^R|\boldsymbol{x}^S, s) = 2s\rho\hat{q}^S(s)\int_{\boldsymbol{x}\in\Lambda_g}\hat{G}(\boldsymbol{x}-\boldsymbol{x}^R, s)\partial_3\hat{G}(\boldsymbol{x}-\boldsymbol{x}^S, s)\mathrm{dA}\,, \tag{13.121}$$

in the case of a perfectly compliant lamina, and (cf. Eqs. (9.11) and (9.17))

$$\hat{p}^r(\boldsymbol{x}^R|\boldsymbol{x}^S, s) = -2s\rho\hat{q}^S(s)\int_{\boldsymbol{x}\in\Lambda_g}\hat{G}(\boldsymbol{x}-\boldsymbol{x}^S, s)\partial_3\hat{G}(\boldsymbol{x}-\boldsymbol{x}^R, s)\mathrm{dA}\,, \tag{13.122}$$

in the case of an immovable disk.

On account of the reciprocity relation of Eq. (13.14), the two expressions can be combined to one expression of the form

$$\begin{aligned}\hat{p}^r(\boldsymbol{x}^R|\boldsymbol{x}^S, s) = \pm\hat{W}(s)\int_{\boldsymbol{x}\in\Lambda_g}[\hat{G}(\boldsymbol{x}-\boldsymbol{x}^R, s)\partial_3\hat{G}(\boldsymbol{x}-\boldsymbol{x}^S, s)\\ +\hat{G}(\boldsymbol{x}-\boldsymbol{x}^S, s)\partial_3\hat{G}(\boldsymbol{x}-\boldsymbol{x}^R, s)]\mathrm{dA}\,.\end{aligned} \tag{13.123}$$

The upper sign holds for the case of a compliant lamina and the lower sign holds for the case of a rigid disk. The high-frequency analysis of Sections

13.2 and 13.3 can be circumvented. We can directly use the representations of the Green's function.

Using Eq. (4.60), the Green's function follows as an inverse spatial Fourier transform with respect to the source coordinates,

$$\hat{G}(\boldsymbol{x}-\boldsymbol{x}^S, s) = \frac{1}{(2\pi)^2} \int_{(s\alpha_1^S, s\alpha_2^S)\in \mathbb{R}^2} \frac{\exp[js\alpha_1^S(x_1-x_1^S) + js\alpha_2^S(x_2-x_2^S) - s\Gamma^S x_3]}{2s\Gamma^S} \mathrm{dA}\,, \tag{13.124}$$

with

$$\Gamma^S = \left[\frac{1}{c^2} + (\alpha_1^S)^2 + (\alpha_2^S)^2\right]^{\frac{1}{2}}\,, \tag{13.125}$$

and as an inverse spatial Fourier transform with respect to the receiver coordinates as

$$\hat{G}(\boldsymbol{x}-\boldsymbol{x}^R, s) = \frac{1}{(2\pi)^2} \int_{(s\alpha_1^R, s\alpha_2^R)\in \mathbb{R}^2} \frac{\exp[js\alpha_1^R(x_1-x_1^R) + js\alpha_2^R(x_2-x_2^R) - s\Gamma^R x_3]}{2s\Gamma^R} \mathrm{dA}\,, \tag{13.126}$$

with

$$\Gamma^R = \left[\frac{1}{c^2} + (\alpha_1^R)^2 + (\alpha_2^R)^2\right]^{\frac{1}{2}}\,. \tag{13.127}$$

Note that we have used $x_3^S = x_3^R = 0$ and $x_3 > 0$.

Using these representations and performing the Fourier transforms with respect to the horizontal source coordinates (Eq. (13.21)) and to the horizontal receiver coordinates (Eq. (13.48)), we directly obtain

$$\overline{\overline{p}}^r(js\boldsymbol{\alpha}^R|js\boldsymbol{\alpha}^S, s) = \mp\frac{\hat{W}(s)}{s}\,\frac{\Gamma^R+\Gamma^S}{4\Gamma^R\Gamma^S}\exp[-s(\Gamma^R+\Gamma^S)h] \int_{(x_1,x_2)\in\Lambda_g} \exp[js(\alpha_1^R+\alpha_1^S)x_1 + js(\alpha_2^R+\alpha_2^S)x_2]\mathrm{dA}\,. \tag{13.128}$$

Subsequently, we take $s \to j\omega$ and introduce the horizontal offset coordinates $x_1^O = x_1^R - x_1^S$, $x_2^O = x_2^R - x_2^S$, and in the spatial Fourier domain the coordinates $p_1 = j\alpha_1^R$, $p_2 = j\alpha_2^R$, $q_1 = j(\alpha_1^R + \alpha_1^S)$, $q_2 = j(\alpha_2^R + \alpha_2^S)$. Then, Eq. (13.128) transfers into

$$\overline{\overline{p}}^r(j\omega \boldsymbol{p}|j\omega(\boldsymbol{q}-\boldsymbol{p}), j\omega) = \mp \frac{\hat{W}(j\omega)}{j\omega} \frac{\varphi}{2\Gamma^R\Gamma^S} \exp(-2j\omega\varphi h)\, \overline{\chi}_{\Lambda_g} . \tag{13.129}$$

where the vertical slowness $\varphi$ is given by Eq. (13.85) and

$$\overline{\chi}_{\Lambda_g}(j\omega q_1, j\omega q_2) = \int_{(x_1,x_2)\in \mathbb{R}^2} \exp[j\omega(q_1 x_1 + q_2 x_2)] \chi_{\Lambda_g}(x_1, x_2) \mathrm{dA} \tag{13.130}$$

denotes the spatial Fourier transform of the characteristic function

$$\chi_{\Lambda_g}(x_1, x_2) = \{1, \frac{1}{2}, 0\} \quad \text{when } (x_1, x_2) \in \{\Lambda_g, \partial\Lambda_g, \Lambda_g'\} . \tag{13.131}$$

The boundary $\partial\Lambda_g$ denotes the edge of the disk and $\Lambda_g'$ denotes the complement of $\Lambda_g \cup \partial\Lambda_g$ in the plane $\{(x_1, x_2) \in \mathbb{R}^2;\ x_3 = h\}$. We subsequently take $\varphi$ as independent variable. Performing the weighted average according to Eqs. (13.91) - (13.92) we arrive at

$$\overline{p}^{wght}(\boldsymbol{q}, \varphi, j\omega) = \mp \overline{\chi}_{\Lambda_g}(j\omega q_1, j\omega q_2) \hat{W}(j\omega) \exp(-2j\omega\varphi h), \quad \boldsymbol{q} \in \mathcal{Q}(\varphi) . \tag{13.132}$$

From Eq. (13.130) it follows that $\overline{\chi}_{\Lambda_g}$ tends to zero when $q_1 \neq 0$, $q_2 \neq 0$ and $\omega$ tends to infinity (Riemann Lebesgue lemma, WHITTAKER and WATSON, 1927, p. 127), and consequently we write

$$0 = \mp \overline{\chi}_{\Lambda_g}(j\omega q_1, j\omega q_2) \hat{W}(j\omega) \exp(-2j\omega\varphi h), \quad \boldsymbol{q} \in \mathbb{R}^2 \backslash \mathcal{Q}(\varphi) . \tag{13.133}$$

Hence, from Eqs. (13.132) and (13.133) we observe that the inverse spatial Fourier transformation can now be carried out, so that we obtain

$$\hat{p}^{image}(x_1, x_2, j\omega; \varphi) = \mp \chi_{\Lambda_g}(x_1, x_2) \hat{W}(j\omega) \exp(-2j\omega\varphi\, h), \quad \omega \to \infty , \tag{13.134}$$

where $\hat{p}^{image}$ is given by Eq. (13.97). Finally we apply an inverse temporal Fourier transform. Since Eq. (13.134) holds for $\omega \to \infty$, we arrive at

$$\lim_{t \downarrow T_h} p^{image}(x_1, x_2, t; \varphi) = \mp \chi_{\Lambda_g}(x_1, x_2)\, W(t - T_h) , \tag{13.135}$$

where the laterally variant vertical travel time $T_h$ is given by

$$T_h = 2\varphi h\,. \tag{13.136}$$

We observe that the space-time function of Eq. (13.135) starts to act at the instant $T_h = 2\varphi h$. This instant is an image of the depth location of the disk. The horizontal extent of the disk is completely determined by its characteristic function.

## 13.8. Two-dimensional case

In the two-dimensional case, the interface $\partial\mathbb{D}_g$ is considered to be independent of $x_2$, hence this interface is described as

$$x_3 = h(x_1)\,. \tag{13.137}$$

We further consider the two-dimensional wavefield to be generated by a monopole line source at $\boldsymbol{x}_T^S = (x_1^S, x_3^S = 0)$. In the $s$-domain the source wavefield at the two-dimensional observation point $\boldsymbol{x}_T = (x_1, x_3)$ is given by

$$\hat{p}^{inc}(\boldsymbol{x}_T|\boldsymbol{x}_T^S, s) = s\rho\hat{q}^S(s)\hat{\mathcal{G}}(\boldsymbol{x}_T-\boldsymbol{x}_T^S, s)\,, \tag{13.138}$$

$$\hat{v}_k^{inc}(\boldsymbol{x}_T|\boldsymbol{x}_T^S, s) = -\hat{q}^S(s)\partial_k\hat{\mathcal{G}}(\boldsymbol{x}_T-\boldsymbol{x}_T^S, s)\,, \tag{13.139}$$

with $\hat{v}_2^{inc} = 0$ and where the two-dimensional Green's function is given by (cf. Eqs. (9.84) and (9.87))

$$\hat{\mathcal{G}}(\boldsymbol{x}_T, s) = \int_{x_2\in\mathbb{R}} \hat{G}(x_1, x_2, x_3)\mathrm{d}x_2\,. \tag{13.140}$$

*The boundary-integral representation*

Carrying out these two-dimensional modifications in the analysis of Section 13.1, Eq. (13.17) has to be modified into

$$\hat{p}^r(\boldsymbol{x}_T^R|\boldsymbol{x}_T^S, s) = \frac{s}{c}\hat{W}(s)\int_{\boldsymbol{x}_T\in\partial\mathbb{D}_g} A\,\hat{\mathcal{G}}(\boldsymbol{x}_T-\boldsymbol{x}_T^R, s)\hat{\mathcal{G}}(\boldsymbol{x}_T-\boldsymbol{x}_T^S, s)\mathrm{d}l\,, \tag{13.141}$$

where $\boldsymbol{x}_T^R = (\boldsymbol{x}_1^R, \boldsymbol{x}_3^R = 0)$ is the two-dimensional line-receiver position. The function $A = A(\boldsymbol{x}_T, \boldsymbol{x}_T - \boldsymbol{x}_T^R, \boldsymbol{x}_T - \boldsymbol{x}_T^S)$ is given by (cf. Eq. (13.20)) by

$$A = -\zeta(\boldsymbol{x}_T|\boldsymbol{x}_T - \boldsymbol{x}_T^S)\,(\nu_1\partial_1 + \nu_3\partial_3)|\boldsymbol{x}_T - \boldsymbol{x}_T^S| \\ -\zeta(\boldsymbol{x}_T|\boldsymbol{x}_T - \boldsymbol{x}_T^R)(\nu_1\partial_1 + \nu_3\partial_3)|\boldsymbol{x}_T - \boldsymbol{x}_T^R|\,. \tag{13.142}$$

Substituting the expression of Eq. (13.140) for the two-dimensional Green's function in Eq. (13.141) and interchanging the order of integration, we obtain

$$\hat{p}^r(\boldsymbol{x}_T^R|\boldsymbol{x}_T^S, s) = \int_{x_2^R \in \mathbb{R}} \int_{x_2^S \in \mathbb{R}} \hat{p}^r(\boldsymbol{x}^R|\boldsymbol{x}^S, s)\mathrm{d}\boldsymbol{x}_2^R \mathrm{d}\boldsymbol{x}_2^S\,, \tag{13.143}$$

where

$$\hat{p}^r(\boldsymbol{x}^R|\boldsymbol{x}^S, s) \\ = \frac{s}{c}\hat{W}(s)\int_{\boldsymbol{x}_T \in \partial\mathbb{D}_g} A\,\hat{G}(\boldsymbol{x}_1 - \boldsymbol{x}_1^R, \boldsymbol{x}_2^R, \boldsymbol{x}_3 - \boldsymbol{x}_3^R, s)\hat{G}(\boldsymbol{x}_1 - \boldsymbol{x}_1^S, \boldsymbol{x}_2^S, \boldsymbol{x}_3 - \boldsymbol{x}_3^S, s)\mathrm{d}l\,. \tag{13.144}$$

We observe that Eq. (13.144) has the same appearance as Eq. (13.17). Moreover, Eq. (13.143) can be regarded as the spatial Fourier transform of $\hat{p}^r(\boldsymbol{x}^R|\boldsymbol{x}^S, s)$ with respect to $\boldsymbol{x}_2^S$ and $\boldsymbol{x}_2^R$, with transform parameters $js\alpha_2^S = 0$ and $js\alpha_2^R = 0$.

*Fourier transform with respect to* $\boldsymbol{x}_1^R$ *and* $\boldsymbol{x}_1^S$

In order to take advantage of the results of Sections 13.2 and 13.3 we only need to transform Eq. (13.143) with respect to the receiver coordinate $\boldsymbol{x}_1^R$ and the source coordinate $\boldsymbol{x}_1^S$. Let us define the Fourier transform with respect to $\boldsymbol{x}_1^R$ and $\boldsymbol{x}_1^S$ as

$$\bar{\bar{p}}^r(js\alpha_1^R|js\alpha_1^S, s) \\ = \int_{x_1^R \in \mathbb{R}} \int_{x_1^S \in \mathbb{R}} \exp(js\alpha_1^R \boldsymbol{x}_1^R + js\alpha_1^S \boldsymbol{x}_1^S)\hat{p}^r(\boldsymbol{x}_T^R|\boldsymbol{x}_T^S, s)\mathrm{d}\boldsymbol{x}_1^R \mathrm{d}\boldsymbol{x}_1^S\,. \tag{13.145}$$

Then, the two-dimensional variant of the analysis of Sections 13.2 and 13.3 becomes (cf. Eq. (13.78))

$$\bar{\bar{p}}^r(js\alpha_1^R|js\alpha_2^R, s) = \frac{\hat{W}(s)}{4sc\Upsilon^R\Upsilon^S}\int_{x_1 \in \mathbb{R}} E\exp[js(\alpha_1^R + \alpha_1^S)\boldsymbol{x}_1 - s(\Upsilon^R + \Upsilon^S)h]\mathrm{d}\boldsymbol{x}_1\,, \tag{13.146}$$

with $E = E(x_1; j\alpha_1^R, j\alpha_1^S)$ given by

$$E = c\zeta(\boldsymbol{x}_T|\frac{-j\alpha_1^S h}{\Upsilon^S}, h)\,[j\alpha_1^S\partial_1 h + \Upsilon^S] + c\zeta(\boldsymbol{x}_T|\frac{-j\alpha_1^R h}{\Upsilon^R}, h)\,[j\alpha_1^R\partial_1 h + \Upsilon^R]\,, \tag{13.147}$$

and

$$\Upsilon^R = \left[\frac{1}{c^2} + (\alpha_1^R)^2\right]^{\frac{1}{2}}, \quad \Upsilon^S = \left[\frac{1}{c^2} + (\alpha_1^S)^2\right]^{\frac{1}{2}}. \tag{13.148}$$

We remark that this result is obtained from Eq. (13.78) by taking $\alpha_2^R = 0$ and $\alpha_2^S = 0$.

*Angular-frequency-domain analysis*

First, we introduce the horizontal offset coordinate $x_1^O$ as

$$x_1^O = x_1^R - x_1^S\,, \tag{13.149}$$

and in the spatial Fourier domain the coordinates $p_1$ and $q_1$ as

$$p_1 = j\alpha_1^R, \quad q_1 = j(\alpha_1^R + \alpha_1^S)\,. \tag{13.150}$$

Substituting these new coordinates in Eq. (13.145) and taking $s = j\omega$, we obtain

$$\bar{\bar{p}}^r(j\omega p_1|j\omega(q_1-p_1), j\omega)$$
$$= \int_{x_1^O\in\mathbb{R}}\int_{x_1^S\in\mathbb{R}} \exp(j\omega p_1 x_1^O + j\omega q_1 x_1^S)\hat{p}^r(x_1^O + x_1^S, 0|\boldsymbol{x}_T^S, s)\mathrm{d}x_1^O\mathrm{d}x_1^S\,, \tag{13.151}$$

while (cf. Eqs. (13.84) and (13.85))

$$\bar{\bar{p}}^r(j\omega p_1|j\omega(q_1-p_1), j\omega)$$
$$= \frac{\hat{W}(j\omega)}{4j\omega c\Upsilon^R\Upsilon^S}\int_{x_1\in\mathbb{R}} E(x_1; p_1, q_1-p_1)\exp(j\omega q_1 x_1 - 2j\omega\varphi h)\mathrm{d}x_1\,, \tag{13.152}$$

where the vertical two-dimensional slowness $\varphi = \varphi(p_1, q_1)$ is given by

$$\varphi = \frac{1}{2}\left[\frac{1}{c^2} - p_1^2\right]^{\frac{1}{2}} + \frac{1}{2}\left[\frac{1}{c^2} - (q_1-p_1)^2\right]^{\frac{1}{2}}. \tag{13.153}$$

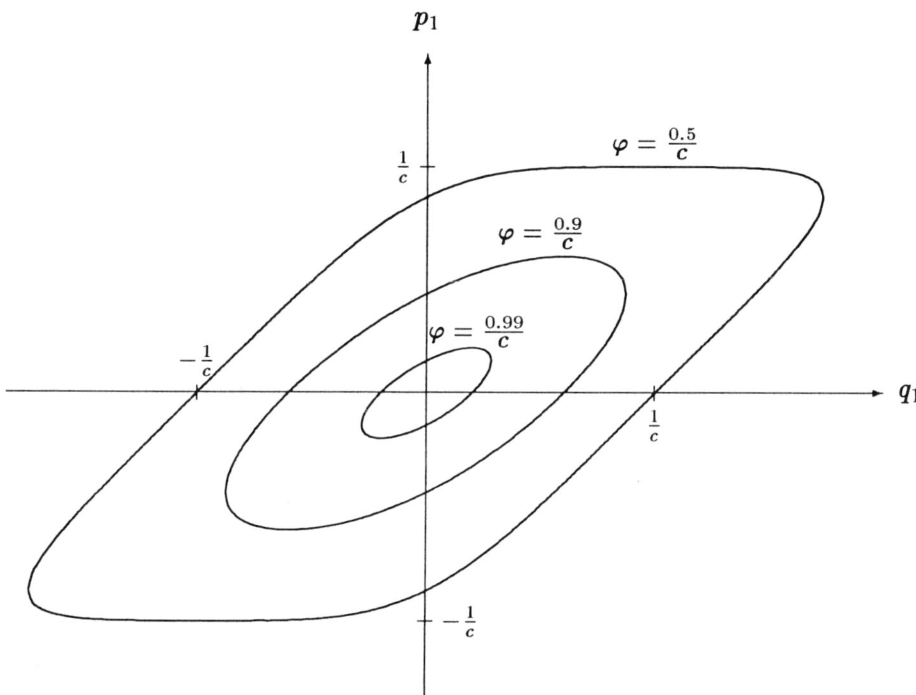


Figure 13.6. The solution $p_1$ as a function of $q_1$ for some values of $\varphi$.

In Section 13.5 we have studied the consequences of taking $\varphi$ as new independent variable. In the present two-dimensional case, by taking $p_2 = q_2 = 0$, we deduce from Eq. (13.88) that two real solutions of $p_1$ exist, viz.,

$$p_1^{\pm} = \frac{1}{2}q_1 \pm \varphi \left( \frac{\frac{1}{c^2} - \frac{1}{4}q_1^2 - \varphi^2}{\frac{1}{4}q_1^2 + \varphi^2} \right)^{\frac{1}{2}} . \tag{13.154}$$

Fig. 13.6 shows the loci in the $(q_1, p_1)$-plane for some values of $\varphi$. For $\varphi = 1/c$ the locus degrades to a point in the origin. The real solutions of Eq. (13.154) only exist for values of $q_1$ in the linear segment $\mathcal{Q}(\varphi)$,

$$\mathcal{Q}(\varphi) = \{q_1 \in \mathbb{R};\ \frac{q_1^2}{4} \leq \frac{1}{c^2} - \varphi^2\} . \tag{13.155}$$

For values of $q_1$ located inside this segment and for values of $\varphi \leq c^{-1}$, real solutions for $p_1$ exist and we define the domain $\mathcal{P} = \mathcal{P}(q_1, \varphi)$ as the pertaining solution space. In this domain we perform the two-dimensional imaging procedure.

*Imaging*

Following a similar procedure as in Section 13.5 we introduce $\varphi$ as new independent variable. From Eq. (13.154) it is clear that the solution space $\mathcal{P}(q_1, \varphi)$ only contains the two values $p_1^+$ and $p_1^-$. Therefore the weighted average of Eq. (13.152) is given by

$$\overline{p}^{wght}(q_1, \varphi, j\omega) = \hat{W}(j\omega) \int_{x_1 \in \mathbb{R}} E^{wght} \exp[j\omega(q_1 x_1 - 2\varphi h)] \mathrm{d}x_1 \,, \quad (13.156)$$

for $q_1 \in \mathcal{Q}(\varphi)$, where

$$\overline{p}^{wght}(q_1, \varphi, j\omega) = \frac{j\omega}{\varphi} \Big[ \Upsilon^R(p_1^+) \Upsilon^S(q_1 - p_1^+) \overline{\overline{p}}^r (j\omega p_1^+ | j\omega(q_1 - p_1^+), j\omega)$$

$$+ \Upsilon^R(p_1^-) \Upsilon^S(q_1 - p_1^-) \, \overline{\overline{p}}^r (j\omega p_1^- | j\omega(q_1 - p_1^-), j\omega) \Big] \,, \quad (13.157)$$

and

$$E^{wght}(x_1; q_1, \varphi) = \frac{1}{4c\varphi} \Big[ E(x_1; p_1^+, q_1 - p_1^+) + E(x_1; p_1^-, q_1 - p_1^-) \Big] \,. \quad (13.158)$$

Observing that the contribution of the integral on the right-hand side of Eq. (13.156), when $\omega$ tends to infinity, comes from the stationary point following from

$$q_1 - 2\varphi \partial_1 h(x_1) = 0 \,, \quad (13.159)$$

Eq. (13.156) is replaced by

$$\overline{p}^{wght}(q_1, \varphi, j\omega) \,,$$

$$= \hat{W}(j\omega) \int_{x_1 \in \mathbb{R}} E^{image}(x_1; \varphi) \exp(j\omega q_1 x_1 - 2j\omega\varphi \, h) \mathrm{d}x_1 \,, \quad (13.160)$$

with

$$E^{image}(x_1; \varphi) = E^{wght}_{|_{q_1 = 2\varphi\partial_1 h}} \,, \quad (13.161)$$

for $\omega \to \infty$ and for $q_1 \in \mathcal{Q}(\varphi)$. The next step is to apply the inverse spatial Fourier transformation yielding

$$\hat{p}^{image}(\boldsymbol{x}_1, j\omega; \varphi) = E^{image}(\boldsymbol{x}_1; \varphi)\, \hat{W}(j\omega) \exp(-2j\omega\varphi\, h), \quad \omega \to \infty\,, \tag{13.162}$$

with

$$\hat{p}^{image}(\boldsymbol{x}_1, j\omega; \varphi) = \frac{\omega}{2\pi} \int_{q_1 \in \mathcal{Q}(\varphi)} \exp(-j\omega q_1 \boldsymbol{x}_1) \overline{p}^{wght}(q_1, \varphi, j\omega) \mathrm{d}q_1\,. \tag{13.163}$$

Equation (13.163) indicates how the recorded data have to be mapped. We subsequently transform Eq. (13.162) back to the time domain. Since Eq. (13.162) holds for $\omega \to \infty$, we observe that

$$\lim_{t \downarrow T_h} p^{image}(\boldsymbol{x}_1, t; \varphi) = E^{image}(\boldsymbol{x}_1, \varphi)\, W(t - T_h)\,, \tag{13.164}$$

where the laterally variant vertical travel time $T_h = T_h(\boldsymbol{x}_1; \varphi)$ is given by

$$T_h(\boldsymbol{x}_1; \varphi) = 2\varphi\, h(\boldsymbol{x}_1)\,. \tag{13.165}$$

We observe that the space-time function of Eq. (13.164) starts to act at the instant $T_h = 2\varphi\, h(\boldsymbol{x}_1)$, so that we have a space-time image of the profile of the interface as the envelope of the arrivals of the reflected causal wavelet function. For a complete image of the profile of the interface it is necessary that the domain $\mathcal{Q} = \mathcal{Q}(\varphi)$ contains all the stationary points given by $q_1 = 2\varphi\partial_1 h$. Hence, from Eq. (13.155) it follows

$$\sup_{x_1 \in \mathbb{R}} (\partial_1 h)^2 \leq \frac{1}{c^2\varphi^2} - 1\,. \tag{13.166}$$

*Computational procedure*

The computational procedure can be summarized as follows:

- Collect the reflected wavefield $p^r(\boldsymbol{x}_1^O + \boldsymbol{x}_1^S, 0|\boldsymbol{x}_T^S, t)$ in the $t$-domain.
- Transform these data to the angular-frequency domain. We then arrive at the data $\hat{p}^r(\boldsymbol{x}_1^O + \boldsymbol{x}_1^S, 0|\boldsymbol{x}_T^S, j\omega)$ in the $(j\omega)$-domain.
- Transform these data to the $(j\omega p_1, j\omega q_1)$-domain using Eq. (13.151). We arrive at $\overline{\overline{p}}^r(j\omega p_1, j\omega(q_1 - p_1), j\omega)$.

- Perform the weighted average (see Eq. (13.157)) of the data for some particular choice of $q_1$ and $\varphi$. The resulting data $\bar{p}^{wght}(q_1, \varphi, \omega)$ are stacked over the domain $\mathcal{Q}(\varphi)$; using Eq. (13.163), we obtain the function $\hat{p}^{image}(x_1, j\omega; \varphi)$.

- Apply an inverse temporal Fourier transformation. We finally obtain a space-time function $p^{image}(x_1, t; \varphi)$ that contains the profile function of the interface in its time delay (see Eq. (13.164)).

Note that all operations can be performed with Fourier transformations.

A numerical example of an imaged boundary is presented by FOKKEMA and VAN DEN BERG (1992).

## 13.9. Imaging of the strip configuration

In order to arrive at an efficient scheme for computing the Fourier transforms of the Radon type (see Eq. (13.151)), we proceed as follows. We compute this transforms for a fixed frequency. Then from inspection of the argument of the exponential function in the transformation kernel it follows that the the desired result is best obtained by using a standard FFT routine, where the products of slowness and frequency ($\omega p_1$, $\omega q_1$) act as transformation parameters. The final result for discrete slowness values is obtained by a linear interpolation. Operating in this way necessarily makes the valid slowness range frequency dependent (FOKKEMA *et al.*, 1992). The integral on the right-hand side of Eq. (13.163) is identified as an inverse Fourier transform of the Radon type by introducing a characteristic function of the set $\mathcal{Q}(\varphi)$ and is computed in a similar way.

To illustrate the imaging procedure, we use the wavefield in the strip configuration in the homogeneous halfspace as it has been computed with the integral-equation method of Chapter 9. The resulting two-dimensional dataset of 255 sources and 255 receiver per source arranged in a split-spread configuration is first deghosted with the method of Section 11.6. Next, this dataset is used as input for the multiple-removal procedure outlined in Section 12.5. These steps have been carried out to mimic a realistic processing sequence. After these preprocessing steps, the imaging procedure discussed in the previous section is applied.

As comparison, we employ the results of the Kirchhoff approximation explained in Section 13.7 for the case of the disk. The imaging result of the strip using the Kirchhoff approximation follows immediately from the analysis of the disk as the two-dimensional equivalent (cf. Eq. (13.135))

$$\lim_{t \downarrow T_h} p^{image}(\boldsymbol{x}_1, t; \varphi) = \mp \chi_{\Lambda_g}(\boldsymbol{x}_1)\, W(t - T_h)\,, \tag{13.167}$$

where the laterally variant vertical travel time $T_h$ is given by

$$T_h = 2\varphi h\,, \tag{13.168}$$

and where $\chi_{\Lambda_g}$ denotes the spatial characteristic function of the strip area

$$\Lambda_g = \{\boldsymbol{x} \in \mathbb{R}^3;\ -a < \boldsymbol{x}_1 < a,\ -\infty < \boldsymbol{x}_2 < \infty,\ \boldsymbol{x}_3 = h\}\,. \tag{13.169}$$

The $-$ sign holds for the case of the compliant strip and the $+$ sign holds for the case of the rigid strip.

In Fig. 13.7 (top) we show the image for the compliant strip based on the Kirchhoff approximation. Here, we only have calculated the analytical expression of the right-hand side of Eq. (13.167). We have chosen $\varphi = 0.9/c$ with $c = 1457$ m/s. In Fig. 13.7 (bottom) we present the image result for the compliant strip based on the complete processing sequence. In comparing these results we conclude that the imaging procedure performs very well at the center, but slightly degrades towards the edge of the strip, as far as the wavelet appearance is concerned. This means that not all energy diffracted from the edges has been migrated. Similar results are shown in Fig. 13.8 for the rigid strip, both for the Kirchhoff approach (top) and the complete processing sequence (bottom).

These results conclude our discussion of the boundary imaging.

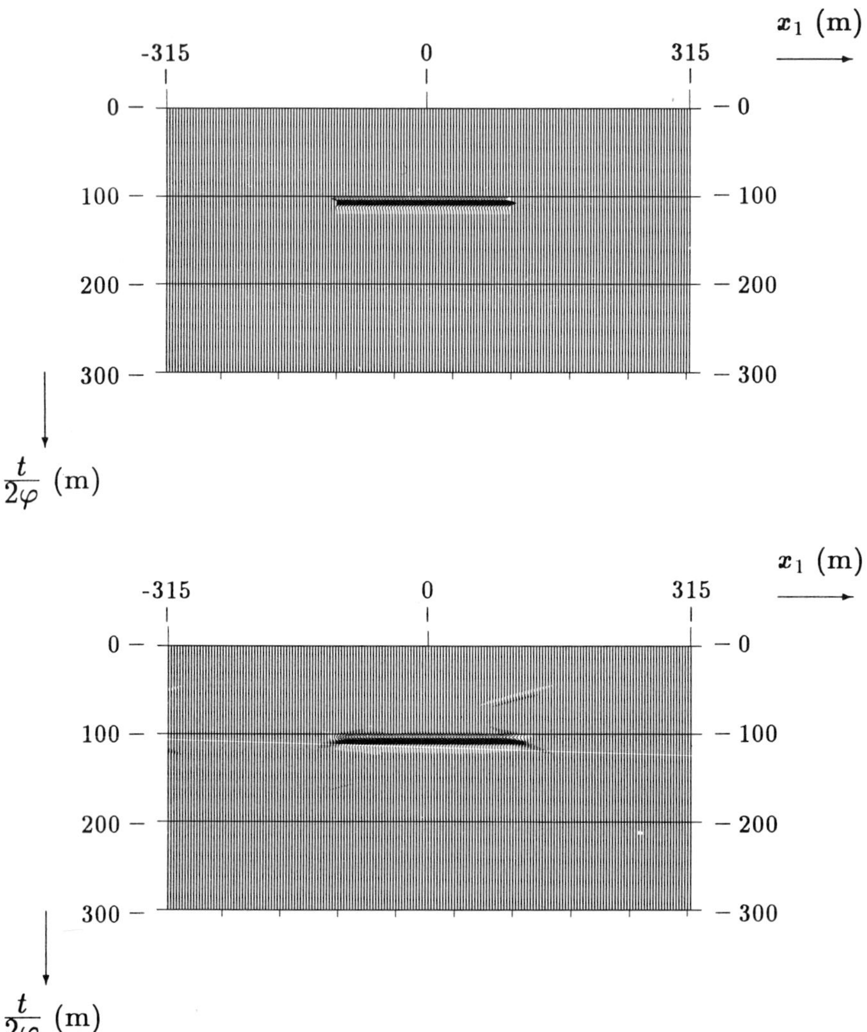


Figure 13.7. The imaged compliant strip based on the Kirchhoff approximation (top) and the imaged compliant strip based on the complete processing sequence (bottom).

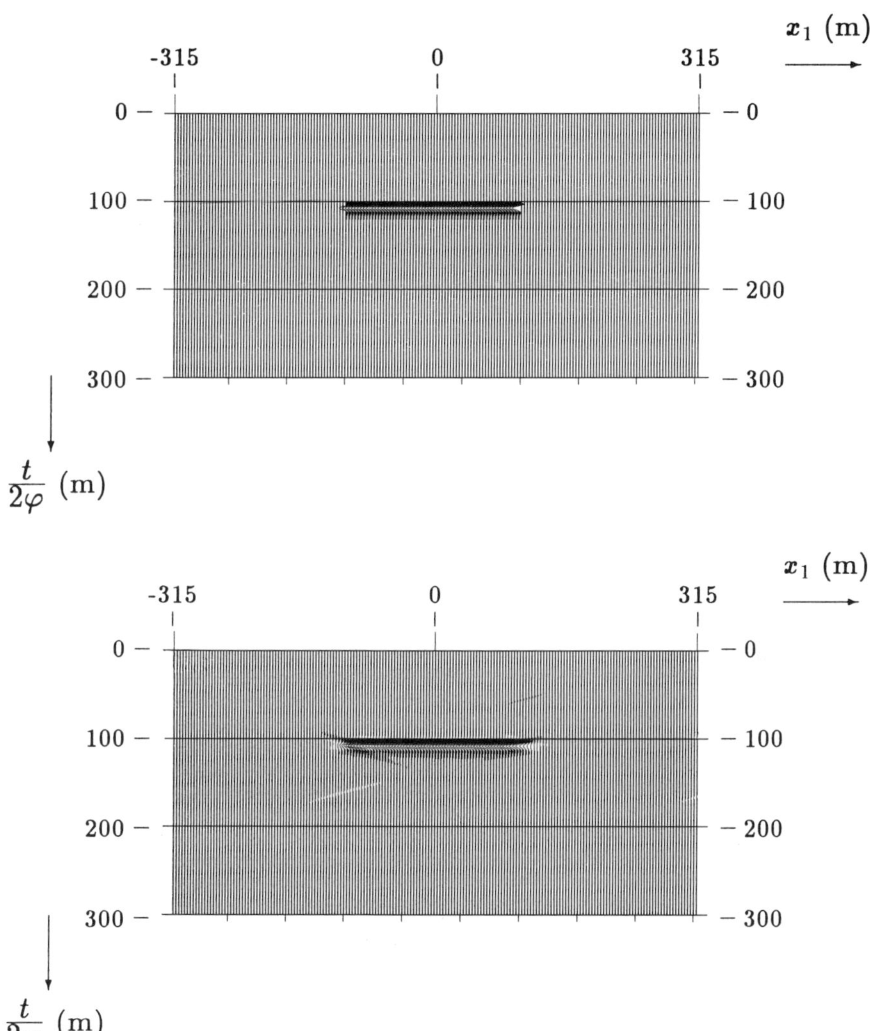


Figure 13.8. The imaged rigid strip based on the Kirchhoff approximation (top) and the imaged rigid strip based on the complete processing sequence (bottom).

# Chapter 14

# Domain Imaging

In this chapter we discuss a method to image a domain discontinuity of the earth parameters. As starting point, we take the domain-integral representation. We employ the first Born approximation for the total wavefield in the contrasting domain. In the wavefield, we separate the horizontal phase contribution from the vertical one by employing a Radon transform with respect to the horizontal source and receiver coordinates only. As next step we switch to the frequency domain (imaginary axis in the complex Laplace-transform domain) and the vertical slowness is taken as independent variable. In the space-time domain we obtain an image of the contrast function as a temporal convolution of the wavelet and the time depth of the contrast function.

## 14.1. The domain-integral representation

We consider the wavefield to be reflected by a contrasting domain of bounded extent, present in a homogeneous embedding of infinite extent. Let $\mathbb{D}_g$ be the bounded domain occupied by the scatterer. The domain exterior to $\mathbb{D}_g$ is denoted by $\mathbb{D}$. In $\mathbb{D}$, the medium is homogeneous with material constants $\rho$ and $\kappa$ (Fig. 14.1). We assume that the contrasting domain has no contrast in the volume density of mass. Then, in $\mathbb{D}_g$ the constant $\rho$ is the volume density of mass and $\kappa^g = \kappa^g(\boldsymbol{x})$ is the compressibility. The impulsive wave motion generated by a point source at $\boldsymbol{x}^S = \{x_1^S, x_2^S, 0\}$ starts to act at $t = 0$. The receiver is located at $\boldsymbol{x}^R = \{x_1^R, x_2^R, 0\}$.

In the $s$-domain, the source wavefield at the observation point $\boldsymbol{x}$ originating from a monopole source at $\boldsymbol{x}^S$ is then denoted as $\{\hat{p}^{inc}, \hat{v}_k^{inc}\}(\boldsymbol{x}|\boldsymbol{x}^S, s)$ and is given by

$$\hat{p}^{inc}(\boldsymbol{x}|\boldsymbol{x}^S, s) = s\rho\hat{q}^S(s)\hat{G}(\boldsymbol{x}-\boldsymbol{x}^S, s)\,, \tag{14.1}$$

$$\hat{v}_k^{inc}(\boldsymbol{x}|\boldsymbol{x}^S, s) = -\hat{q}^S(s)\partial_k\hat{G}(\boldsymbol{x}-\boldsymbol{x}^S, s)\,, \tag{14.2}$$

where the Green's function

$$\hat{G}(\boldsymbol{x}, s) = \frac{\exp(-\frac{s}{c}|\boldsymbol{x}|)}{4\pi|\boldsymbol{x}|}, \quad c = (\rho\kappa)^{-\frac{1}{2}}\,. \tag{14.3}$$

The wavefield in $\mathbb{D}$ reflected by the contrasting domain $\mathbb{D}_g$ is given by $\{\hat{p}^r, \hat{v}_k^r\}(\boldsymbol{x}|\boldsymbol{x}^S, s)$. The reflected wavefield can be considered as the wavefield generated by secondary volume sources (monopole sources) in the domain $\mathbb{D}_g$ and it can be expressed in terms of these volume sources as (cf. Eq. (8.16))

$$\hat{p}^r(\boldsymbol{x}^R|\boldsymbol{x}^S, s) = \int_{\boldsymbol{x}\in\mathbb{D}_g} \hat{G}^q\, s(\kappa-\kappa^g)\hat{p}(\boldsymbol{x}|\boldsymbol{x}^S, s)\mathrm{dV}\,, \tag{14.4}$$

where $\hat{p}$ denotes the total wavefield in the contrasting domain and where the Green's state $\hat{G}^q$ is given by

$$\hat{G}^q(\boldsymbol{x}^R|\boldsymbol{x}, s) = s\rho\hat{G}(\boldsymbol{x}-\boldsymbol{x}^R, s)\,. \tag{14.5}$$

Using the first Born approximation for the total acoustic pressure in the contrasting domain (cf. Eq. (8.24))

$$\hat{p}(\boldsymbol{x}|\boldsymbol{x}^S, s) \cong \hat{p}^{inc}(\boldsymbol{x}|\boldsymbol{x}^S, s)\,, \tag{14.6}$$

together with Eqs. (14.1) and (14.3), the integral representation of Eq. (14.4) becomes

$$\hat{p}^r(\boldsymbol{x}^R|\boldsymbol{x}^S, s) = \frac{s^2}{c^2}\hat{W}(s)\int_{\boldsymbol{x}\in\mathbb{D}_g} \chi(\boldsymbol{x})\,\hat{G}(\boldsymbol{x}-\boldsymbol{x}^R, s)\hat{G}(\boldsymbol{x}-\boldsymbol{x}^S, s)\mathrm{dV}\,, \tag{14.7}$$

where $\hat{W}(s) = s\rho\hat{q}^S(s)$ is the source wavelet and $\chi$ is the contrast function

$$\chi(\boldsymbol{x}) = 1 - \frac{\kappa^g}{\kappa} = 1 - (\frac{c}{c^g})^2, \quad c^g = (\rho\kappa^g)^{-\frac{1}{2}}\,. \tag{14.8}$$

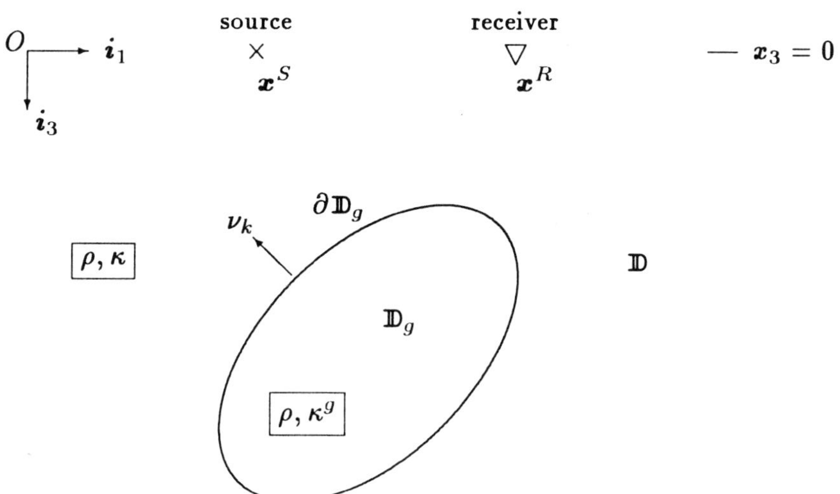


Figure 14.1. The contrasting-domain configuration.

Using Eq. (4.60), the Green's function follows as an inverse spatial Fourier transform with respect to the source coordinates,

$$\hat{G}(\boldsymbol{x}-\boldsymbol{x}^S,s)=\frac{1}{(2\pi)^2}\int_{(s\alpha_1^S,s\alpha_2^S)\in\mathbb{R}^2}\frac{\exp[js\alpha_1^S(x_1-x_1^S)+js\alpha_2^S(x_2-x_2^S)-s\Gamma^S x_3]}{2s\Gamma^S}\mathrm{dA}\,, \tag{14.9}$$

with

$$\Gamma^S=\left[\frac{1}{c^2}+(\alpha_1^S)^2+(\alpha_2^S)^2\right]^{\frac{1}{2}}\,, \tag{14.10}$$

and as an inverse spatial Fourier transform with respect to the receiver coordinates as

$$\hat{G}(\boldsymbol{x}-\boldsymbol{x}^R,s)=\frac{1}{(2\pi)^2}\int_{(s\alpha_1^R,s\alpha_2^R)\in\mathbb{R}^2}\frac{\exp[js\alpha_1^R(x_1-x_1^R)+js\alpha_2^R(x_2-x_2^R)-s\Gamma^R x_3]}{2s\Gamma^R}\mathrm{dA}\,, \tag{14.11}$$

with

$$\Gamma^R = \left[\frac{1}{c^2} + (\alpha_1^R)^2 + (\alpha_2^R)^2\right]^{\frac{1}{2}} . \tag{14.12}$$

Note that we have used $x_3^S = x_3^R = 0$ and $x_3 > 0$ in $\mathbb{D}_g$.

We consider a seismic experiment as the collection of measurements of the seismic signal at different receiver positions. This procedure is repeated for different source positions. This means that the values of $\hat{p}^r(\boldsymbol{x}^R|\boldsymbol{x}^S, s)$ are the (to the $s$-domain transformed) recorded signals for all $\boldsymbol{x}^R$ and $\boldsymbol{x}^S$. Our objective is to use Eq. (14.7) for the reconstruction of the contrast function $\chi(\boldsymbol{x})$. To meet our objective we apply a spatial Fourier transformation with respect to the horizontal source coordinates followed by a spatial Fourier transformation with respect to the horizontal receiver coordinates.

## 14.2. Fourier transform of source and receiver coordinates

We first define a Fourier transform with respect to the horizontal source coordinates, viz.

$$\bar{p}^r(\boldsymbol{x}^R|js\boldsymbol{\alpha}^S, s) = \int_{(x_1^S, x_2^S)\in\mathbb{R}^2} \exp(js\alpha_1^S x_1^S + js\alpha_2^S x_2^S)\hat{p}^r(\boldsymbol{x}^R|\boldsymbol{x}^S, s)\mathrm{dA} , \tag{14.13}$$

where $\boldsymbol{\alpha}^S = \{\alpha_1^S, \alpha_2^S\}$ is a two-dimensional vector, while $s\alpha_1^S$ and $s\alpha_2^S$ are real transform parameters. We subsequently define a Fourier transform with respect to the horizontal receiver coordinates, viz.

$$\bar{\bar{p}}^r(js\boldsymbol{\alpha}^R|js\boldsymbol{\alpha}^S, s) = \int_{(x_1^R, x_2^R)\in\mathbb{R}^2} \exp(js\alpha_1^R x_1^R + js\alpha_2^R x_2^R)\bar{p}^r(\boldsymbol{x}^R|js\boldsymbol{\alpha}^S, s)\mathrm{dA} , \tag{14.14}$$

where $\boldsymbol{\alpha}^R = \{\alpha_1^R, \alpha_2^R\}$ is a two-dimensional vector, while $s\alpha_1^R$ and $s\alpha_2^R$ are real transform parameters.

Substituting the expressions for the Green's functions of Eqs. (14.9) and (14.11) in Eq. (14.7), the result of the Fourier transforms of Eqs. (14.13) and

(14.14) becomes

$$\overline{\overline{p}}^r(js\boldsymbol{\alpha}^R|js\boldsymbol{\alpha}^S,s)=\frac{\hat{W}(s)}{4c^2\Gamma^R\Gamma^S}\int_{\boldsymbol{x}\in\mathbb{D}_g}$$

$$\chi(\boldsymbol{x})\exp[js(\alpha_1^R+\alpha_1^S)x_1+js(\alpha_2^R+\alpha_2^S)x_2-s(\Gamma^R+\Gamma^S)x_3]\mathrm{dV}\,. \tag{14.15}$$

With this result we finish our analysis for arbitrary values of $s$ ($\mathrm{Re}(s)>0$). We want to carry out the further analysis along the same lines as in Chapter 13. We therefore switch to the angular-frequency domain.

## 14.3. Angular-frequency-domain analysis

We introduce the horizontal offset coordinates $x_1^O$ and $x_2^O$ as

$$x_1^O=x_1^R-x_1^S,\quad x_2^O=x_2^R-x_2^S\,, \tag{14.16}$$

and in the spatial Fourier domain the coordinates $\boldsymbol{p}=\{p_1,p_2\}$ and $\boldsymbol{q}=\{q_1,q_2\}$ as

$$\boldsymbol{p}=j\boldsymbol{\alpha}^R,\quad \boldsymbol{q}=j(\boldsymbol{\alpha}^R+\boldsymbol{\alpha}^S)\,. \tag{14.17}$$

Substituting these new coordinates in Eqs. (14.13) and (14.14) and taking $s=j\omega$, we obtain

$$\overline{\overline{p}}^r(j\omega\boldsymbol{p}|j\omega(\boldsymbol{q}-\boldsymbol{p}),j\omega)=\int_{(x_1^S,x_2^S)\in\mathbb{R}^2}\exp(j\omega q_1x_1^S+j\omega q_2x_2^S)\mathrm{dA}$$

$$\int_{(x_1^O,x_2^O)\in\mathbb{R}^2}\exp(j\omega p_1x_1^O+j\omega p_2x_2^O)\hat{p}^r(x_1^O+x_1^S,x_2^O+x_2^S,0|\boldsymbol{x}^S,j\omega)\mathrm{dA}\,. \tag{14.18}$$

Equation (14.18) constitutes the temporal Fourier transform of the Radon type with respect to the source coordinates and the offset coordinates. In view of Eqs. (14.13) and (14.14), we may replace Eq. (14.15) by

$$\overline{\overline{p}}^{r}(j\omega\boldsymbol{p}|j\omega(\boldsymbol{q}-\boldsymbol{p}),j\omega)$$

$$= \frac{\hat{W}(j\omega)}{4c^2\Gamma^R\Gamma^S}\int_{\boldsymbol{x}\in\mathbb{R}^3}\chi(\boldsymbol{x})\exp(j\omega q_1x_1 + j\omega q_2x_2 - 2j\omega\varphi x_3)\mathrm{dV}\,. \tag{14.19}$$

Here, we have introduced the vertical slowness $\varphi = \varphi(\boldsymbol{p},\boldsymbol{q})$ as

$$\varphi = \frac{1}{2}\left[\frac{1}{c^2} - p_1^2 - p_2^2\right]^{\frac{1}{2}} + \frac{1}{2}\left[\frac{1}{c^2} - (q_1-p_1)^2 - (q_2-p_2)^2\right]^{\frac{1}{2}}\,. \tag{14.20}$$

In Eq. (14.19) we have extended the integration over the contrasting domain $\mathbb{D}_g$ to $\mathbb{R}^3$. This is allowed because the contrast function $\chi$ is zero outside the contrasting domain.

## 14.4. Imaging

After introducing $\varphi$ as new independent variable, we have observed that the solutions $\boldsymbol{p}$ of Eq. (14.20), for given values of $\boldsymbol{q}$ and $\varphi$, are located in the domain $\mathcal{P}(\boldsymbol{q},\varphi)$ (see Section 13.5). Of all these solutions, we take some average solution. For a particular value of $\boldsymbol{q}$ and $\varphi$, we take a weighted averaging of the quantity of Eq. (14.19) over the domain $\mathcal{P} = \mathcal{P}(\boldsymbol{q},\varphi)$. Therefore, we multiply both sides of Eq. (14.19) with the factor $8\varphi c^2\Gamma^R\Gamma^S$ and integrate both sides of the result over the domain $\mathcal{P}$. This leads to

$$\overline{p}^{wght}(\boldsymbol{q},\varphi,j\omega) = 2\varphi\hat{W}(j\omega)\int_{\boldsymbol{x}\in\mathbb{R}^3}\chi(\boldsymbol{x})\exp[j\omega(q_1x_1 + q_2x_2 - 2\varphi x_3)]\mathrm{dV}\,,$$

$$\boldsymbol{q}\in\mathcal{Q}(\varphi)\,, \tag{14.21}$$

where

$$\overline{p}^{wght}(\boldsymbol{q},\varphi,\omega) = 8\varphi c^2\frac{\displaystyle\int_{\boldsymbol{p}\in\mathcal{P}(\boldsymbol{q},\varphi)}\Gamma^R(\boldsymbol{p})\Gamma^S(\boldsymbol{q}-\boldsymbol{p})\overline{\overline{p}}^{r}(j\omega\boldsymbol{p},j\omega(\boldsymbol{q}-\boldsymbol{p}),j\omega)\mathrm{dA}}{\displaystyle\int_{\boldsymbol{p}\in\mathcal{P}(\boldsymbol{q},\varphi)}\mathrm{dA}}\,. \tag{14.22}$$

We stress that Eq. (14.21) holds for $\boldsymbol{q} \in \mathcal{Q}(\varphi)$. Now, outside this domain we assume that the right-hand side of Eq. (14.21) vanishes. After the Born approximation this is the second approximation we make. In order to maximize the domain of validity of this approximation we take the domain $\mathcal{Q}(\varphi)$ as large as possible. From Eq. (13.89) it follows that this is arrived at by taking $\varphi$ as small as possible. For high frequencies we observe that the right-hand side of Eq. (14.21) has its main contribution around the point $q_1 = q_2 = 0$. This is an interior point of $\mathcal{Q}(\varphi)$ and the assumption that the right-hand side of Eq. (14.21) vanishes outside $\mathcal{Q}(\varphi)$ is realistic. Hence

$$0 = 2\varphi\hat{W}(j\omega)\int_{\boldsymbol{x}\in\mathbb{R}^3}\chi(\boldsymbol{x})\exp[j\omega(q_1x_1 + q_2x_2 - 2\varphi x_3)]\mathrm{dV}\,, \qquad \boldsymbol{q} \in \mathbb{R}^2\backslash\mathcal{Q}(\varphi),\ \ \omega \to \infty\,. \tag{14.23}$$

The combination of Eqs. (14.21) and (14.23) defines a spatial Fourier transform with respect to the coordinates $x_1$ and $x_2$ with transform parameters $j\omega q_1$ and $j\omega q_2$. The inverse spatial Fourier transform yields

$$\hat{p}^{image}(x_1, x_2, j\omega; \varphi) = 2\varphi\int_{x_3\in\mathbb{R}}\chi(\boldsymbol{x})\,\hat{W}(j\omega)\exp(-2j\omega\varphi x_3)\mathrm{d}x_3\,, \tag{14.24}$$

with

$$\hat{p}^{image}(x_1, x_2, j\omega; \varphi) = \left(\frac{\omega}{2\pi}\right)^2\int_{\boldsymbol{q}\in\mathcal{Q}(\varphi)}\exp(-j\omega q_1x_1 - j\omega q_2x_2)\bar{p}^{wght}(\boldsymbol{q}, \varphi, j\omega)\mathrm{dA}\,. \tag{14.25}$$

Equation (14.25) indicates how the recorded data have to be mapped. We subsequently transform Eq. (14.24) back to the time domain; this leads to the result

$$2\varphi\int_{x_3\in\mathbb{R}}\chi(x_1, x_2, x_3)\,W(t - 2\varphi x_3)\,\mathrm{d}x_3 = p^{image}(x_1, x_2, t; \varphi)\,. \tag{14.26}$$

Introducing the new integration variable $t' = 2\varphi x_3$, we obtain

$$\int_{t'\in\mathbb{R}}\chi(x_1, x_2, \frac{t'}{2\varphi})\,W(t - t')\,\mathrm{d}t' = p^{image}(x_1, x_2, t; \varphi)\,. \tag{14.27}$$

We observe that the temporal convolution of the wavelet and the contrast function at the depth of $t/2\varphi$ is equal to $p^{image}$.

*Constant contrast in* $\mathbb{D}_g$

In the special case that the contrast in $\mathbb{D}_g$ is constant, we can image the boundary of $\partial\mathbb{D}_g$ as follows. First we multiply the data in the frequency domain with the factor $j\omega$. This also happened in the procedure of the boundary imaging of Chapter 13. Then, Eq. (14.24) is replaced by

$$2j\omega\varphi \int_{x_3 \in \mathbb{R}} \chi(\boldsymbol{x})\, \hat{W}(j\omega) \exp(-2j\omega\varphi x_3) \mathrm{d}x_3 = \partial\hat{p}^{image}(x_1, x_2, j\omega; \varphi)\,, \tag{14.28}$$

with

$$\partial\hat{p}^{image}(x_1, x_2, j\omega; \varphi) = j\omega\hat{p}^{image}(x_1, x_2, j\omega; \varphi)\,. \tag{14.29}$$

Next we apply an inverse temporal Fourier transform

$$2\varphi\partial_t \int_{x_3 \in \mathbb{R}} \chi(\boldsymbol{x}) W(t - 2\varphi x_3)\, \mathrm{d}x_3 = \partial p^{image}(x_1, x_2, t; \varphi)\,. \tag{14.30}$$

Since the medium in the contrasting domain is homogeneous, we may express the contrast function in the characteristic function of the domain as

$$\chi(\boldsymbol{x}) = \left[1 - \left(\frac{c}{c^g}\right)^2\right] \chi_{\mathbb{D}_g}(\boldsymbol{x})\,. \tag{14.31}$$

With this expression, we can calculate the integral at the left-hand side of Eq. (14.30) as

$$\begin{aligned} 2\varphi\partial_t & \int_{x_3 \in \mathbb{R}} \chi(\boldsymbol{x})\, W(t - 2\varphi x_3)\, \mathrm{d}x_3 \\ &= \partial_t \int_{t' \in \mathbb{R}} \chi(x_1, x_2, \frac{t'}{2\varphi})\, W(t - t')\, \mathrm{d}t' \\ &= \left[1 - \left(\frac{c}{c^g}\right)^2\right] \partial_t \int_{t' \in \mathbb{R}} \chi_{\mathbb{D}_g}(x_1, x_2, \frac{t - t'}{2\varphi})\, W(t')\, \mathrm{d}t' \\ &= \left[1 - \left(\frac{c}{c^g}\right)^2\right] \mathrm{sign}(-\nu_3) W(t - 2\varphi x_3)\big|_{\boldsymbol{x} \in \partial\mathbb{D}_g}\,, \end{aligned} \tag{14.32}$$

where $\nu_3 = \nu_3(\boldsymbol{x})$ is the vertical component of the normal vector on $\partial\mathbb{D}_g$. The latter is directed away from $\mathbb{D}_g$ (see Fig. 14.1). Hence, we finally obtain

$$\partial p^{image}(x_1, x_2, t; \varphi) = \left[1 - \left(\frac{c}{c^g}\right)^2\right] \mathrm{sign}(-\nu_3) W(t - 2\varphi x_3)\big|_{\boldsymbol{x} \in \partial\mathbb{D}_g}\,. \tag{14.33}$$

This result is equivalent to the boundary imaging of Eq. (13.98). The multiplication of the data with $j\omega$ leads to an extra factor in the averaging procedure of Eq. (14.22). Apart from a constant factor, the domain averaging procedure is then equivalent to the boundary weighting procedure of Eq. (13.91).

## 14.5. Two-dimensional case

In the two-dimensional case, the configuration $\partial\mathbb{D}_g$ is considered to be independent of $x_2$. We further consider the two-dimensional wavefield to be generated by a monopole line source at $\boldsymbol{x}_T^S = (x_1^S, x_3^S = 0)$. In the $s$-domain the source wavefield at the two-dimensional observation point $\boldsymbol{x}_T = (x_1, x_3)$ is given by

$$\hat{p}^{inc}(\boldsymbol{x}_T|\boldsymbol{x}_T^S, s) = s\rho\hat{q}^S(s)\hat{\mathcal{G}}(\boldsymbol{x}_T - \boldsymbol{x}_T^S, s)\,, \tag{14.34}$$

$$\hat{v}_k^{inc}(\boldsymbol{x}_T|\boldsymbol{x}_T^S, s) = -\hat{q}^S(s)\partial_k\hat{\mathcal{G}}(\boldsymbol{x}_T - \boldsymbol{x}_T^S, s)\,, \tag{14.35}$$

with $\hat{v}_2^{inc} = 0$ and where the two-dimensional Green's function is given by (cf. Eqs. (9.84) and (9.87))

$$\hat{\mathcal{G}}(\boldsymbol{x}_T, s) = \int_{x_2\in\mathbb{R}} \hat{G}(x_1, x_2, x_3)\mathrm{d}x_2\,. \tag{14.36}$$

*The domain-integral representation*

In the two-dimensional case, the representation for the reflected wavefield becomes (cf. Eqs. (14.4) and (14.5))

$$\hat{p}^r(\boldsymbol{x}_T^R|\boldsymbol{x}_T^S, s) = s^2\rho\int_{\boldsymbol{x}_T\in\mathbb{D}_g}(\kappa - \kappa^g)\hat{\mathcal{G}}(\boldsymbol{x}_T - \boldsymbol{x}_T^R, s)\hat{p}(\boldsymbol{x}_T|\boldsymbol{x}_T^S, s)\mathrm{dA}\,, \tag{14.37}$$

where $\boldsymbol{x}_T^R = (x_1^R, x_3^R = 0)$ is the two-dimensional line-receiver position. We now use the first Born approximation for the total acoustic pressure in the two-dimensional contrasting domain $\mathbb{D}_g$,

$$\hat{p}(\boldsymbol{x}_T|\boldsymbol{x}_T^S, s) \cong \hat{p}^{inc}(\boldsymbol{x}_T|\boldsymbol{x}_T^S, s)\,. \tag{14.38}$$

Substituting the expression of Eq. (14.36) for the two-dimensional Green's function in Eq. (14.37) and interchanging the order of integration, we obtain

$$\hat{p}^r(\boldsymbol{x}_T^R|\boldsymbol{x}_T^S, s) = \int_{x_2^R \in \mathbb{R}} \int_{x_2^S \in \mathbb{R}} \hat{p}^r(\boldsymbol{x}^R|\boldsymbol{x}^S, s) \mathrm{d}x_2^R \mathrm{d}x_2^S , \tag{14.39}$$

where

$$\hat{p}^r(\boldsymbol{x}^R|\boldsymbol{x}^S, s) = \frac{s^2}{c^2}\hat{W}(s) \int_{\boldsymbol{x}_T \in \mathbb{D}_g} \chi(\boldsymbol{x}_T)\, \hat{G}(x_1 - x_1^R, x_2^R, x_3 - x_3^R, s)\hat{G}(x_1 - x_1^S, x_2^S, x_3 - x_3^S, s) \mathrm{dA} . \tag{14.40}$$

We observe that Eq. (14.40) has the same appearance as Eq. (14.7). Moreover, Eq. (14.39) can be regarded as the spatial Fourier transform of the quantity $\hat{p}^r(\boldsymbol{x}^R|\boldsymbol{x}^S, s)$ with respect to $x_2^S$ and $x_2^R$, with transform parameters $js\alpha_2^S = 0$ and $js\alpha_2^R = 0$.

*Fourier transform with respect to $x_1^R$ and $x_1^S$*

In order to take advantage of the results of Section 14.2 we only need to transform Eq. (14.39) with respect to the receiver coordinate $x_1^R$ and the source coordinate $x_1^S$. Let us define the Fourier transform with respect to $x_1^R$ and $x_1^S$ as

$$\overline{\overline{p}}^r(js\alpha_1^R|js\alpha_1^S, s) = \int_{x_1^R \in \mathbb{R}} \int_{x_1^S \in \mathbb{R}} \exp(js\alpha_1^S x_1^S + js\alpha_1^R x_1^R)\hat{p}^r(\boldsymbol{x}_T^R|\boldsymbol{x}_T^S, s) \mathrm{d}x_1^R \mathrm{d}x_1^S . \tag{14.41}$$

Then, the two-dimensional variant of Eq. (14.15) becomes

$$\overline{\overline{p}}^r(js\alpha_1^R|js\alpha_2^R, s) = \frac{\hat{W}(s)}{4c^2\Upsilon^R\Upsilon^S} \int_{\boldsymbol{x}_T \in \mathbb{D}_g} \chi(\boldsymbol{x}_T) \exp[js(\alpha_1^R + \alpha_1^S)x_1 - s(\Upsilon^R + \Upsilon^S)x_3] \mathrm{dA} , \tag{14.42}$$

with

$$\Upsilon^R = \left[\frac{1}{c^2} + (\alpha_1^R)^2\right]^{\frac{1}{2}} , \quad \Upsilon^S = \left[\frac{1}{c^2} + (\alpha_1^S)^2\right]^{\frac{1}{2}} . \tag{14.43}$$

*Angular-frequency-domain analysis*

First, we introduce the horizontal offset coordinate $x_1^O$ as

$$x_1^O = x_1^R - x_1^S , \tag{14.44}$$

and in the spatial Fourier domain the coordinates $p_1$ and $q_1$ as

$$p_1 = j\alpha_1^R, \quad q_1 = j(\alpha_1^R + \alpha_1^S) . \tag{14.45}$$

Substituting these new coordinates in Eq. (14.41) and taking $s = j\omega$, we obtain

$$\bar{\bar{p}}^r(j\omega p_1|j\omega(q_1-p_1), j\omega)$$

$$= \int_{x_1^O \in \mathbb{R}} \int_{x_1^S \in \mathbb{R}} \exp(j\omega p_1 x_1^O + j\omega q_1 x_1^S)\, \hat{p}^r(x_1^O + x_1^S, 0|x_T^S, s) \mathrm{d}x_1^O \mathrm{d}x_1^S , \tag{14.46}$$

while (cf. Eqs. (14.19) and (14.20))

$$\bar{\bar{p}}^r(j\omega p_1|j\omega(q_1-p_1), j\omega)$$

$$= \frac{\hat{W}(j\omega)}{4c^2\Upsilon^R\Upsilon^S} \int_{x_T \in \mathbb{R}^2} \chi(x_T) \exp(j\omega q_1 x_1 - 2j\omega\varphi x_3) \mathrm{dA} , \tag{14.47}$$

where the vertical two-dimensional slowness function $\varphi = \varphi(p_1, q_1)$ is given by

$$\varphi = \frac{1}{2}\left[\frac{1}{c^2} - p_1^2\right]^{\frac{1}{2}} + \frac{1}{2}\left[\frac{1}{c^2} - (q_1 - p_1)^2\right]^{\frac{1}{2}} . \tag{14.48}$$

In Eq. (14.47) we have extended the integration over the contrasting domain $\mathbb{D}_g$ to $\mathbb{R}^2$. This is allowed because the contrast function is zero outside the contrasting domain. In Section 13.5 we have studied the consequences of taking $\varphi$ as new independent variable. In the present two-dimensional case, by taking $p_2 = q_2 = 0$, we deduce from Eq. (13.88) that two real solutions of $p_1$ exist, viz.,

$$p_1^{\pm} = \frac{1}{2}q_1 \pm \varphi \left( \frac{\frac{1}{c^2} - \frac{1}{4}q_1^2 - \varphi^2}{\frac{1}{4}q_1^2 + \varphi^2} \right)^{\frac{1}{2}} . \tag{14.49}$$

The real solutions of Eq. (14.49) only exist for values of $q_1$ in the linear segment $\mathcal{Q}(\varphi)$,

$$\mathcal{Q}(\varphi) = \{q_1 \in \mathbb{R};\ \frac{q_1^2}{4} \leq \frac{1}{c^2} - \varphi^2\}\,. \tag{14.50}$$

For values of $q_1$ located inside this segment and for values of $\varphi \leq c^{-1}$, real solutions for $p_1$ exist and we define the domain $\mathcal{P} = \mathcal{P}(q_1, \varphi)$ as the pertaining solution space. In this domain we perform the two-dimensional imaging procedure.

*Imaging*

Following a similar procedure as in Section 14.4 we introduce $\varphi$ as new independent variable. From Eq. (14.49) it is clear that the solution space $\mathcal{P}(q_1, \varphi)$ only contains the two values $p_1^+$ and $p_1^-$. Therefore the weighted average of Eq. (14.47) is given by

$$\overline{p}^{wght}(q_1, \varphi, j\omega) = 2\varphi\hat{W}(j\omega) \int_{\boldsymbol{x}_T \in \mathbb{R}^2} \chi(\boldsymbol{x}_T) \exp[j\omega(q_1 x_1 - 2\varphi x_3)]\mathrm{dA}\,, \tag{14.51}$$

for $q_1 \in \mathcal{Q}(\varphi)$, where

$$\overline{p}^{wght}(q_1, \varphi, j\omega) = 4\varphi c^2 \Big[\Upsilon^R(p_1^+)\Upsilon^S(q_1 - p_1^+)\,\overline{\overline{p}}^r(j\omega p_1^+ | j\omega(q_1 - p_1^+), j\omega)$$

$$+\Upsilon^R(p_1^-)\Upsilon^S(q_1 - p_1^-)\,\overline{\overline{p}}^r(j\omega p_1^- | j\omega(q_1 - p_1^-), j\omega)\Big]\,. \tag{14.52}$$

Similar arguments as used in Section 14.4 lead to the following representation for the imaged pressure wavefield

$$\hat{p}^{image}(\boldsymbol{x}_1, j\omega; \varphi) = 2\varphi \int_{x_3 \in \mathbb{R}} \chi(\boldsymbol{x}_T)\, \hat{W}(j\omega) \exp(-2j\omega\varphi x_3)\mathrm{d}x_3\,, \tag{14.53}$$

with

$$\hat{p}^{image}(\boldsymbol{x}_1, j\omega; \varphi) = \frac{\omega}{2\pi} \int_{q_1 \in \mathcal{Q}(\varphi)} \exp(-j\omega q_1 x_1)\overline{p}^{wght}(q_1, \varphi, j\omega)\mathrm{d}q_1\,. \tag{14.54}$$

Equation (14.54) indicates how the recorded data have to be mapped. We subsequently transform Eq. (14.53) back to the time domain; this leads to the result

$$2\varphi \int_{x_3 \in \mathbb{R}} \chi(x_1, x_3)\, W(t - 2\varphi x_3)\, \mathrm{d}x_3 = p^{image}(x_1, t; \varphi)\,. \tag{14.55}$$

Introducing the new integration variable $t' = 2\varphi x_3$, we obtain

$$\int_{t'\in\mathbb{R}} \chi(x_1, \frac{t'}{2\varphi})\, W(t-t')\,\mathrm{d}t' = p^{image}(x_1, t; \varphi)\,. \qquad (14.56)$$

We observe that the temporal convolution of the wavelet and the contrast function at the depth of $t/2\varphi$ is equal to $p^{image}$.

*Computational procedure*

The computational procedure can be summarized as follows:

- Collect the reflected wavefield $p^r(x_1^O + x_1^S, 0|\boldsymbol{x}^S, t)$ in the $t$-domain.
- Transform these data to the angular-frequency domain. We then arrive at the data $\hat{p}^r(x_1^O + x_1^S, 0|\boldsymbol{x}^S, j\omega)$ in the $(j\omega)$-domain.
- Transform these data to the $(j\omega p_1, j\omega q_1)$-domain using Eq. (14.46). We arrive at $\bar{\bar{p}}^r(j\omega p_1, j\omega(q_1 - p_1), j\omega)$.
- Perform the weighted average (see Eq. (14.52)) of the data for some particular choice of $q_1$ and $\varphi$. The resulting data $\bar{p}^{wght}(q_1, \varphi, \omega)$ are stacked over the domain $\mathcal{Q}(\varphi)$; using Eq. (14.54), we obtain the function $\hat{p}^{image}(x_1, j\omega; \varphi)$.
- Apply an inverse temporal Fourier transform. We finally obtain a space-time function $p^{image}(x_1, t; \varphi)$ that is a convolution of the wavelet and the time depth of the contrast function (see Eq. (14.56)).

Note that all operations can be performed with Fourier transformations.

*Constant contrast in* $\mathbb{D}_g$

In the special case that the contrast in $\mathbb{D}_g$ is constant, we can image the boundary of $\partial\mathbb{D}_g$ as follows. First we multiply the data in the frequency domain with the factor $j\omega$. This also happened in the procedure of the boundary imaging of Chapter 13. Then, Eq. (14.53) is replaced by

$$2j\omega\varphi \int_{x_3\in\mathbb{R}} \chi(\boldsymbol{x}_T)\, \hat{W}(j\omega) \exp(-2j\omega\varphi x_3)\mathrm{d}x_3 = \partial\hat{p}^{image}(x_1, j\omega; \varphi)\,, \quad (14.57)$$

with

$$\partial \hat{p}^{image}(\boldsymbol{x}_1, j\omega; \varphi) = j\omega \hat{p}^{image}(\boldsymbol{x}_1, j\omega; \varphi) . \tag{14.58}$$

Evaluation the left-hand side of Eq. (14.57), see Eq. (14.32), leads to

$$\partial p^{image}(\boldsymbol{x}_1, t; \varphi) = \left[1 - (\frac{c}{c^g})^2\right] \operatorname{sign}(-\nu_3) W(t - 2\varphi \boldsymbol{x}_3)\big|_{\boldsymbol{x}_T \in \partial \mathbb{D}_g} . \tag{14.59}$$

where $\nu_3 = \nu_3(\boldsymbol{x}_T)$ is the vertical component of the normal vector on $\partial \mathbb{D}_g$. The latter is directed away from $\mathbb{D}_g$ (see Fig. 14.1). This result is equivalent to the two-dimensional boundary imaging of Eq. (13.164). The multiplication of the data with $j\omega$ leads to an extra factor in the averaging procedure of Eq. (14.52).

## 14.6. Two-dimensional scattering by a circular cylinder

In order to generate synthetic data for a domain scatterer, we consider the scattering problem by a homogeneous, circular cylinder (Fig. 14.2). The cylinder has a radius $a$. The medium in the interior of the cylinder is characterized by the constants $\rho$ and $\kappa^g$. The center of the cylinder is at $\boldsymbol{x}_1 = 0$, $\boldsymbol{x}_3 = d$. The incident wavefield is generated by a monopole line source at $\boldsymbol{x}_T^S = (\boldsymbol{x}_1^S, \boldsymbol{x}_3^S = 0)$. From Eqs. (14.34) and (9.96) it follows that this incident wavefield is given by

$$\hat{p}^{inc}(\boldsymbol{x}_T|\boldsymbol{x}_T^S, s) = \frac{\hat{W}(s)}{2\pi} K_0(\frac{s}{c}|\boldsymbol{x}_T - \boldsymbol{x}_T^S|) . \tag{14.60}$$

In order to solve our scattering problem at hand, we introduce local polar coordinates adapted to the geometry of the circular cylinder,

$$\boldsymbol{x}_1 = r\cos(\phi) , \quad \boldsymbol{x}_3 = d + r\sin(\phi) , \quad -\pi \le \phi < \pi . \tag{14.61}$$

Similarly for the source coordinates, we introduce

$$\boldsymbol{x}_1^S = r^S \cos(\phi^S) , \quad \boldsymbol{x}_3^S = d + r^S \sin(\phi^S) , \quad -\pi \le \phi^S < \pi . \tag{14.62}$$

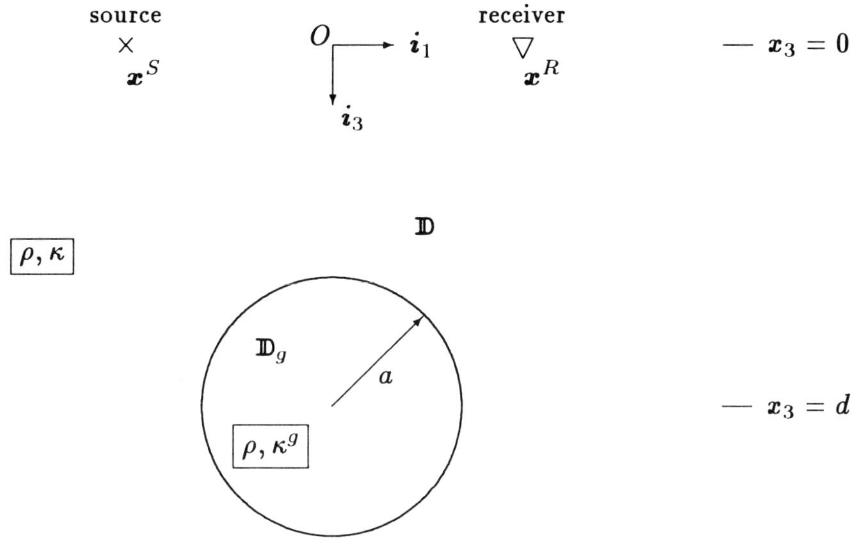


Figure 14.2. The circular-cylinder configuration.

Then, using the addition theorem for modified Bessel functions (WATSON, 1944, p. 361) we observe that the incident acoustic pressure is represented as an infinite series, in which the spatial dependencies of $\boldsymbol{x}_T$ and $\boldsymbol{x}_T^S$ are degenerated, viz.,

$$\hat{p}^{inc}(\boldsymbol{x}_T|\boldsymbol{x}_T^S, s) = \begin{cases} \dfrac{\hat{W}(s)}{2\pi} \displaystyle\sum_{m=-\infty}^{\infty} I_m(\frac{s}{c}r)\, K_m(\frac{s}{c}r^S) \cos[m(\phi-\phi^S)]\,, \\ \qquad\qquad r \leq r^S\,, \\ \dfrac{\hat{W}(s)}{2\pi} \displaystyle\sum_{m=-\infty}^{\infty} K_m(\frac{s}{c}r)\, I_m(\frac{s}{c}r^S) \cos[m(\phi-\phi^S)]\,, \\ \qquad\qquad r \geq r^S\,. \end{cases} \tag{14.63}$$

Here, $I_m(\cdot)$ is the modified Bessel function of the first kind and $m^{\text{th}}$ order, while $K_m(\cdot)$ is the modified Bessel function of the second kind and $m^{\text{th}}$ order. Note that, according to the first expression of Eq. (14.63), the incident wavefield in $r = 0$ is bounded; according to the second expression of Eq. (14.63), the incident wavefield satisfies the causality condition when $r \to \infty$. For points on the surface of the cylinder with radius $a$, the first expression of Eq. (14.63) applies. Taking into account the causality condition for the reflected wavefield, we write the reflected wavefield outside the scattering cylinder as

$$\hat{p}^r(\boldsymbol{x}_T|\boldsymbol{x}_T^S, s) = \frac{\hat{W}(s)}{2\pi} \sum_{m=-\infty}^{\infty} \hat{A}_m \, K_m(\frac{s}{c}r) \, K_m(\frac{s}{c}r^S) \cos[m(\phi - \phi^S)] \,. \quad (14.64)$$

The wavefield inside the cylinder has to be bounded at $r = 0$, hence, we write this interior wavefield as

$$\hat{p}^i(\boldsymbol{x}_T|\boldsymbol{x}_T^S, s) = \frac{\hat{W}(s)}{2\pi} \sum_{m=-\infty}^{\infty} \hat{B}_m \, I_m(\frac{s}{c^g}r) \, K_m(\frac{s}{c}r^S) \cos[m(\phi - \phi^S)] \,. \quad (14.65)$$

The unknown expansion factors $\hat{A}_m$ and $\hat{B}_m$ follow from the boundary conditions at $r = a$, viz.,

$$\lim_{r \downarrow a} [\hat{p}^{inc}(\boldsymbol{x}_T|\boldsymbol{x}_T^S, s) + \hat{p}^r(\boldsymbol{x}_T|\boldsymbol{x}_T^S, s)] = \lim_{r \uparrow a} \hat{p}^i(\boldsymbol{x}_T|\boldsymbol{x}_T^S, s) \,, \quad (14.66)$$

$$\lim_{r \downarrow a} \nu_k [\hat{v}_k^{inc}(\boldsymbol{x}_T|\boldsymbol{x}_T^S, s) + \hat{v}_k^r(\boldsymbol{x}_T|\boldsymbol{x}_T^S, s)] = \lim_{r \uparrow a} \nu_k \hat{v}_k^i(\boldsymbol{x}_T|\boldsymbol{x}_T^S, s) \,, \quad (14.67)$$

where $\nu_k$ is the normal vector in the radial direction of the cylinder. Using Eq. (7.70), that expresses the particle velocity in terms of the pressure, the resulting two boundary conditions can be solved for the reflected and interior wavefield expansion factors. The results are

$$\hat{A}_m = - \frac{\frac{s}{c^g} \partial I_m(\frac{s}{c^g}a) \, I_m(\frac{s}{c}a) - \frac{s}{c} \partial I_m(\frac{s}{c}a) \, I_m(\frac{s}{c^g}a)}{\frac{s}{c^g} \partial I_m(\frac{s}{c^g}a) \, K_m(\frac{s}{c}a) - \frac{s}{c} \partial K_m(\frac{s}{c}a) \, I_m(\frac{s}{c^g}a)} \,, \quad (14.68)$$

where $\partial I_m(\cdot)$ and $\partial K_m(\cdot)$ denote the derivatives of the functions $I_m$ and $K_m$ with respect to their arguments. Substituting these reflection factors in the expression for the reflected wavefield of Eq. (14.64) and taking $\boldsymbol{x}_T = \boldsymbol{x}_T^R$, we are able to compute the synthetic data pertaining to the present example of a domain scatterer. The various expressions to compute the modified Bessel functions can be found in ABRAMOWITZ and STEGUN, 1968, pp. 374-388).

*The numerical procedure*

As a scattering configuration we consider the situation shown in Fig. 14.2. The center of the circular cylinder is located at $x_1 = 0$ m and $x_3 = 150$ m, while the radius of the cylinder is 50 m. The acoustic wave speed of the fluid embedding amounts to 1457 m/s. We consider a cylinder of low contrast ($c^g = 1800$ m/s) and high contrast ($c^g = 2500$ m/s). The sources and receivers are located at $x_3 = 0$. The shape of the input source wavelet $W(t)$ in the time domain with a sample rate of 2 ms is shown in Fig. 9.10.

The series in the representation of the reflected wavefield of Eq. (14.64) is truncated and 61 terms are taken into account ($-30 \leq m \leq 30$), independent of the frequency. In our numerical computations we take

$$s = 4 + j\omega \,. \tag{14.69}$$

Note that we have taken a real part that is independent of $\omega$. This means that the temporal forward and inverse Fourier transformations can be employed. Apart from a factor $\exp(4t)$, these transforms are standard Fourier transforms and are computed with FFT routines. We employ a 512-points FFT routine with a temporal sample interval of 2 ms. Hence, the interval 0 - 0.51 s is the time interval in which we model and present the causal wave motion.

To simulate a two-dimensional seismic experiment we have computed the synthetic seismograms for 255 different source positions, starting at $x_1^S = -444.5$ m and ending at $x_1^S = 444.5$ m with an increment of 3.5 m. We have used 255 receiver positions per source, arranged in a symmetrical split-spread fashion, starting at $x_1^R = x_1^S - 444.5$ m and ending at $x_1^R = x_1^S + 444.5$ m, where $x_1^S$ is the horizontal source position.

Fig. 14.3 shows the reflected wavefield from a cylinder with low contrast for the lateral source positions $x_1^S = 0$ m and $x_1^S = -105$ m, respectively. Fig. 14.4 shows the pertaining results for a cylinder with high contrast. We observe that the upper curve in all these figures represent the reflection from the upper part of the boundary of the cylinder, while the lower curve represents the reflection from the lower part of the cylinder boundary, after transmission through the interior medium with wave speed $c_g$. The shape and the position of the upper curve are almost independent of the contrast of the cylinder, while the shape and the position of the lower curve clearly depend on the interior wave speed.

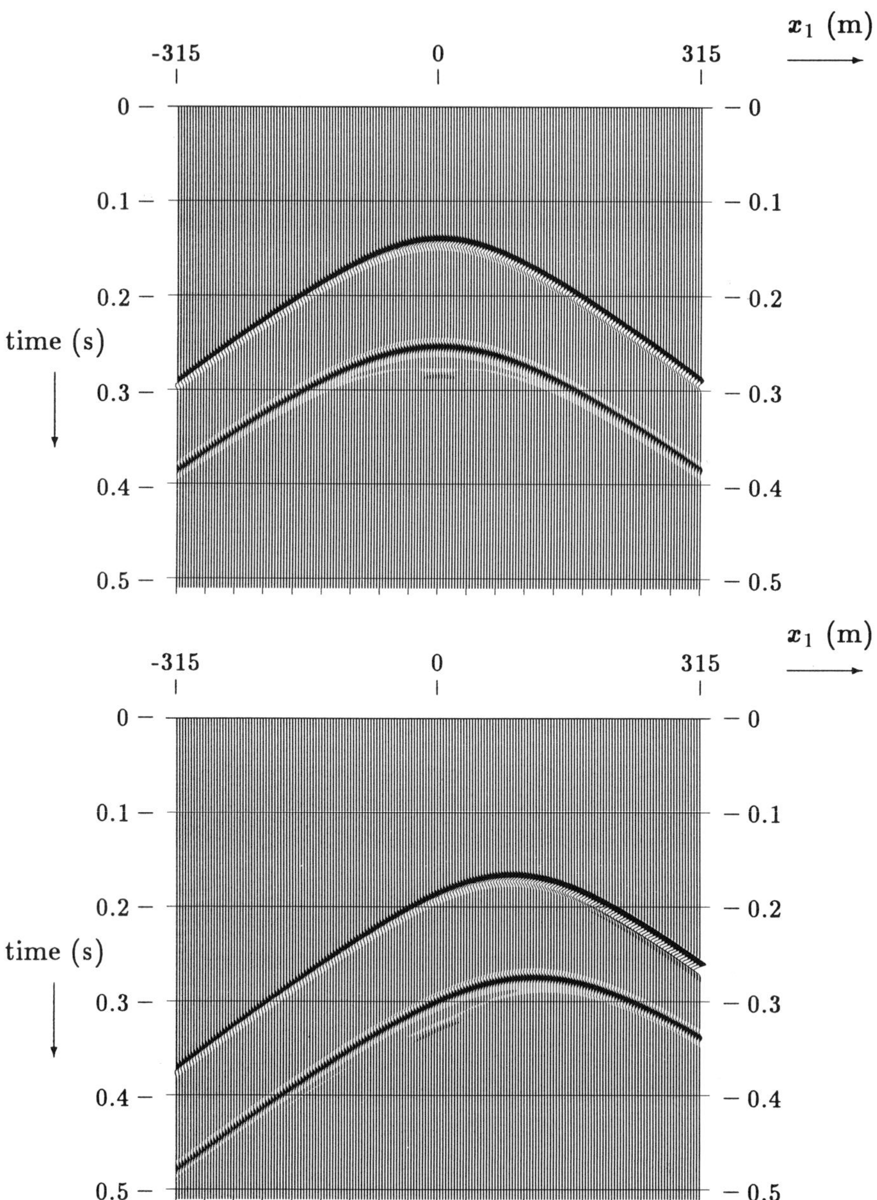


Figure 14.3. The reflected wavefield from a circular cylinder with low contrast ($c^g$ = 1800 m/s). The source is located at $x_1^S$ = 0 m (top) and $x_1^S$ = −105 m (bottom), respectively.

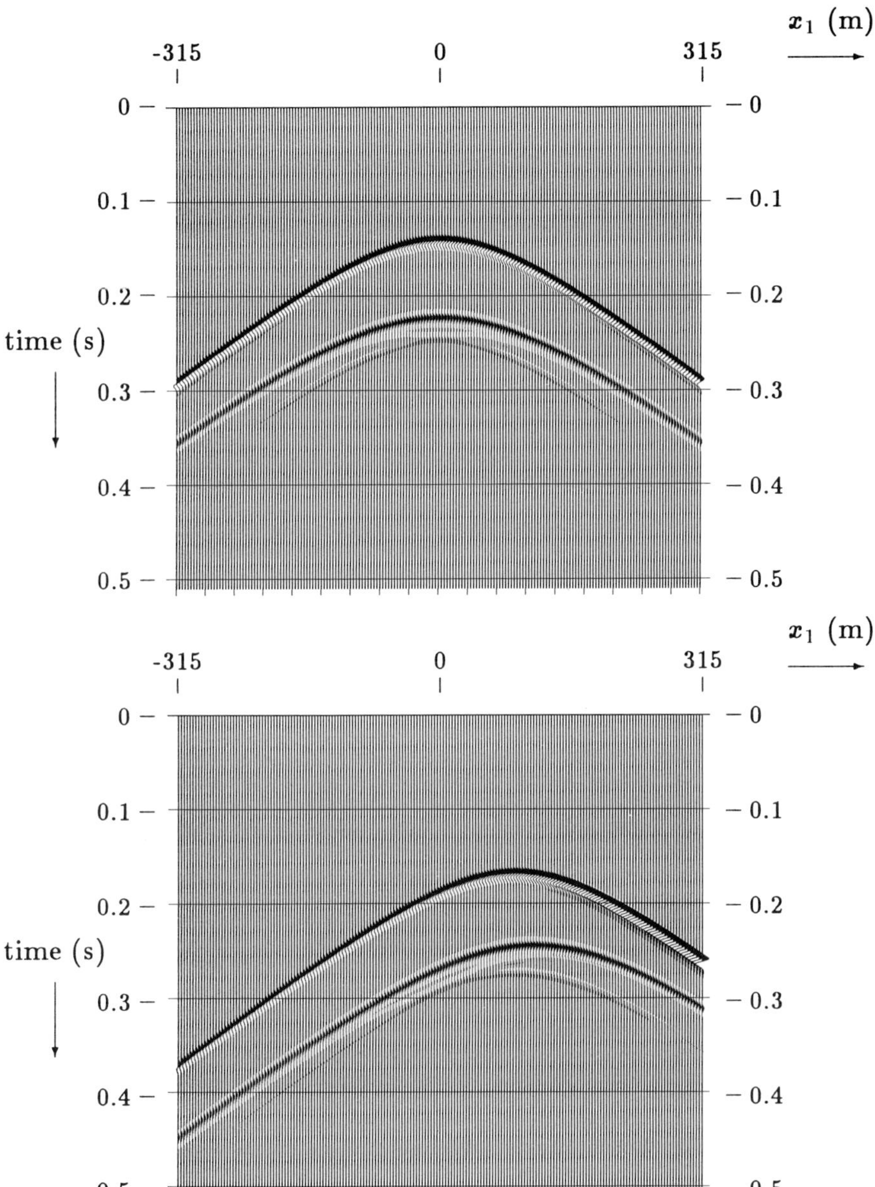


Figure 14.4. The reflected wavefield from a circular cylinder with high contrast ($c^g = 2500$). The source located at $x_1^S = 0$ m (top) and $x_1^S = -105$ m (bottom), respectively.

## 14.7. Imaging of the circular cylinder

To illustrate the domain imaging of the circular cylinder, we use the synthetic wavefield computed in the previous section. To image the circular cylinder, we follow the steps in the computational procedure outlined in Section 14.5.

As comparison, we employ the results of the Born approximation, when the contrast is constant. For the present case of the circular cylinder, it follows that Eq. (14.53) may be calculated as

$$\hat{p}^{image}(x_1, j\omega; \varphi) = \frac{1}{j\omega}\hat{W}(j\omega)\left[1 - (\frac{c}{c^g})^2\right]\chi_{cyl}(x_1)$$
$$\left\{\exp\left[-2j\omega\varphi\left(d - (a^2 - x_1^2)^{\frac{1}{2}}\right)\right] - \exp\left[-2j\omega\varphi\left(d + (a^2 - x_1^2)^{\frac{1}{2}}\right)\right]\right\}, \tag{14.70}$$

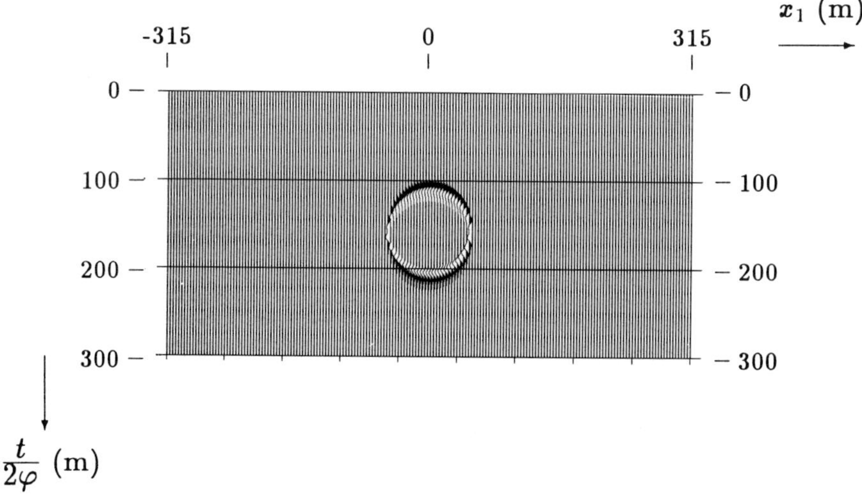


Figure 14.5. The ideally imaged circular cylinder.

where

$$\chi_{cyl}(\boldsymbol{x}_1) = \{1, \frac{1}{2}, 0\} \text{ when } \{|\boldsymbol{x}_1| < a, |\boldsymbol{x}_1| = a, |\boldsymbol{x}_1| > a\}. \qquad (14.71)$$

The time-domain counterpart of Eq. (14.70) yields the ideal image of the circular cylinder. This result for $p^{image}$ is shown in Fig. 14.5. Note that $p^{image}$ represents the superposition of two time retardations of the integrated wavelet. When $|\boldsymbol{x}_1| \to a$, we observe that $p^{image}$ tends to zero. Hence, a vertical boundary cannot be imaged.

In the next figures we show the image of the circular cylinder. The results are based on the processing sequence discussed in Section 14.5. In the imaging process, the choice of the vertical slowness parameter $\varphi$ is important. As we have shown in Section 14.4 the objective is to maximize the domain $\mathcal{Q}(\varphi)$ by taking $\varphi$ as small as possible. However, on the practical side, the smallest value of $\varphi$ that we can take, is constrained by the availability of the data in the domain $\mathcal{Q}$. In Fig. 14.6, we show the results for the cylinder of low contrast, where we have taken $\varphi = 0.9/c$ and $\varphi = 0.7/c$, respectively. Clearly, the results for $\varphi = 0.7/c$ reveals more curvature of the boundary of the contrasting domain than the results for $\varphi = 0.9/c$. The artifacts in the images are caused by the discretization of Eq. (14.54). Similar results are shown in Fig. 14.7 for the circular cylinder of high contrast. Again the imaged curvature depends on the chosen values of $\varphi$. Apart from these phenomena, the imaged circular cylinder becomes egg-shaped. This is caused by the failure of the Born approximation in reliably representing the interior wavefield. The upper part of the boundary of the contrasting domain is correctly positioned, while the lower part of the boundary is deformed. Improvement could be obtained by considering an adaptive version of an inhomogeneous background medium, but then the resulting imaging analysis has to be modified completely. This modification will lead to a domain-imaging procedure based on the generalized Radon transform (BEYLKIN, 1985). The same applies for the boundary-imaging procedure with an inhomogeneous background (BLEISTEIN, 1987). It is remarked that it does not make sense to complicate the imaging procedure too much. We suggest to switch to a full seismic-inversion process, where we can use the results from the imaging step as a priori information. This is discussed in the next chapter.

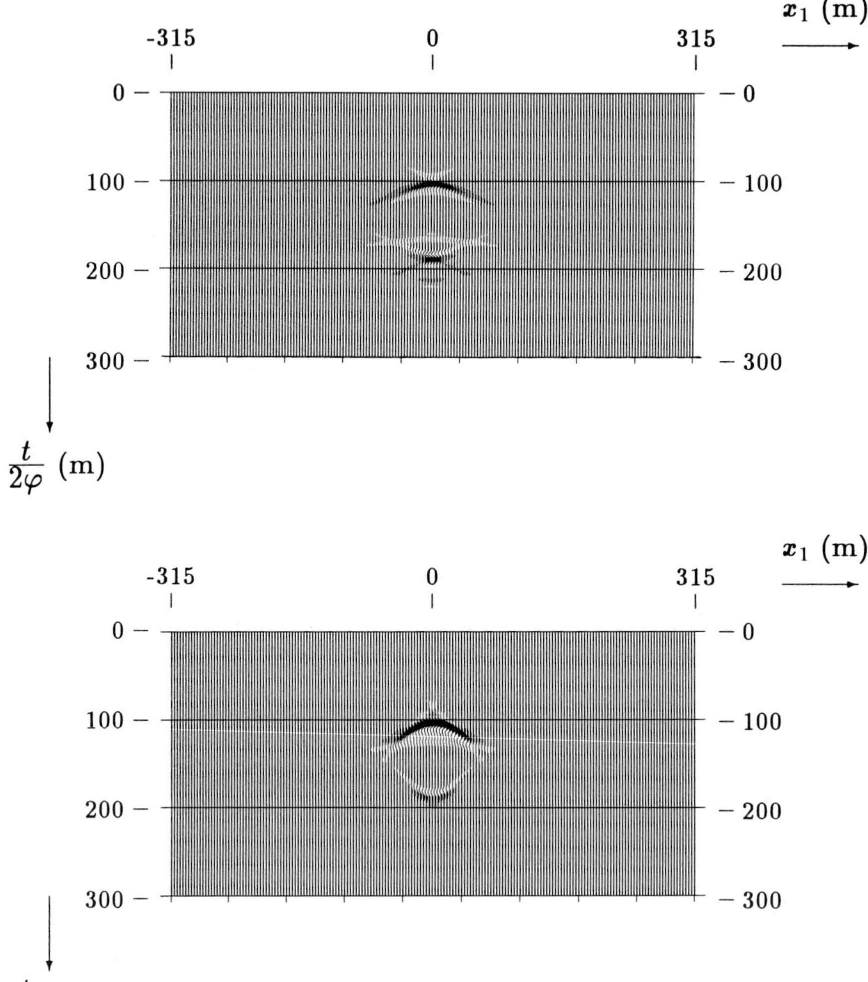


Figure 14.6. The imaged circular cylinder of low contrast ($c^g$ = 1800 m/s), for $\varphi = 0.9/c$ (top) and $\varphi = 0.7/c$ (bottom), respectively.

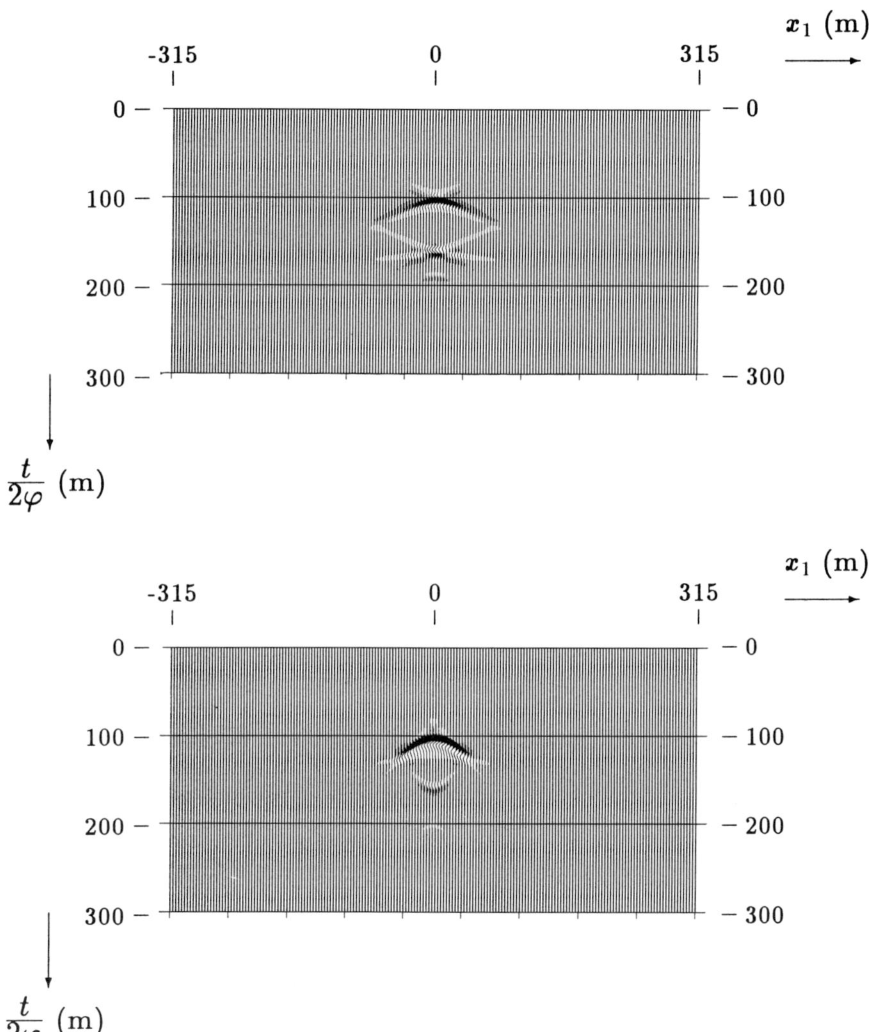


Figure 14.7. The imaged circular cylinder of high contrast ($c^g = 2500$ m/s), for $\varphi = 0.9/c$ (top) and $\varphi = 0.7/c$ (bottom), respectively.

# Chapter 15

# Seismic Inversion

In a seismic inversion problem, the aim is to reconstruct the distribution of constitutive parameters of a domain, whose interior is inaccessible to direct measurements, by probing it from the outside. To this end, the domain is considered as a contrasting domain in a known background configuration. The probing is carried out by irradiating the object by known acoustic sources, while the resulting wavefield is detected at a number of receiver positions. In seismic imaging, the approximate location of the subsurface scatterers is determined. Then, seismic inversion as a following-up process uses this result in its quest for the more precise information about the spatial dependent constitutive parameters (CLAERBOUT, 1985, 1992).

Application of both the field reciprocity and the power reciprocity theorem leads to various mathematical formulations of the inversion problem at hand (BLOK and ZEYLMANS, 1987). A common feature is the interaction between the actual state and a computational state. The analysis of this chapter confines to the application of the field reciprocity theorem. As point of departure, the domain-integral representation derived in Chapter 8 is employed.

As a consequence of the non-destructive nature of the measurements, the data are confined outside the scattering object. This means that the resulting integral equation lacks data support inside the scattering domain. This makes the inverse problem essentially ill posed. Our inversion strategy is to enforce the consistency of the domain-integral equation inside the scattering object. The problem of data uncertainty and data redundancy (TARANTOLA, 1987) is not addressed in this chapter.

## 15.1. The domain-integral representation

We consider the wavefield from a monopole source at $\boldsymbol{x}^S$ to be reflected by a contrasting domain of bounded extent, present in an inhomogeneous embedding of infinite extent, the background medium, with material parameters $\rho(\boldsymbol{x})$ and $\kappa(\boldsymbol{x})$. Let $\mathbb{D}_g$ be the bounded domain occupied by the scatterer. The domain exterior to $\mathbb{D}_g$ is denoted by $\mathbb{D}$ (Fig. 15.1). We assume that the contrasting domain has no contrast in the volume density of mass. Then, in $\mathbb{D}_g$ $\rho(\boldsymbol{x})$ is the volume density of mass and $\kappa^g(\boldsymbol{x})$ is the compressibility. The acoustic wave motion generated by a point source at $\boldsymbol{x}^S = \{x_1^S, x_2^S, 0\}$ starts to act at $t = 0$. The receiver is located at $\boldsymbol{x}^R = \{x_1^R, x_2^R, 0\}$.

In the $s$-domain, the source wavefield at the observation point $\boldsymbol{x}$ originating from a monopole source at $\boldsymbol{x}^S$ is then denoted as $\{\hat{p}^{inc}, \hat{v}_k^{inc}\}(\boldsymbol{x}|\boldsymbol{x}^S, s)$. The incident pressure wavefield is given by

$$\hat{p}^{inc}(\boldsymbol{x}|\boldsymbol{x}^S, s) = \hat{q}^S(s)\hat{G}^q(\boldsymbol{x}|\boldsymbol{x}^S, s)\,, \tag{15.1}$$

where $\hat{G}^q$ is the Green's state of the background medium (cf. Eq. (7.8)). The wavefield in $\mathbb{D}$ reflected by the contrasting domain $\mathbb{D}_g$ is given by $\{\hat{p}^r, \hat{v}_k^r\}(\boldsymbol{x}|\boldsymbol{x}^S, s)$. The reflected wavefield can be considered as the wavefield generated by secondary volume sources (monopole sources) in the domain $\mathbb{D}_g$ and it can be expressed in terms of these volume sources as (cf. Eq. (8.16))

$$\hat{p}^r(\boldsymbol{x}^R|\boldsymbol{x}^S, s) = \int_{\boldsymbol{x}\in\mathbb{D}_g} \hat{G}^q(\boldsymbol{x}^R|\boldsymbol{x}, s)s(\kappa - \kappa^g)\hat{p}(\boldsymbol{x}|\boldsymbol{x}^S, s)\mathrm{dV}\,, \tag{15.2}$$

where $\hat{p}$ denotes the total wavefield in the contrasting the domain and where $\hat{G}^q$ is the Green's state of the background medium.

The inverse problem consists of determining the compressibility $\kappa - \kappa^g$ in $\mathbb{D}_g$ from measurements of $\hat{p}^r$ in $\mathbb{D}$. If we assume that the total wavefield $\hat{p}$ in $\mathbb{D}_g$ is known, then, in principle, the compressibility $\kappa - \kappa^g$ can be found

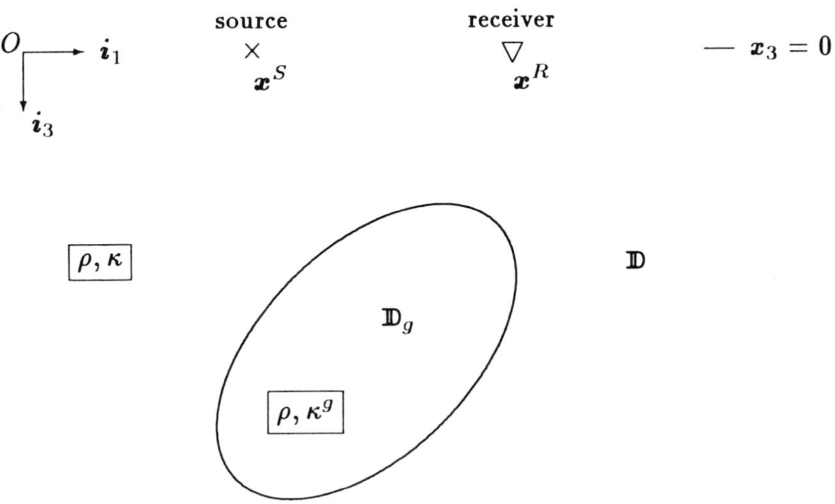


Figure 15.1. The inaccessible domain $\mathbb{D}_g$.

by matching the right-hand side of the integral representation of Eq. (15.2) to the data measured at all receiver locations, for all source positions and all $s$ values. However this is an ill-posed problem, because in Eq. (15.2) the equality sign applies at the receiver locations in $\mathbb{D}$, while the unknown has to be determined in $\mathbb{D}_g$. Moreover, the total wavefield $\hat{p}$ is also unknown in $\mathbb{D}_g$. If we assume that the compressibility $\kappa - \kappa^g$ is known, the latter total wavefield in $\mathbb{D}_g$ follows from the integral equation (cf. Eq. (8.18))

$$\hat{p}(\boldsymbol{x}|\boldsymbol{x}^S, s) - \int_{\boldsymbol{x}' \in \mathbb{D}_g} \hat{G}^q(\boldsymbol{x}|\boldsymbol{x}', s) s(\kappa - \kappa^g)\hat{p}(\boldsymbol{x}'|\boldsymbol{x}^S, s)\mathrm{dV} \tag{15.3}$$
$$= \hat{p}^{inc}(\boldsymbol{x}|\boldsymbol{x}^S, s), \quad \boldsymbol{x} \in \mathbb{D}_g \, .$$

Hence, we observe that the compressibility $\kappa - \kappa^g$ can be determined from Eq. (15.2) using Eq. (15.3) as a constraint. In fact, we have a non-linear problem for the solution of $\kappa - \kappa^g$, since the total wavefield in Eq. (15.2) depends on $\kappa - \kappa^g$ through Eq. (15.3). There are several methods to solve this non-linear problem.

The Born approximation is well known as a tool in attempts to solve inverse problems (see e.g., Clayton and Stolt, 1981, Cohen *et al.*, 1986). Application of this method to the present problem boils down to the following procedure. One tries to determine an unknown compressibility from measurements of a reflected wavefield in some measurement domain exterior to the scattering object. The essence of this approach involves making a guess of the wavefield in the object. In the Born approximation, the wavefield is guessed to be the incident wavefield. Then, the compressibility contrast is determined by minimizing the discrepancy between the right-hand side of Eq. (15.2) and the measured data at the receiver locations on the left-hand side of Eq. (15.2).

In the iterative Born method (see e.g., Devaney, 1982, Tijhuis 1987, Habashy and Mittra, 1987), the compressibility is updated iteratively as follows. With the compressibility found from the Born approximation, the integral equation (15.3) is solved in order to update the wavefield in $\mathbb{D}_g$. This wavefield is used again in the integral representation of Eq. (15.2) to determine a new compressibility by minimizing the discrepancy between the right-hand side of Eq. (15.2) and the measured data at the receiver locations. This iterative process is continued until the defect in matching the measured data is reduced to an acceptable level. Essentially, the updating involves a linearization of the non-linear dependence of the wavefield on the compressibility. In this method, the background medium and its Green's function remain the same in all iterations.

A variant is the distorted-wave Born method (see e.g., Beylkin and Oristaglio, 1985, Tobocman, 1986, Chew and Wang, 1990). In a particular iteration of this method an update of the compressibility is obtained. In the next iteration, the background medium is determined using the updated compressibility. Hence, the Green's function is updated in each iteration. In seismics this is known as the updating of the background model or macro model.

The main disadvantage of all these methods is that finding the compressibility from the integral representation of Eq. (15.2) remains an ill-posed problem and regularizing techniques should be used. Since the application of the latter techniques in the present inversion problem seems to be more *art* than *science*, we shall propose a method where the integral equation (15.3) itself serves as a regularizer. In this method we minimize the discrepancy in the satisfaction of both Eq. (15.2) and (15.3) simultaneously. This procedure retains the non-linear relation between the unknown compressibility and the unknown acoustic wavefield in the object.

## 15.2. Simultaneous minimization

Introduce the (compressibility) contrast $\kappa$ by

$$\kappa(\boldsymbol{x}) = \kappa(\boldsymbol{x}) - \kappa^g(\boldsymbol{x}) . \tag{15.4}$$

Then, the integral representation of Eq. (15.2) can be written as

$$\hat{p}^r(\boldsymbol{x}^R|\boldsymbol{x}^S, s) = \int_{\boldsymbol{x}\in\mathbb{D}_g} \hat{G}^q(\boldsymbol{x}^R|\boldsymbol{x}, s) s\kappa(\boldsymbol{x})\hat{p}(\boldsymbol{x}|\boldsymbol{x}^S, s)\mathrm{dV} , \tag{15.5}$$

where $\boldsymbol{x}^R \in \mathbb{D}_R$ and $\mathbb{D}_R$ is the domain where the receivers are located. The domain where the sources are located is denoted as $\mathbb{D}_S$. In our seismic situation these two domains coincide, viz., the plane $x_3 = 0$. The integral equation (15.3) can be written as

$$\hat{p}(\boldsymbol{x}|\boldsymbol{x}^S, s) - \int_{\boldsymbol{x}'\in\mathbb{D}_g} \hat{G}^q(\boldsymbol{x}|\boldsymbol{x}', s) s\kappa(\boldsymbol{x}')\hat{p}(\boldsymbol{x}'|\boldsymbol{x}^S, s)\mathrm{dV} = \hat{p}^{inc}(\boldsymbol{x}|\boldsymbol{x}^S, s) , \tag{15.6}$$

where $\boldsymbol{x} \in \mathbb{D}_g$ and $\mathbb{D}_g$ is the domain of the unknown compressibility contrast.

We assume that the (discrete) values of $s$ ($\mathrm{Re}(s) > 0$) are located in the domain ($s \in \mathcal{C}^+$).

Next, we introduce the operator

$$M\{\kappa\hat{p}\}(\boldsymbol{x}^R|\boldsymbol{x}^S, s) = \int_{\boldsymbol{x}\in\mathbb{D}_g} \hat{G}^q(\boldsymbol{x}^R|\boldsymbol{x}, s) s\kappa(\boldsymbol{x})\hat{p}(\boldsymbol{x}|\boldsymbol{x}^S, s)\mathrm{dV} , \tag{15.7}$$

that maps $\kappa\hat{p}$ from the domain $\{\boldsymbol{x}|\boldsymbol{x}^S, s\} \in \mathbb{D}_g \times \mathbb{D}_S \times \mathcal{C}^+$ to the domain $\{\boldsymbol{x}^R|\boldsymbol{x}^S, s\} \in \mathbb{D}_R \times \mathbb{D}_S \times \mathcal{C}^+$. We further define an inner product and norm on the domain $\mathbb{D}_R \times \mathbb{D}_S \times \mathcal{C}^+$ as

$$\langle \hat{u}, \hat{v} \rangle_R = \int_{\boldsymbol{x}^R\in\mathbb{D}_R} \mathrm{dA} \int_{\boldsymbol{x}^S\in\mathbb{D}_S} \sum_{s\in\mathcal{C}^+} \hat{u}(\boldsymbol{x}^R|\boldsymbol{x}^S, s)\hat{v}^\star(\boldsymbol{x}^R|\boldsymbol{x}^S, s)\mathrm{dA} \tag{15.8}$$

and

$$\|\hat{u}\|_R = \left[ \int_{\boldsymbol{x}^R \in \mathbb{D}_R} \mathrm{dA} \int_{\boldsymbol{x}^S \in \mathbb{D}_S} \sum_{s \in \mathcal{C}^+} |\hat{u}(\boldsymbol{x}^R, \boldsymbol{x}^S, s)|^2 \mathrm{dA} \right]^{\frac{1}{2}} . \tag{15.9}$$

Subsequently, we introduce the operator

$$L\{\kappa, \hat{p}\}(\boldsymbol{x}|\boldsymbol{x}^S, s) = \hat{p}(\boldsymbol{x}|\boldsymbol{x}^S, s) - K\{\kappa \hat{p}\}(\boldsymbol{x}|\boldsymbol{x}^S, s) , \tag{15.10}$$

where

$$K\{\kappa \hat{p}\}(\boldsymbol{x}|\boldsymbol{x}^S, s) = \int_{\boldsymbol{x}' \in \mathbb{D}_g} \hat{G}^q(\boldsymbol{x}|\boldsymbol{x}', s) s \kappa(\boldsymbol{x}') \hat{p}(\boldsymbol{x}'|\boldsymbol{x}^S, s) \mathrm{dV} , \tag{15.11}$$

that maps $\kappa \hat{p}$ from the domain $\{\boldsymbol{x}'|\boldsymbol{x}^S, s\} \in \mathbb{D}_g \times \mathbb{D}_S \times \mathcal{C}^+$ to the domain $\{\boldsymbol{x}|\boldsymbol{x}^S, s\} \in \mathbb{D}_g \times \mathbb{D}_S \times \mathcal{C}^+$. We also define an inner product and norm on the domain $\mathbb{D}_g \times \mathbb{D}_S \times \mathcal{C}^+$ as

$$\langle \hat{u}, \hat{v} \rangle_g = \int_{\boldsymbol{x} \in \mathbb{D}_g} \mathrm{dV} \int_{\boldsymbol{x}^S \in \mathbb{D}_S} \sum_{s \in \mathcal{C}^+} \hat{u}(\boldsymbol{x}|\boldsymbol{x}^S, s) \hat{v}^*(\boldsymbol{x}|\boldsymbol{x}^S, s) \mathrm{dA} \tag{15.12}$$

and

$$\|\hat{u}\|_g = \left[ \int_{\boldsymbol{x} \in \mathbb{D}_g} \mathrm{dV} \int_{\boldsymbol{x}^S \in \mathbb{D}_S} \sum_{s \in \mathcal{C}^+} |\hat{u}(\boldsymbol{x}|\boldsymbol{x}^S, s)|^2 \mathrm{dA} \right]^{\frac{1}{2}} . \tag{15.13}$$

The seismic inversion problem is that of finding $\kappa(\boldsymbol{x})$ and $\hat{p}(\boldsymbol{x}|\boldsymbol{x}^S, s)$, for $\boldsymbol{x}$ in $\mathbb{D}_g$, from the two equations

$$M\{\kappa \hat{p}\}(\boldsymbol{x}^R|\boldsymbol{x}^S, s) = \hat{p}^r(\boldsymbol{x}^R|\boldsymbol{x}^S, s) , \quad \boldsymbol{x}^R \in \mathbb{D}_R, \quad \boldsymbol{x}^S \in \mathbb{D}_S, \quad s \in \mathcal{C}^+, \tag{15.14}$$

$$L\{\kappa, \hat{p}\}(\boldsymbol{x}|\boldsymbol{x}^S, s) = \hat{p}^{inc}(\boldsymbol{x}|\boldsymbol{x}^S, s) , \quad \boldsymbol{x} \in \mathbb{D}_g, \quad \boldsymbol{x}^S \in \mathbb{D}_S, \quad s \in \mathcal{C}^+ . \tag{15.15}$$

We observe that this is a non-linear problem for the compressibility contrast $\kappa(\boldsymbol{x})$ and the wavefield $\hat{p}(\boldsymbol{x}|\boldsymbol{x}^S, s)$.

One way to solve this non-linear problem is to linearize and consider them either as the linear operator equation

$$\mathcal{L}_{\{\hat{p}\}}\{\kappa\} = \begin{pmatrix} M\{\kappa \hat{p}\} \\ L\{\kappa, \hat{p}\} \end{pmatrix} = \begin{pmatrix} \hat{p}^r \text{ in } \mathbb{D}_R \times \mathbb{D}_S \times \mathcal{C}^+ \\ \hat{p}^{inc} \text{ in } \mathbb{D}_g \times \mathbb{D}_S \times \mathcal{C}^+ \end{pmatrix} , \tag{15.16}$$

from which, for fixed values of the pressure wavefield, the compressibility contrast may be determined, or as the linear operator equation

$$\mathcal{L}_{\{\kappa\}}\{\hat{p}\} = \begin{pmatrix} M\{\kappa\hat{p}\} \\ L\{\kappa,\hat{p}\} \end{pmatrix} = \begin{pmatrix} \hat{p}^r \text{ in } \mathbb{D}_R \times \mathbb{D}_S \times \mathcal{C}^+ \\ \hat{p}^{inc} \text{ in } \mathbb{D}_g \times \mathbb{D}_S \times \mathcal{C}^+ \end{pmatrix}, \tag{15.17}$$

from which, for fixed values of the contrast, the wavefield may be determined. However, we postpone this linearization procedure and specifically we will seek $\kappa(\boldsymbol{x})$ and $\hat{p}(\boldsymbol{x}|\boldsymbol{x}^S, s)$ simultaneously to minimize the functional

$$F = \frac{\|\hat{p}^r - M\{\kappa\hat{p}\}\|_R^2}{\|\hat{p}^r\|_R^2} + \frac{\|\hat{p}^{inc} - L\{\kappa,\hat{p}\}\|_g^2}{\|\hat{p}^{inc}\|_g^2}. \tag{15.18}$$

We choose to put the two terms in Eq. (15.18) on equal footing by normalizing them in the sense that they are both equal to one when $\hat{p} = 0$.

## 15.3. Inversion algorithm

A first inspection of the minimization of Eq. (15.18) might suggest a substantial numerical effort. However, an adequate iterative scheme may overcome this problem. There is no need to compute the wavefield function to a higher order degree of accuracy than the degree of accuracy of the approximated contrast. We therefore propose an iterative inversion algorithm based upon the ideas of conjugate-gradient methods of Chapter 2. Let us remind that in Chapter 2 we have sought a solution of a linear operator equation for an unknown wavefield. In the present problem, we have to update the wavefield and the contrast simultaneously. Specifically, we propose the iterative construction of $\kappa_N = \kappa_N(\boldsymbol{x})$ and $\hat{p}_N = \hat{p}_N(\boldsymbol{x}|\boldsymbol{x}^S, s)$ as follows

$$\kappa_0 = \kappa^{initial}, \tag{15.19}$$

$$\kappa_N = \kappa_{N-1} + \beta_N \phi_N, \tag{15.20}$$

$$\hat{r}_N^R = \hat{p}^r - M\{\kappa_N \hat{p}_N\}, \quad \boldsymbol{x}^R \in \mathbb{D}_R, \quad \boldsymbol{x}^S \in \mathbb{D}_S, \quad s \in \mathcal{C}^+, \tag{15.21}$$

and

$$\hat{p}_0 = \hat{p}^{initial}, \tag{15.22}$$

$$\hat{p}_N = \hat{p}_{N-1} + \hat{\alpha}_N \hat{\psi}_N, \tag{15.23}$$

$$\hat{r}_N^g = \hat{p}^{inc} - L\{\kappa_N, \hat{p}_N\}, \quad \boldsymbol{x} \in \mathbb{D}_g, \quad \boldsymbol{x}^S \in \mathbb{D}_S, \quad s \in \mathcal{C}^+, \tag{15.24}$$

where $\beta_N$ is a real constant and $\hat{\alpha}_N$ is a complex constant, which are chosen at each step to minimize (cf. Eq. (15.18))

$$F_N = \frac{\|\hat{r}_N^R\|_R^2}{\|\hat{p}^r\|_R^2} + \frac{\|\hat{r}_N^g\|_g^2}{\|\hat{p}^{inc}\|_g^2} . \tag{15.25}$$

Here, the residual errors can recursively be written as

$$\hat{r}_N^R = \hat{r}_{N-1}^R - \beta_N M\{\phi_N \hat{p}_{N-1}\} - \hat{\alpha}_N M\{\kappa_{N-1}\hat{\psi}_N\} - \beta_N \hat{\alpha}_N M\{\phi_N \hat{\psi}_N\} , \tag{15.26}$$

and

$$\hat{r}_N^g = \hat{r}_{N-1}^g + \beta_N K\{\phi_N \hat{p}_{N-1}\} - \hat{\alpha}_N L\{\kappa_{N-1}, \hat{\psi}_N\} + \beta_N \hat{\alpha}_N K\{\phi_N \hat{\psi}_N\}, \tag{15.27}$$

where the real correction function $\phi_N$ and the complex correction function $\hat{\psi}_N$ are chosen appropriately.

The minimization of the quantity $F_N$ of Eq. (15.25), using Eqs. (15.26) and (15.27), leads to a non-linear problem for the variables $\beta_N$ and $\hat{\alpha}_N$ at each step, which can be solved by an appropriate numerical algorithm, e.g., the Fletcher-Reeves-Polak-Ribière conjugate-gradient method (PRESS *et al.*, 1986, p. 305).

### *The gradient directions*

In order to guarantee improvement in the iterative scheme, we require improvement in the independent directions varying either real $\beta_N$ or complex $\hat{\alpha}_N$. This is easily accomplished by first taking $\hat{\alpha}_N$ is zero. Then, Eqs. (15.26) and (15.27) simplify to

$$\hat{r}_N^R = \hat{r}_{N-1}^R - \beta_N M\{\phi_N \hat{p}_{N-1}\} \tag{15.28}$$

and

$$\hat{r}_N^g = \hat{r}_{N-1}^g + \beta_N K\{\phi_N \hat{p}_{N-1}\} . \tag{15.29}$$

Taking into account that at the $N^{\text{th}}$ step $\hat{r}_{N-1}^R$ and $\hat{r}_{N-1}^g$ are known, minimization of $F_N$ leads to the relation (cf. Eqs. (2.15) - (2.16))

$$\text{Re}\left[\frac{\langle \hat{r}_N^R, M\{\phi_N \hat{p}_{N-1}\}\rangle_R}{\|\hat{p}^r\|_R^2} - \frac{\langle \hat{r}_N^g, K\{\phi_N \hat{p}_{N-1}\}\rangle_g}{\|\hat{p}^{inc}\|_g^2}\right] = 0 , \tag{15.30}$$

where we have used that $\beta_N$ is real. After substituting Eqs. (15.28) and (15.29) in Eq. (15.30) we obtain

$$\beta_N = \operatorname{Re}\left[\frac{\dfrac{\langle \hat{r}^R_{N-1}, M\{\phi_N \hat{p}_{N-1}\}\rangle_R}{\|\hat{p}^r\|^2_R} - \dfrac{\langle \hat{r}^g_{N-1}, K\{\phi_N \hat{p}_{N-1}\}\rangle_g}{\|\hat{p}^{inc}\|^2_g}}{\dfrac{\|M\{\phi_N \hat{p}_{N-1}\}\|^2_R}{\|\hat{p}^r\|^2_R} + \dfrac{\|K\{\phi_N \hat{p}_{N-1}\}\|^2_g}{\|\hat{p}^{inc}\|^2_g}}\right] . \tag{15.31}$$

Let us define the adjoint operator $M^*$ through

$$\langle M\hat{u}, \hat{v}\rangle_R = \langle \hat{u}, M^*\hat{v}\rangle_g \,, \tag{15.32}$$

from which it follows that $M^*$ is an operator that maps a function from the domain $\mathbb{D}_R \times \mathbb{D}_S \times \mathcal{C}^+$ to the domain $\mathbb{D}_g \times \mathbb{D}_S \times \mathcal{C}^+$, viz.,

$$M^*\hat{v} = \int_{\boldsymbol{x}^R \in \mathbb{D}_R} \hat{G}^{q\star}(\boldsymbol{x}^R|\boldsymbol{x}, s)s^\star\, \hat{v}(\boldsymbol{x}^R|\boldsymbol{x}^S, s)\mathrm{dA} \,. \tag{15.33}$$

Let us further define the adjoint operator $K^*$ through

$$\langle K\hat{u}, \hat{v}\rangle_g = \langle \hat{u}, K^*\hat{v}\rangle_g \,, \tag{15.34}$$

from which it follows that $K^*$ is an operator that maps a function from the domain $\mathbb{D}_g \times \mathbb{D}_S \times \mathcal{C}^+$ to the domain $\mathbb{D}_g \times \mathbb{D}_S \times \mathcal{C}^+$, viz.,

$$K^*\hat{v} = \int_{\boldsymbol{x}' \in \mathbb{D}_g} \hat{G}^{q\star}(\boldsymbol{x}'|\boldsymbol{x}, s)s^\star\, \hat{v}(\boldsymbol{x}'|\boldsymbol{x}^S, s)\mathrm{dV} \,. \tag{15.35}$$

Now improvement ($\beta_N \neq 0$) is obtained if we require that the real part of the numerator of Eq. (15.31) does not vanish. Using the definition of the adjoint operators (cf. Eqs. (15.32) and (15.34)) we require that

$$\operatorname{Re}\left[\frac{\langle \hat{p}^\star_{N-1} M^* \hat{r}^R_{N-1}, \phi_N\rangle_g}{\|\hat{p}^r\|^2_R} - \frac{\langle \hat{p}^\star_{N-1} K^* \hat{r}^g_{N-1}, \phi_N\rangle_g}{\|\hat{p}^{inc}\|^2_g}\right] \neq 0 \,. \tag{15.36}$$

We further define the inner product and norm on $\mathbb{D}_g$ as

$$\langle \hat{u}, \hat{v}\rangle = \int_{\boldsymbol{x} \in \mathbb{D}_g} \hat{u}(\boldsymbol{x})\hat{v}^\star(\boldsymbol{x})\mathrm{dV} \tag{15.37}$$

and

$$\|\hat{u}\| = \int_{\boldsymbol{x} \in \mathbb{D}_g} |\hat{u}(\boldsymbol{x})|^2\mathrm{dV} \,. \tag{15.38}$$

Then, we rewrite the improvement condition of Eq. (15.36) as

$$\mathrm{Re}\left[\frac{\langle \int_{\boldsymbol{x}^S \in \mathbb{D}_S} \sum_{s \in \mathcal{C}+} \hat{p}^{\star}_{N-1} M^* \hat{r}^R_{N-1} \mathrm{dA}, \phi_N \rangle}{\|\hat{p}^r\|^2_R} - \frac{\langle \int_{\boldsymbol{x}^S \in \mathbb{D}_S} \sum_{s \in \mathcal{C}+} \hat{p}^{\star}_{N-1} K^* \hat{r}^g_{N-1} \mathrm{dA}, \phi_N \rangle}{\|\hat{p}^{inc}\|^2_g}\right] \neq 0 . \tag{15.39}$$

This improvement condition will be satisfied if we take the (real) direction $\phi_N$ to be

$$\phi_N(\boldsymbol{x}) = \mathcal{L}^*_{\{\hat{p}_{N-1}\}} \left\{ \begin{array}{c} \hat{r}^R_{N-1} \\ \hat{r}^g_{N-1} \end{array} \right\} , \quad \boldsymbol{x} \in \mathbb{D}_g , \tag{15.40}$$

in which

$$\mathcal{L}^*_{\{\hat{p}_{N-1}\}} \left\{ \begin{array}{c} \hat{r}^R_{N-1} \\ \hat{r}^g_{N-1} \end{array} \right\} = \int_{\boldsymbol{x}^S \in \mathbb{D}_S} \sum_{s \in \mathcal{C}+} \mathrm{Re}\left[\frac{\hat{p}^{\star}_{N-1} M^* \hat{r}^R_{N-1}}{\|\hat{p}^r\|^2_R} - \frac{\hat{p}^{\star}_{N-1} K^* \hat{r}^g_{N-1}}{\|\hat{p}^{inc}\|^2_g}\right] \mathrm{dA} . \tag{15.41}$$

This operator is the adjoint of the operator defined in Eq. (15.16). These operators are the linear operators, assuming that the wavefields do not change.

As next step we take $\beta_N$ is zero. Then, Eqs. (15.26) and (15.27) simplify to

$$\hat{r}^R_N = \hat{r}^R_{N-1} - \hat{\alpha}_N M\{\kappa_{N-1} \hat{\psi}_N\} \tag{15.42}$$

and

$$\hat{r}^g_N = \hat{r}^g_{N-1} - \hat{\alpha}_N L\{\kappa_{N-1}, \hat{\psi}_N\} . \tag{15.43}$$

Taking into account that at the $N^{\mathrm{th}}$ step $\hat{r}^R_{N-1}$ and $\hat{r}^g_{N-1}$ are known, minimization of $F_N$ leads to the relation (cf. Eqs. (2.15) - (2.16))

$$\frac{\langle \hat{r}^R_N, M\{\kappa_{N-1} \hat{\psi}_N\} \rangle_R}{\|\hat{p}^r\|^2_R} + \frac{\langle \hat{r}^g_N, L\{\kappa_{N-1}, \hat{\psi}_N\} \rangle_g}{\|\hat{p}^{inc}\|^2_g} = 0 . \tag{15.44}$$

After substituting Eqs. (15.42) and (15.43) in Eq. (15.44) we obtain

$$\hat{\alpha}_N = \frac{\dfrac{\langle \hat{r}^R_{N-1}, M\{\kappa_{N-1}\hat{\psi}_N\}\rangle_R}{\|\hat{p}^r\|^2_R} + \dfrac{\langle \hat{r}^g_{N-1}, L\{\kappa_{N-1}, \hat{\psi}_N\}\rangle_g}{\|\hat{p}^{inc}\|^2_g}}{\dfrac{\|M\{\kappa_{N-1}\hat{\psi}_N\}\|^2_R}{\|\hat{p}^r\|^2_R} + \dfrac{\|L\{\kappa_{N-1}, \hat{\psi}_N\}\|^2_g}{\|\hat{p}^{inc}\|^2_g}} . \tag{15.45}$$

Now improvement ($\hat{\alpha}_N \neq 0$) is obtained if we require that the numerator of Eq. (15.45) does not vanish. Using the definition of the adjoint operators (cf. Eqs. (15.32) and (15.34)) we require that

$$\frac{\langle \kappa_{N-1} M^* \hat{r}^R_{N-1}, \hat{\psi}_N\rangle_g}{\|\hat{p}^r\|^2_R} + \frac{\langle \hat{r}^g_{N-1} - \kappa_{N-1} K^* \hat{r}^g_{N-1}, \hat{\psi}_N\rangle_g}{\|\hat{p}^{inc}\|^2_g} \neq 0 . \tag{15.46}$$

This improvement condition will be satisfied if we take the (complex) direction $\hat{\psi}_N$ to be

$$\hat{\psi}_N(\boldsymbol{x}|\boldsymbol{x}^S, s) = \mathcal{L}^*_{\{\kappa_{N-1}\}} \left\{ \begin{array}{c} \hat{r}^R_{N-1} \\ \hat{r}^g_{N-1} \end{array} \right\} , \quad \boldsymbol{x} \in \mathbb{D}_g, \ \boldsymbol{x}^S \in \mathbb{D}_S, \ s \in \mathcal{C}^+ , \tag{15.47}$$

in which

$$\mathcal{L}^*_{\{\kappa_{N-1}\}} \left\{ \begin{array}{c} \hat{r}^R_{N-1} \\ \hat{r}^g_{N-1} \end{array} \right\} = \frac{\kappa_{N-1} M^* \hat{r}^R_{N-1}}{\|\hat{p}^r\|^2_R} + \frac{\hat{r}^g_{N-1} - \kappa_{N-1} K^* \hat{r}^g_{N-1}}{\|\hat{p}^{inc}\|^2_g} . \tag{15.48}$$

This operator is the adjoint of the operator defined in Eq. (15.17). These operators are the linear operators, assuming that the contrast does not change.

*The conjugate-gradient directions*

Assuming that the wavefields $\hat{p}$ do not change ($\hat{\alpha}_N = 0$), the determination of the contrast by minimizing $F_N$ is a linear problem. In this case, an improved direction for $N = 2, \cdots$, is the conjugate-gradient direction (see Chapter 2, the case where $LT$ is selfadjoint, $L \equiv \mathcal{L}_{\{\hat{p}_{N-1}\}}$, $T \equiv \mathcal{L}^*_{\{\hat{p}_{N-1}\}}$)

$$\phi_N = \mathcal{L}^*_{\{\hat{p}_{N-1}\}} \left\{ \begin{array}{c} \hat{r}^R_{N-1} \\ \hat{r}^g_{N-1} \end{array} \right\} + \frac{\|\phi_N\|^2}{\|\phi_{N-1}\|^2} \phi_{N-1} . \tag{15.49}$$

This direction is similar to the Fletcher-Reeves direction for non-linear problems. If we still assume that in all previous iterations, $n = 1, \cdots, N$, the wavefields have not changed, the directions satisfy the orthogonality relation (cf. Eq. (2.87))

$$\langle \phi_n, \phi_m \rangle = 0, \quad \text{for } n \neq m. \tag{15.50}$$

Then, Eq. (15.49) is identical to

$$\phi_N = \mathcal{L}^*_{\{\hat{p}_{N-1}\}} \left\{ \begin{array}{c} \hat{r}^R_{N-1} \\ \hat{r}^g_{N-1} \end{array} \right\} + \frac{\langle \phi_N, \phi_N - \phi_{N-1} \rangle}{\|\phi_{N-1}\|^2} \phi_{N-1} \,. \tag{15.51}$$

This is similar to the Polak-Ribière direction for non-linear problems. For a linear problem, the Polak-Ribière choice is identical to the Fletcher-Reeves choice; however, for non-linear problems, the two choices behave differently and numerical evidence suggests that the Polak-Ribière choice is preferable. This may be argued as follows. Suppose the conjugate gradient algorithm is making little progress, and $\phi_N \cong \phi_{N-1}$. In these circumstances the Polak-Ribière direction becomes the gradient and the scheme seems to restart automatically.

Similarly, assuming that the contrast does not change ($\beta_N = 0$), the determination of the wavefield by minimizing $F_N$ is a linear problem. Then, an improved direction for $N = 2, \cdots$, is the Fletcher-Reeves direction (see Chapter 2, the case where $LT$ is selfadjoint, $L \equiv \mathcal{L}_{\{\kappa_{N-1}\}}$, $T \equiv \mathcal{L}^*_{\{\kappa_{N-1}\}}$)

$$\hat{\psi}_N = \mathcal{L}^*_{\{\kappa_{N-1}\}} \left\{ \begin{array}{c} \hat{r}^R_{N-1} \\ \hat{r}^g_{N-1} \end{array} \right\} + \frac{\|\hat{\psi}_N\|^2_g}{\|\hat{\psi}_{N-1}\|^2_g} \hat{\psi}_{N-1} \,. \tag{15.52}$$

If we still assume that in all previous iterations, $n = 1, \cdots, N$, the contrast has not changed, the directions satisfy the orthogonality relation (cf. Eq. (2.87))

$$\langle \hat{\psi}_n, \hat{\psi}_m \rangle = 0, \quad \text{for } n \neq m. \tag{15.53}$$

Then, Eq. (15.52) is identical to Polak-Ribière direction

$$\hat{\psi}_N = \mathcal{L}^*_{\{\kappa_{N-1}\}} \left\{ \begin{array}{c} \hat{r}^R_{N-1} \\ \hat{r}^g_{N-1} \end{array} \right\} + \frac{\langle \hat{\psi}_N, \hat{\psi}_N - \hat{\psi}_{N-1} \rangle_g}{\|\hat{\psi}_{N-1}\|^2_g} \hat{\psi}_{N-1} \,. \tag{15.54}$$

For a linear problem, the Polak-Ribière choice is identical to the Fletcher-Reeves choice; however, for non-linear problems, the two choices behave differently and numerical evidence suggests that the Polak-Ribière choice is preferable.

We finally note that the linearization of the non-linear inversion problem is incorporated in the choice of the directions $\phi_N$ and $\hat{\psi}_N$. As soon they are determined, the constants $\beta_N$ and $\hat{\alpha}_N$ follow from the simpler non-linear problem of minimizing $F_N$ of Eq. (15.25), using Eqs. (15.26) - (15.27). It is now not difficult to find the the constants $\beta_N$ and $\hat{\alpha}_N$ by a standard conjugate-gradient algorithm, e.g., the Fletcher-Reeves-Polak-Ribière scheme (PRESS *et al.*, 1986, p. 305).

The present algorithm has been shown to be very effective in the case that one is dealing with one parameter $s$ and that sources and receivers are located around a finite object to be probed (KLEINMAN and VAN DEN BERG, 1992a, 1992b). The application of this inversion algorithm to the seismic situation has to be investigated. With this challenging research problem we conclude this chapter and this book.

# Bibliography


M. Abramowitz and I.A. Stegun (1968), *Handbook of Mathematical Functions*, Dover, New York, 1046 pp.

K. Aki and P.G. Richards (1980), *Quantitative Seismology, Theory and Methods*, vol. I, Freeman, New York, 557 pp.

R.H.T. Bates and D.J.N. Wall (1977), Null field approach to scalar diffraction, I. General Methods, II. Approximate methods, *Phil. Trans. R. Soc. Lond.*, **A287**, pp. 45-95.

M. Båth and A.J. Berkhout (1984), *Mathematical Aspects of Seismology*, Geophysical Press, second edition, London, 484 pp.

A.J. Berkhout (1987), *Applied Seismic Wave Theory*, Elsevier, Amsterdam, 377 pp.

G. Beylkin (1985), Imaging of discontinuities in the inverse scattering problem by inversion of a causal generalized Radon transform, *J. Math. Phys.*, **26**, pp. 99-108.

G. Beylkin and M.L. Oristaglio (1985), Distorted-wave Born and distorted-wave Rytov approximation, *Optics Communications*, **53**, pp. 213-216.

N. Bleistein (1987), On the imaging of reflectors in the earth, *Geophysics*, **52**, 931-942.

H. Blok and M.C.S. Zeylmans (1987), Reciprocity and the formulation of inverse profiling problems, *Radio Science*, **22**, pp. 1137-1147.

N.N. Bojarski (1983), Generalized reaction principles and reciprocity theorems for the wave equations and the relationship between the time-advanced and time-retarded fields, *J. Acoust. Soc. Am.*, **74**, pp. 281-285.

M. Born and E. Wolf (1965), *Principles of Optics*, Pergamon Press, London, 808 pp.

V. Červený (1989), Seismic ray theory, in: D.E. James (Ed.), *Encyclopedia of Solid Earth Geophysics*, Van Nostrand Reinhold Co., New York, pp. 1098-1118.

C.H. Chapman (1981), Generalized Radon transforms and slant stacks, *Geophysic. J. Roy. Astr. Soc.*, **66**, pp. 445-453.

W.C. Chew and Y.M. Wang (1990), Reconstruction of two-dimensional permittivity distribution using the distorted Born iterative method, *IEEE Trans. Medical Imaging*, **9**, pp. 218-225.

J.F. Claerbout (1985), *Imaging the Earth's Interior*, Blackwell, Oxford, 398 pp.

J.F. Claerbout (1992), *Earth Soundings Analysis; Processing versus Inversion*, Blackwell, Oxford, 320 pp.

R.W. Clayton and R.H. Stolt (1981), A Born-WKBJ inversion method for acoustic reflection data, *Geophysics*, **46**, pp. 1559-1567.

J.K. Cohen, F.G. Hagin and N. Bleistein (1986), Three-dimensional Born inversion with an arbitrary reference, *Geophysics*, **51**, pp. 1552-1558.

S.R. Deans (1983), *The Radon Transform and Some of its Applications*, John Wiley & Sons, New York, 289 pp.

A.T. de Hoop (1960), A modification of Cagniard's method for solving seismic pulse problems, *Applied Scientific Research*, **B8**, pp. 349-356.

A.T. de Hoop (1988), Time-domain reciprocity theorems for acoustic wave fields in fluids with relaxation, *J. Acoust. Soc. Am.*, **84**, pp. 1877-1882.

A.T. de Hoop (1991), Convergence criterion for the time-domain iterative Born approximation to scattering by an inhomogeneous, dispersive object, *J. Optical. Soc. Am.*, **A8**, pp. 1256-1260.

A.J. Devaney (1982), A filtered backprojection algorithm for diffraction tomography, *Ultrasonic Imaging*, **4**, pp. 336-360.

J.T. Fokkema, P.M. van den Berg and M. Vissinga (1992), On the computation of Radon transforms of seismic data, *Journal of Seismic Exploration*, **1**, pp. 93-105.

J.T. Fokkema and P.M. van den Berg (1992), Reflector imaging, *Geophys. J. Int.*, **110**, pp. 191-200.

F.G. Friedlander (1958), *Sound Pulses*, Cambridge University Press, London, 202 pp.

G.H. Golub and C.F. van Loan (1983), *Matrix Computations*, John Hopkins University Press, Baltimore, MD, 476 pp.

G.H. Golub and D.P. O'Leary (1989), Some history of the conjugate gradient and Lanczos algorithms: 1948–1976, *SIAM Review*, **31**, pp. 50-102.

T.M. Habashy and R. Mittra (1987), On some inverse methods in electromagnetics, *J. of Electromagn. Waves Appl.*, **1**, pp. 25-58.

S. Hanish (1981), *A Treatise on Acoustic Radiation*, vol. I, Naval Research Laboratory, Washington, D.C., 575 pp.

R.F. Harrington (1968), *Field Computation by Moment Methods*, Macmillan, New York, 229 pp.

H. Hönl, A.W. Maue, and K. Westpfahl (1961), Theorie der Beugung, in: S. Flügge, *Handbuch der Physik*, **XXV/1**, Kristaloptik-Beugung, 592 pp.

H. Jeffreys (1974), *Cartesian Tensors*, Cambridge University Press, London, 93 pp.

D.S. Jones (1952), Diffraction by an edge and by a corner, *Quart. J. Mech. Appl. Math.*, **5**, pp. 363-378.

L.V. Kantorovich and V.I.R. Krylov (1964), *Approximate Methods of Higher Analysis*, Interscience Publishers, New York, 681 pp.

L.E. Kinsler, A.R. Frey, A.B. Coppens and J.V. Sanders (1982), *Fundamentals of Acoustics*, third edition, John Wiley & Sons, 480 pp.

R.E. Kleinman and P.M. van den Berg (1991), Iterative methods for solving integral equations, Chapter 3 in: T.K. Sarkar (Ed.), *Application of Conjugate Gradient Method to Electromagnetics and Signal Analysis*, PIER 5, Elsevier, Amsterdam, pp. 67-102.

R.E. Kleinman and P.M. van den Berg (1992a), A modified gradient method for two-dimensional problems in tomography, *Journal of Computational and Applied Mathematics*, **42**, pp. 17-35.

R.E. Kleinman and P.M. van den Berg (1992b), An extended range modified gradient technique for profile inversion, *Proceedings of the 1992 URSI International Symposium on Electromagnetic Theory*, Sydney, August 17-20, 1992, pp. 251-253.

H. Lamb (1925), *The Dynamical Theory of Sound*, Dover, New York, 1960, 307 pp.

S.W. Lee, J. Boersma, C.L. Law and G.A. Deschamps (1980), Singularity in Green's function and its numerical evaluation, *IEEE Trans. Antennas*

*Propagat.*, **AP-28**, pp. 311-317.

A.E.H. Love (1944), *A Treatise on the Mathematical Theory of Elasticity*, Macmillan, London, fourth edition; Dover, New York, 1944, 643 pp.

W.P. Mason (Ed.) (1964 - ), *Physical Acoustics*, Academic Press, New York.

D.W. McCowan and H. Brysk (1982), Cartesian and cylindrical slant stacks, in: P.L. Stoffa (ed.), *Tau-p: a plane wave approach to the analysis of seismic data*, Kluwer, Dordrecht, pp. 1-33.

P.M. Morse and H. Feshbach (1953), *Methods of Theoretical Physics*, McGraw-Hill, New York, part I and part II, 1978 pp.

P.M. Morse and K.V. Ingard (1968), *Theoretical Acoustics*, McGraw-Hill, New York, 972 pp.

A.V. Oppenheim, A.S. Willsky and J.T. Young (1983), Prentice Hall, London, 796 pp.

G.W. Oseen (1915), Über die Wechselwirkung zwischen zwei elektrischen Dipolen und über die Drehung der Polarisationsebene in Kristallen und Flüssigkeiten, *Annalen der Physik*, **48**, pp. 1-56.

W.H. Press, B.P. Flannery, S.A. Teukolsky and W.T. Vetterling (1986), *Numerical Recipes, The Art of Scientific Computing*, Cambridge University Press, London, 818 pp.

A. Ralston and P. Rabinowitz (1978), *A First Course in Numerical Analysis*, McGraw-Hill, London, 556 pp.

Lord Rayleigh (J.W. Strutt, 1894), *Theory of Sound*, Macmillan, London, second edition; Dover, New York, 1945, vol. I, 480 pp., vol. II, 504 pp.

S.N. Rschevkin (1963), *A Course of Lectures on Theory of Sound*, Pergamon Press, London, 464 pp.

W.A. Schneider (1978), Integral formulation for migration in two and three dimensions, *Geophysics*, **43**, pp. 49-76.

A. Tarantola (1987), *Inverse Problem Theory*, Elsevier, Amsterdam, 613 pp.

E.C. Titchmarsh (1939), *The Theory of Functions*, Oxford University Press, London, second edition, 454 pp.

E.C. Titchmarsh (1948), *Introduction to the Theory of Fourier Integrals*, Oxford University Press, London, second edition, 394 pp.

A.G. Tijhuis (1987), *Electromagnetic Inverse Profiling*, VNU Science Press, Utrecht, 465 pp.

W. Tobocman (1986), Iterative inverse scattering method employing Gram-Schmidt orthogonalization, *J. Comput. Phys.*, **64**, pp. 230-245.

C.J. Tranter (1966), *Integral Transforms in Mathematical Physics*, third edition, Methuen & Co; Chapman and Hall, London, 1971, 139 pp.

E. Skudrzyk (1971), The Foundations of Acoustics, Springer-Verlag, New York, 790 pp.

R.G. van Borselen, J. Thorbecke, J.T. Fokkema and P.M. van den Berg (1991), Surface-related multiple elimination based on reciprocity, *61st Annual International SEG Meeting, Expanded Abstracts*, Society of Exploration Geophysicists, Houston, pp. 1339-1342.

P.M. van den Berg (1984), Iterative computational techniques in scattering based upon the integrated square error criterion, *IEEE Trans. Antennas Propagat.*, **AP-32**, pp. 1063-1071.

P.M. van den Berg (1991), Iterative schemes based on minimization of a uniform error criterion, Chapter 2 in: T.K. Sarkar (Ed.), *Application of Conjugate Gradient Method to Electromagnetics and Signal Analysis*, PIER 5, Elsevier, Amsterdam, pp. 27-65.

J.H.M.T. van der Hijden (1987), *Propagation of Transient Elastic Waves in Stratified Anisotropic media*, North Holland, Amsterdam, 288 pp.

Balth. van der Pol and H. Bremmer (1950), *Operational Calculus*, Cambridge University Press, London, 415 pp.

D.J. Verschuur, A.J. Berkhout and C.P.A. Wapenaar (1992), Adaptive surface-related multiple elimination, *Geophysics*, **57**, pp. 1166-1177.

D.V. Widder (1946), *The Laplace Transform*, Princeton University Press, Princeton, 406 pp.

E.T. Whittaker and G.N. Watson (1927), *A Course of Modern Analysis*, Cambridge University Press, London, fourth edition, 608 pp.

C.P.A. Wapenaar and A.J. Berkhout (1989), *Elastic Wave Field Extrapolation*, Elsevier, Amsterdam, 468 pp.

P.C. Waterman (1969), New formulation of acoustic scattering, *J. Acoust. Soc. Am.*, **45**, pp. 1417-1429.

G.N. Watson (1944), *A Treatise on the Theory of Bessel Functions*, Cambridge University Press, London, second edition, 804 pp.

A. Ziolkowski (1991), Why don't measure seismic signatures?, *Geophysics*, **56**, pp. 190-201.

# Index

## — R —



## — S —



## — T —